JAGUAR XJS
Gold Portfolio
1975-1990

Compiled by
R.M. Clarke

ISBN 1 85520 0198

Distributed by
Brooklands Book Distribution Ltd.
'Holmerise', Seven Hills Road,
Cobham, Surrey, England

Printed in Hong Kong

BROOKLANDS BOOKS

BROOKLANDS BOOK SERIES
AC Ace & Aceca 1953-1983
Alfa Romeo Alfasud 1972-1984
Alfa Romeo Alfetta Coupes GT GTV GTV6 1974-1987
Alfa Romeo Guilia Berlinas 1962-1976
Alfa Romeo Giulia Coupes 1963-1976
Alfa Romeo Spider 1966-1987
Allard Gold Portfolio 1937-1958
Alvis Gold Portfolio 1919-1969
American Motors Muscle Cars 1966-1970
Aston Martin Gold Portfolio 1972-1985
Austin Seven 1922-1982
Austin A30 & A35 1951-1962
Austin Healey 3000 1959-1967
Austin Healey 100 & 3000 Col No.1
Austin Healey 'Frogeye' Sprite Col No.1 1958-1961
Austin Healey Sprite 1958-1971
Avanti 1962-1983
BMW Six Cylinder Coupes 1969-1975
BMW 1600 Col. 1 1966-1981
BMW 2002 1968-1976
Bristol Cars Gold Portfolio 1946-1985
Buick Automobiles 1947-1960
Buick Muscle Cars 1965-1970
Buick Riviera 1963-1978
Cadillac Automobiles 1949-1959
Cadillac Automobiles 1960-1969
Cadillac Eldorado 1967-1978
High Performance Capris Gold Portfolio 1969-1987
Chevrolet Camaro & Z-28 1973-1981
High Performance Camaros 1982-1988
Chevrolet Camaro Col No.1 1967-1973
Camaro Muscle Cars 1966-1972
Chevrolet 1955-1957
Chevrolet Impala & SS 1958-1971
Chevrolet Muscle Cars 1966-1971
Chevelle and SS 1964-1972
Chevy EL Camino & SS 1959-1987
Chevy II Nova & SS 1962-1973
Chrysler 300 1955-1970
Citroen Traction Avant 1934-1957
Citroen DS & ID 1955-1975
Citroen 2CV 1949-1980
Cobras & Replicas 1962-1983
Corvair 1959-1968
High Performance Corvettes 1983-1989
Datsun 240Z 1970-1973
Datsun 280Z & ZX 1975-1983
De Tomaso Collection No.1 1962-1981
Dodge Charger 1966-1974
Dodge Muscle Cars 1967-1970
Excalibur Collection No.1 1952-1981
Ferrari Cars 1946-1956
Ferrari Cars 1973-1977
Ferrari Dino 1965-1974
Ferrari Dino 308 1974-1979
Ferrari 308 & Mondial 1980-1984
Ferrair Collection No.1 1960-1970
Fiat-Bertone X1/9 1973-1988
Fiat Pininfarina 124 + 2000 Spider 1968-1985
Ford Automobiles 1949-1959
Ford GT40 Gold Portfolio 1964-1987
Ford Fairlane 1955-1970
Ford Falcon 1960-1970
High Perfomance Mustangs 1982-1988
Ford Cortina 1600E & GT 1967-1970
Ford RS Escorts 1968-1980
High Performance Escorts Mk1 1968-1974
High Performance Escorts Mk II 1975-1980
Honda CRX 1983-1987
Hudson & Railton 1934-1940
Jaguar Cars 1957-1961
Jaguar Cars 1961-1964
Jaguar Mk2 1959-1969
Jaguar E-Type Gold Portfolio 1961-1971
Jaguar E-Type 1966-1971
Jaguar E-Type V-12 1971-1975
Jaguar XKE Collection No.1 1961-1974
Jaguar XJ6 1968-1972
Jaguar XJ6 Series II 1973-1979
Jaguar XJ6 & XJ12 Series III 1979-1985
Jaguar XJ12 1972-1980
Jaguar XJS Gold Portfolio 1975-1990
Jaguar XK120.XK140.XK150 Gold Portfolio 1948-1960
Jensen Cars 1946-1967
Jensen Cars 1967-1979
Jensen Healey 1972-1976
Jensen Interceptor Gold Portfolio 1966-1986
Lamborghini Cars 1964-1970
Lamborghini Cars 1970-1975
Lamborghini Countach Col No.1 1971-1982
Lamborghini Countach & Urraco 1974-1980
Lamborghini Countach & Jalpa 1980-1985
Lancia Stratos 1972-1985
Land Rover 1948-1973 - A Collection
Land Rover Series II & IIa 1958-1971
Land Rover Series III 1971-1985
Land Rover 90 & 110 1983-1989
Lincoln Gold Portfolio 1949-1960
Lincoln Continental 1961-1969
Lotus and Caterham Seven Gold Portfolio 1957-1989
Lotus Elan Gold Portfolio 1962-1974
Lotus Elan Collection No.2 1963-1972
Lotus Elite 1957-1964
Lotus Elite & Eclat 1974-1981
Lotus Turbo Esprit 1980-1986
Lotus Europa 1966-1975
Lotus Europa Collection No.1 1966-1974
Lotus Seven Collection No.1 1957-1982
Marcos Cars 1960-1988
Maserati 1965-1970
Maserati 1970-1975
Mazda RX-7 Collection No.1 1978-1981
Mercedes 190 & 300SL 1954-1963
Mercedes 230/250/280SL 1963-1971
Mercedes Benz SLs & SLCs Gold Portfolio 1971-1989
Mercedes Benz Cars 1949-1954
Mercedes Benz Cars 1954-1957
Mercedes Benz Cars 1957-1961

Mercedes Benz Competion Cars 1950-1957
Mercury Muscle Cars 1966-1971
Metropolitan 1954-1962
MG TC 1945-1949
MG TD 1949-1953
MG TF 1953-1955
MG Cars 1959-1962
MGA Roadsters 1955-1962
MGA Collection No.1 1955-1982
MGB Roadsters 1962-1980
MGB GT 1965-1980
MG Midget 1961-1980
Mini Moke 1964-1989
Mini Muscle Cars 1961-1979
Mopar Muscle Cars 1964-1967
Mopar Muscle Cars 1968-1971
Morgan Three-Wheeler Gold Portfolio 1910-1952
Morgan Cars 1960-1970
Morgan Cars Gold Portfolio 1968-1989
Morris Minor Collection No.1
Mustang Muscle Cars 1967-1971
Oldsmobile Automobiles 1955-1963
Old's Cutlass & 4-4-2 1964-1972
Oldsmobile Muscle Cars 1964-1971
Oldsmobile Toronado 1966-1978
Opel GT 1968-1973
Packard Gold Portfolio 1946-1958
Pantera Gold Portfolio 1970-1989
Plymouth Barracuda 1964-1974
Plymouth Muscle Cars 1966-1971
Pontiac Tempest & GTO 1961-1965
Pontiac GTO 1964-1970
Pontiac Firebird 1967-1973
Pontiac Firebird and Trans-Am 1973-1981
High Performance Firebirds 1982-1988
Pontiac Fiero 1984-1988
Pontiac Muscle Cars 1966-1972
Porsche 356 1952-1965
Porsche Cars in the 60's
Porsche Cars 1960-1964
Porsche Cars 1964-1968
Porsche Cars 1968-1972
Porsche Cars 1972-1975
Porsche Turbo Collection No.1 1975-1980
Porsche 911 1965-1969
Porsche 911 1970-1972
Porsche 911 1973-1977
Porsche 911 Carrera 1973-1977
Porsche 911 Turbo 1975-1984
Porsche 911 SC 1978-1983
Porsche 914 Gold Portfolio 1969-1976
Porsche 914 Collection No.1 1969-1983
Porsche 924 Gold Portfolio 1975-1988
Porsche 928 1977-1989
Porsche 944 1981-1985
Range Rover Gold Portfolio 1970-1988
Reliant Scimitar 1964-1986
Riley 11/2 & 21/2 Litre Gold Portfolio 1945-1955
Rolls Royce Silver Cloud 1955-1965
Rolls Royce Silver Shadow 1965-1981
Rover P4 1949-1959
Rover P4 1955-1964
Rover 3 & 3.5 Litre 1958-1973
Rover 2000 + 2200 1963-1977
Rover 3500 1968-1977
Rover 3500 & Vitesse 1976-1986
Saab Sonett Collection No.1 1966-1974
Saab Turbo 1976-1983
Shelby Mustang Muscle Cars 1965-1970
Stubebaker Gold Portfolio 1947-1966
Stubebaker Hawks & Larks 1956-1963
Sunbeam Tiger & Alpine Gold Portfolio 1959-1967
Thunderbird 1955-1957
Thunderbird 1958-1963
Thunderbird 1964-1976
Toyota MR2 1984-1988
Triumph 2000. 2.5. 2500 1963-1977
Triumph GT6 1966-1974
Triumph Spitfire 1962-1980
Triumph Spitfire Col No.1 1962-1982
Triumph Stag 1970-1980
Triumph Stag Collection No.1 1970-1984
Triumph TR2 & TR3 1952-60
Triumph TR4-TR5-TR250 1961-1968
Triumph TR6 1969-1976
Triumph TR6 Collection No.1 1969-1983
Triumph TR7 & TR8 1975-1982
Triumph Vitesse & Herald 1959-1971
TVR Gold Portfolio 1959-1988
Volkswagen Cars 1936-1956
VW Beetle Collection No.1 1970-1982
VW Golf GTi 1976-1986
VW Karmann Ghia 1955-1982
VW Kubelwagen 1940-1975
VW Scirocco 1974-1981
VW Bus. Camper. Van 1954-1967
VW Bus. Camper. Van 1968-1979
VW Bus. Camper. Van 1979-1989
Volvo 120 1956-1970
Volvo 1800 1960-1973

BROOKLAND ROAD & TRACK SERIES
Road & Track on Alfa Romeo 1949-1963
Road & Track on Alfa Romeo 1964-1970
Road & Track on Alfa Romeo 1971-1976
Road & Track on Alfa Romeo 1977-1989
Road & Track on Aston Martin 1962-1984
Road & Track on Auburn Cord and Duesenburg 1952-1984
Road & Track on Audi & Auto Union 1952-1980
Road & Track on Audi 1980-1986
Road & Track on Austin Healey 1953-1970
Road & Track on BMW Cars 1966-1974
Road & Track on BMW Cars 1975-1978
Road & Track on BMW Cars 1979-1983
Road & Track on Cobra, Shelby & GT40 1962-1983
Road & Track on Corvette 1953-1967
Road & Track on Corvette 1968-1982
Road & Track on Corvette 1982-1986
Road & Track on Datsun Z 1970-1983

Road & Track on Ferrari 1950-1968
Road & Track on Ferrari 1968-1974
Road & Track on Ferrari 1975-1981
Road & Track on Ferrari 1981-1984
Road & Track on Fiat Sports Cars 1968-1987
Road & Track on Jaguar 1950-1960
Road & Track on Jaguar 1961-1968
Road & Track on Jaguar 1968-1974
Road & Track on Jaguar 1974-1982
Road & Track on Jaguar 1983-1989
Road & Track on Lamborghini 1964-1985
Road & Track on Lotus 1972-1981
Road & Track on Maserati 1952-1974
Road & Track on Maserati 1975-1983
Road & Track on Mazda RX7 1978-1986
Road & Track on Mercedes 1952-1962
Road & Track on Mercedes 1963-1970
Road & Track on Mercedes 1971-1979
Road & Track on Mercedes 1980-1987
Road & Track on MG Sports Cars 1949-1961
Road & Track on MG Sprots Cars 1962-1980
Road & Track on Mustang 1964-1977
Road & Track on Peugeot 1955-1986
Road & Track on Pontiac 1960-1983
Road & Track on Porsche 1961-1967
Road & Track on Porsche 1968-1971
Road & Track on Porsche 1972-1975
Road & Track on Porsche 1975-1978
Road & Track on Porsche 1979-1982
Road & Track on Porsche 1982-1985
Road & Track on Porsche 1985-1988
Road & Track on Rolls Royce & B'ley 1950-1965
Road & Track on Rolls Royce & B'ley 1966-1984
Road & Track on Saab 1955-1985
Road & Track on Toyota Sports & GT Cars 1966-1984
Road & Track on Triumph Sports Cars 1953-1967
Road & Track on Triumph Sports Cars 1967-1974
Road & Track on Triumph Sports Cars 1974-1982
Road & Track on Volkswagen 1951-1968
Road & Track on Volkswagen 1968-1978
Road & Track on Volkswagen 1978-1985
Road & Track on Volvo 1957-1974
Road & Track on Volvo 1975-1985
Road & Track - Henry Manney at Large and Abroad

BROOKLAND CAR AND DRIVER SERIES
Car and Driver on BMW 1955-1977
Car and Driver on BMW 1977-1985
Car and Driver on Cobra, Shelby & Ford GT 40 1963-198
Car and Driver on Corvette 1956-1967
Car and Driver on Corvette 1968-1977
Car and Driver on Corvette 1978-1982
Car and Driver on Corvette 1983-1988
Car and Driver on Datsun Z 1600 & 2000 1966-1984
Car and Driver on Ferrari 1955-1962
Car and Driver on Ferrari 1963-1975
Car and Driver on Ferrari 1976-1983
Car and Driver on Mopar 1956-1967
Car and Driver on Mopar 1968-1975
Car and Driver on Mustang 1964-1972
Car and Driver on Pontiac 1961-1975
Car and Driver on Porsche 1955-1962
Car and Driver on Porsche 1963-1970
Car and Driver on Porsche 1970-1976
Car and Driver on Porsche 1977-1981
Car and Driver on Porsche 1982-1986
Car and Driver on Saab 1956-1985
Car and Driver on Volvo 1955-1986

BROOKLANDS PRACTICAL CLASSICS SERIES
PC on Austin A40 Restoration
PC on Land Rover Restoration
PC on Metalworking in Restoration
PC on Midget/Sprite Restoration
PC on Mini Cooper Restoration
PC on MGB Restoration
PC on Morris Minor Restoration
PC on Sunbeam Rapier Restoration
PC on Triumph Herald/Vitesse
PC on Triumph Spitfire Restoration
PC on VW Beetle Restoration
PC on 1930s Car Restoration

BROOKLANDS MOTOR & THOROGHBRED & CLASSIC CAR SERIES
Motor & T & CC on Ferrari 1966-1976
Motor & T & CC on Ferrari 1976-1984
Motor & T & CC on Lotus 1979-1983

BROOKLANDS MILITARY VEHICLES SERIES
Allied Mil. Vehicles No.1 1942-1945
Allied Mil. Vehicles No.2 1941-1946
Dodge Mil. Vehicles Col. 1 1940-1945
Military Jeeps 1941-1945
Off Road Jeeps 1944-1971
Hail to the Jeep
US Military Vehicles 1941-1945
US Army Military Vehicles WW2-TM9-2800

BROOKLAND HOT ROD RESTORATION SERIES
Auto Restoration Tips & Techniques
Basic Bodywork Tips & Techniques
Basic Painting Tips & Techniques
Camaro Restoration Tips & Techniques
Custom Painting Tips & Techniques
How to Build a Street Rod
Mustang Restoration Tips & Techniques
Performance Tuning - Chevrolets of the '60s
Performance Tuning - Ford of the '60s
Performance Tuning - Mopars of the '60s
Performance Tuning - Pontiacs of the '60s

BROOKLANDS BOOKS

CONTENTS

Page	Title	Publication	Date		Year
5	XJS	Autocar	Sept.	13	1975
12	On the Road	Thoroughbred & Classic Cars	Oct.		1975
14	Jaguar XJS Road Test	Car and Driver	Jan.		1976
19	Jaguar XJS – New from Britain	Road & Track	Oct.		1975
22	Quietly, Cat, Quietly	Modern Motor	Feb.		1976
26	Jaguar XJS Road Test	Motor	Feb.	21	1976
32	Jaguar XJS Road Test	Road Test	Feb.		1976
37	Jaguar Journey	Motor Sport	April		1976
40	Move Over Merc, Comparison Test	Motor	May	1	1976
45	Jaguar XJS: Nothing but the Best	Autosport	April	1	1976
48	XJS	Thoroughbred & Classic Cars	Jan.		1977
50	Jaguar XJS Automatic Road Test	Autocar	May	28	1977
56	Two Jaguar Coupes Track Test	Road & Track	May		1977
62	The Top Cat! Road Test	Modern Motor	Oct.		1977
66	Cat Tales in the Sunset Road Test	Motor Trend	June		1979
70	Cat Amongst the Pigeons! Road Test	Motor Manual	Oct.		1979
75	Jaguar XJS Short Take	Car and Driver	Feb.		1981
76	Come to the Cabriolet	What Car?	Sept.		1980
78	Jaguar XJS Road Test	Motor	Oct.	25	1980
82	Jaguar XJS Road Test	Road & Track	May		1981
85	Fireball – The Development of the Decade	Asian Auto	July		1981
88	The Big Cat's Back Track Test	Autosport	Feb.	25	1982
90	State of the Art Road Test	Autosport	Feb.	25	1982
92	Jaguar XJS	Motor Trend	Feb.		1982
94	Jaguar XJS HE Road Test	Autocar	April	24	1982
100	Jaguar XJS Road Test	Road & Track	Dec.		1982
103	Tuning Topics Road Test	Motor Sport	May		1982
107	Jaguar XJS HE Automatic Road Test	Car	Jan.		1983
110	Tail-Piece – XJS Eventer Estate	What Car?	Jan.		1983
112	Jaguar XJS 3.6C Road Test	Motor	March	3	1984
116	Jaguar XJSC 3.6	Motor Sport	July		1984
118	Go Cat Go!	Motor	Aug.	11	1984
120	Cool Cat – Jaguar's Roofless Restyle Road Test	Drive	Jan.		1985
124	Coventry's Future Classic Road Test	Autosport	Sept.	12	1985
126	TWR Jaguar Sport	Road & Track Specials	Aug.		1985
128	The Cat's Whiskers – Jaguar XJS HE Cabriolet Test Update	Autocar	Sept.	25	1985
131	Jaguar XJSC Cabriolet Road Test	Car and Driver	Aug.		1986
136	Cat Fight – Jaguar v Jaguar Comparison Test	Motor	Aug.	30	1986
139	Jaguar XJSC Cabriolet	Motor Trend	Nov.		1986
143	A Classic Opportunity Buying Secondhand	Autocar	April	8	1987
147	Jaguar XJS 3.6 Auto Test Update	Autocar	Oct.	7	1987
150	Lister-Jaguar XJS NAS	Road & Track	Nov.		1987
153	Untamed Road Test	Fast Lane	Jan.		1988
157	XJS Convertible a Snip at £40,000	Motor	March	5	1988
158	Prestige Performance Comparison Test	What Car?	March		1988
164	XJS V12 Convertible Test Extra	Autocar	April	27	1988
168	XJS	Road & Track Specials			1988
172	Tom's Cat	Autocar & Motor	Aug.	23	1989
175	Jaguar XJ-S Convertible Road Test	Motor Trend	Dec.		1989
180	The Fall and Rise of the XJS	Motor	Aug.	6	1988

ACKNOWLEDGEMENTS

The XJS made its debut in September 1975 to a fanfare of publicity. It came as a surprise to many as it was a break with tradition. Jaguar up to then had been famous for value-for-money saloons and brisk Le Mans winning sportscars. The much loved E-Type, which fitted into the sportscar mould had gone out of production only six months earlier and was now being replaced by a much heavier and plusher 2+2 GT.

It came into being at a difficult time. Jaguars build quality in the late seventies was not high and the XJS along with other cars from Browns Lane gained a reputation for poor reliability. The V12 engine had always been a thirsty unit even when fitted in the E-Type and was especially so now when married to this weighty GT body. The outcome was that in 1980 sales sank to below 2000 for the twelve months, and production had to be temporarily halted.

The tide then turned for Jaguar. With the appointment of a new chief executive the finish and reliability of the cars improved beyond belief and confidence in the marque returned. The mileage handicap was dealt with in 1981 with the introduction of the HE engine and the result was that sales in 1982 were 250% greater than those of a year earlier.

The XJS has never looked back. It had been a late developer. Sales in 1987 were nearing 8 times those of 1981 and it is confidently expected that this year over 10,000 units will be sold.

Brooklands books are works of reference and readers will be able to trace the downs and ups of the XJS from the broad selection of international contemporary reports contained herein.

We are as always indebted to the publishers of the worlds leading motoring journals for allowing us to include their copyright informative road tests and other stories in our series. We are sure that XJS owners will wish to join with us in thanking the management of Asian Auto, Autocar, Autosport, British Car, Car South Africa, Car and Driver, Classic Car, Drive, Fast Lane, Modern Motor, Motor, Motor Sport, Motor Trend, Road & Track, Road & Track Specials, Road Test and What Car? for their continued support.

R.M. Clarke

XJ-S

A new concept in Jaguar motoring

by Michael Scarlett   Pictures by Peter Cramer   Drawings by Vic Berris

What is the XJ-S?
THE SHORT ANSWER, in old-fashioned terms, is a fixed-head sports coupé. One must then add "of great refinement, and high performance", which follows when one learns that it is based to a large extent on the Jaguar XJ12 (or XJ 5·3 as it is now called), but it is lower and a little lighter. It shares many saloon features, and judging from some pre-announcement driving Jaguar kindly allowed us, an XJ12 driver put into the driving seat of the XJ-S would certainly notice more performance, a slight and valued improvement in the saloon's already superb handling, and different visibility – the XJ-S tends towards American and Italian fashion in super-fast cars where the faster the car, the less rearward view is deemed necessary – but he would find it more refined if anything, and as gentle to drive.

It is the most expensive Jaguar ever offered (the price appears elsewhere in this issue) – what some would call Jaguar's "flagship", and others a prestige Jaguar. But it is not now a *replacement* for the great E-type, even in that very refined sports-car's fixed-head coupé form. Coupé production of E-types stopped in 1973, and the roadster last February (*Autocar*

In spite of the old E-type 2+2's sleeker-looking shape, its drag coefficient (according to the MIRA tunnel) is not so good as that for the XJ-S. When frontal area figures are taken into account in spite of the new car's greater area, the XJ-S should have less drag.

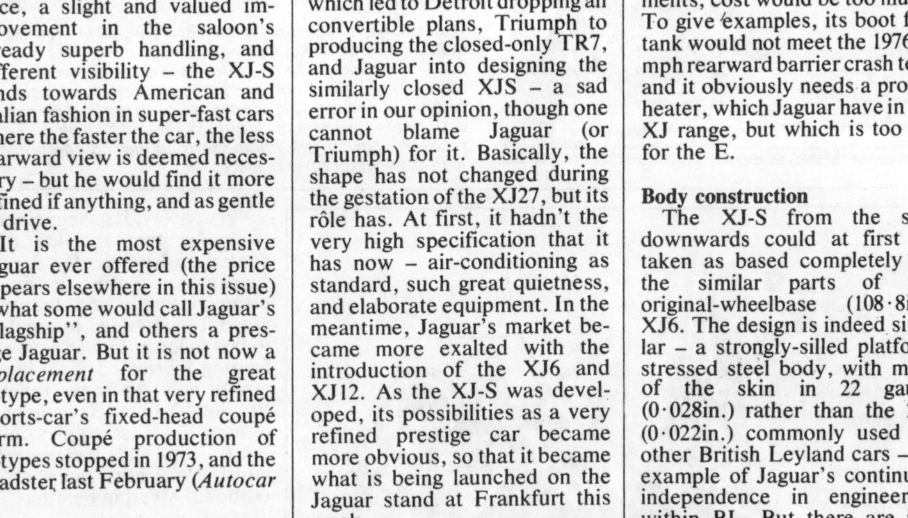

described a drive in the very last E-type V12 roadster round Scotland in last week's issue); in fact, as far as its principal market was concerned, North America, there were no 1975 E-type roadsters.

Originally, in 1970, when the design was first put on paper seriously, the XJ-S (XJ27 in Jaguar drawing office terminology) was to have been an E-type replacement. There was then the widely-held misunderstanding about United States Federal regulations outlawing the open car in years to come, which led to Detroit dropping all convertible plans, Triumph to producing the closed-only TR7, and Jaguar into designing the similarly closed XJS – a sad error in our opinion, though one cannot blame Jaguar (or Triumph) for it. Basically, the shape has not changed during the gestation of the XJ27, but its rôle has. At first, it hadn't the very high specification that it has now – air-conditioning as standard, such great quietness, and elaborate equipment. In the meantime, Jaguar's market became more exalted with the introduction of the XJ6 and XJ12. As the XJ-S was developed, its possibilities as a very refined prestige car became more obvious, so that it became what is being launched on the Jaguar stand at Frankfurt this week.

It is unusual to harp on the death of a new car's predecessor, but both because the E-type was such a notable and delightful machine, and because of the widespread expectation that the XJ-S was to be an "F-type" – which it clearly is not – one may wonder why the E-type roadster at least could not continue alongside its very different new brother, with both 12- and six-cylinder versions. The answer given is, chiefly that to re-design the E-type and improve its few latter-day failings, bringing it up to contemporary standards of controls and equipment, and to meet coming American crash safety requirements, cost would be too much. To give examples, its boot fuel tank would not meet the 1976 30 mph rearward barrier crash test, and it obviously needs a proper heater, which Jaguar have in the XJ range, but which is too big for the E.

Body construction
The XJ-S from the sills downwards could at first be taken as based completely on the similar parts of the original-wheelbase (108·8in.) XJ6. The design is indeed similar – a strongly-silled platform stressed steel body, with most of the skin in 22 gauge (0·028in.) rather than the 24g (0·022in.) commonly used on other British Leyland cars – an example of Jaguar's continued independence in engineering within BL. But there are important differences. For a start, the wheelbase is 102in., 6·8in. shorter. In front of the dashboard bulkhead and at the sides of the engine room the structure is effectively triangulated forwards, more so than the saloon, both for stiffness and to help towards good crash absorption; the triangulation is as near as possible in line with the front screen pillars, making the best use of the inherent increase in section strength given by the cabin. The lower parts of the pillars are wide, and ahead of the radiator area, there are a horizontal cross tube and a diagonal one bolted across to give extra bracing to the sides. Behind the back seat pan, there are detail differences too.

Standard on all cars, whether destined for America or Europe, are impact-absorbing 5 mph bumpers which use deforming-plastic Menasco (American) piston-type shock "absorbers". The wide doors have side-intrusion resisting barriers on reinforced hinges. Some of the strength required for the rear-end barrier tests is evident in the sides of the deep, roomy boot. For the first time on any Jaguar, the 20gal fuel tank spans the rear suspension arch ahead of the boot, instead of being a pannier type, one in each wing as on the XJ saloons. There are obvious crash safety advantages of this layout. The tank itself has a sump below it for the fuel feed, to avoid surge problems with the width of the tank proper. The spare wheel is placed vertically, across the car and against the tank, next to the battery, which is on top of the tank sump and fuel pump. The spare has a cover to protect luggage. An interesting little detail in the floor of the boot is the pair of flexible louvres, in action rather like the self-bailers of a sailing dinghy, and intended to remove any stale air from the boot.

The XJ saloons are exceptionally quiet, even by the standards of supposedly better,

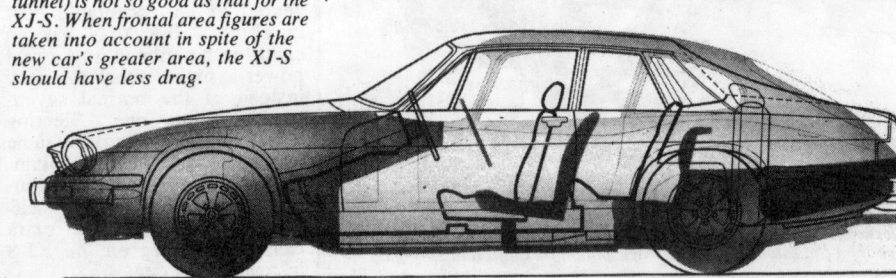

XJ-S

more exclusive cars of much higher price. The XJ-S is certainly as quiet, and perhaps even quieter, due to the meticulous attention to details of sound damping. There isn't space in this description to list every place where sound and vibration damping has been applied; a better understanding is obtained only by walking down the XJ-S production assembly line at the Browns Lane factory. Elaborate foam type moulded materials are found in all manner of places to deaden noise. Under the bonnet, the damping materials lining bulkhead and parts of the sides are faced with reflective aluminium to cut down heat absorption. As on the Series 2 saloons, inter-bulkhead connections are made by plug and socket, and not by inefficient grommeted holes. The XJ-S has positive air extraction from the inside, cockpit air passing out via the tops of the rear seat side pockets, into the large rear quarter panels and out of the vents in the latter; foam is applied to the insides of these collection chambers to keep the air exhaust quiet. Two more little details may help show the sort of attention that has been paid to noise suppression. The fuel pipe in the boot connecting the high pressure pump to the rest of the fuel system forwards is much longer than might seem to be necessary, coiled and carried within a larger foam tube in order to cut down transmission of pump noise. At the other end of the car, under the bonnet, the drain pipes for the air intake plenum chamber are light alloy tubes broken by rubber sections to avoid carrying engine and road noise.

Running gear

Suspension, naturally, is that used on the other Jaguars; subframe mounted, all-independent, with anti-dive angled wishbones in front and the unique three-link, modified-wishbone-geometry system at the back, using at each side, one front coil-spring-damper unit and two rear ones. Compared with the heavier XJ 5·3, spring rates are 90lb per in. front (XJ 5·3 92lb per in.) and 125lb per in. rear (XJ 154lb per in.). Anti-roll bars are used at *both* ends, ⅞in. dia. front (as now on XJ 5·3 – it was ¾in. dia) and ⁹⁄₁₆in. dia rear, where the saloon has none. Jaguar have taken great trouble to select what they consider to be exactly the right springs, roll bars and damper settings to give the best compromise between saloon ride and sporting handling – something they seem, on past XJ saloon evidence, to know more about than any other manufacturer in the world.

Very wisely, Jaguar engineers recognize that the life of a sealed-and-lubricated-for-life suspension bearing is by no means infinite, and that provided it is attended to regularly (6,000 miles in Jaguar's case), a bearing lubricated via a grease nipple is a lot more reliable. Previous XJ models have had a total of 17 grease nipples on the suspension and steering, but now there are two more, on the outboard steering bearings.

For better steering response – a small criticism of the XJ6 and XJ 5·3 is that the steering could easily be a little higher geared at an acceptable cost in effort – the XJ-S has an eight-tooth pinion in its Adwest power-assisted steering rack instead of the normal seven-tooth saloon one. Steering wheel diameter is the same as the saloon, at 15½in. Another important contribution towards better self-centring and feel is extra caster – 3½ deg on the XJ-S instead of 2½ deg.

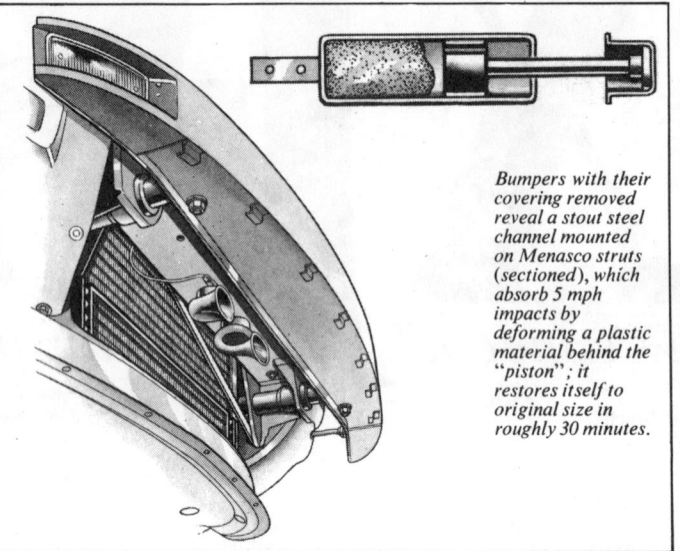

Bumpers with their covering removed reveal a stout steel channel mounted on Menasco struts (sectioned), which absorb 5 mph impacts by deforming a plastic material behind the "piston"; it restores itself to original size in roughly 30 minutes.

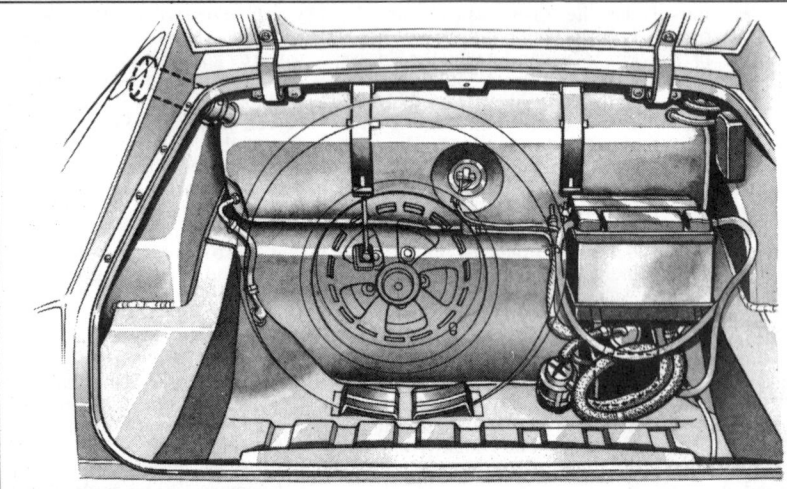

Boot details, with vertical spare wheel ghosted to show the saddle-mounted tank, its anti-surge sump alongside the battery, the long fuel pipe, coiled and encased to avoid noise transmission from the 28 psi Lucas immersed-motor fuel pump, and the carefully controlled strength of the sides, designed to cope with the latest US Federal 30 mph rearwards barrier test. Photo shows finished appearance, with spare wheel cover in place.

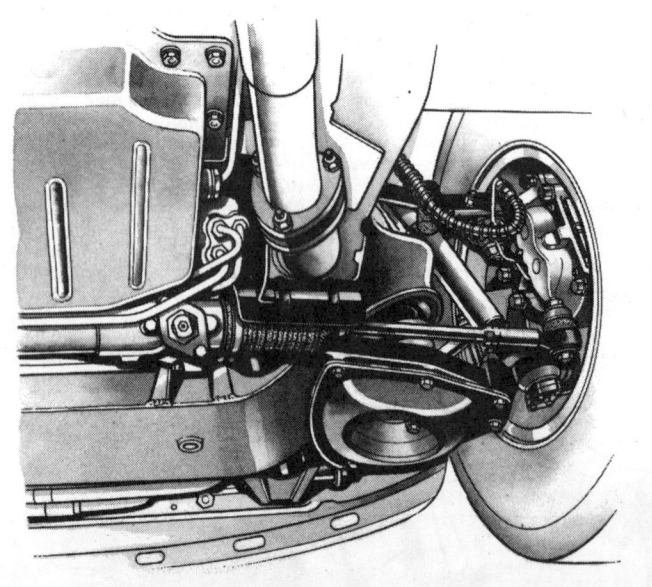

Underneath view of front suspension and steering, which has more caster than the saloons, a bigger rack pinion and different springs.

AUTOCAR, week ending 13 September 1975

XJ-S

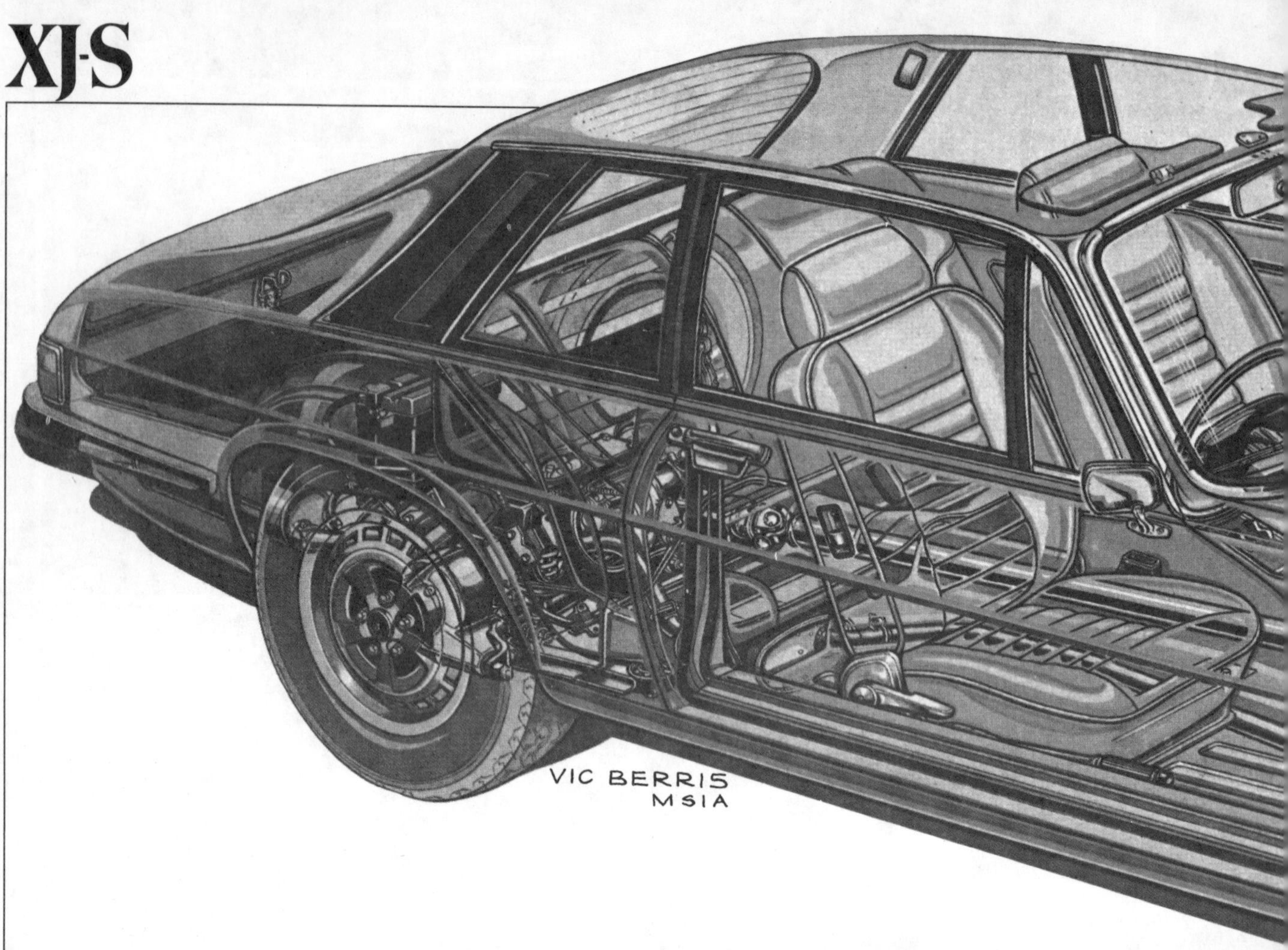

VIC BERRIS
MSIA

For instant recognition, the warning lights are colour-coded—red for major faults such as brake failure or loss of oil pressure, and amber for less urgent matters such as the seat belt reminder

Minor instruments grouped between the speedometer and rev counter, read vertically, with horizontal pointers. Rotary control for the lamps, on the right of the steering column, is matched by the ignition and starter switch on the left

Vic Berris's cutaway drawing of the Jaguar XJ-S shows the neater (if no less complex-looking) engine in its fuel injection form, the single-piece prop shaft, XJ 12 suspension and the cast wheels; also details such as the wing-mounted brake vacuum reservoir, corrosion-resistant front length of exhaust system with stainless steel aft of the first silencers, and the handbrake, on the driver's door sill and which when not in use, on or off, can be pushed down out of the way. The beard spoiler assists both lower drag and better cooling at speed

A release catch on the side of the seat squab allows it to be tipped forward, giving quite easy access to the rear compartment. The seats are generously proportioned, and shaped to accommodate two adults in comfort

XJ-S

Readers will recall that, when the XJ6 was announced in 1968, Jaguar gave great credit to Dunlop for developing and providing the SP Sport E70 VR 15in. rayon-braced tyres which obviously played an important part in the praise the car has always received for its superb road manners. A further development has endowed the XJ-S with a similar tyre from Dunlop, the Formula 70 SP Super Sport, 205 70 VR 15 (XJ 5·3 size) but with a steel-braced, block tread pattern. It is claimed that the tyre gives good wet road performance, quick response, and low wear. The speed rating is effectively higher, really VR+, thanks to a cooler-running tread. A low-pressure, die-cast, aluminium-alloy (LM25) road wheel is used, heat treated for better elongation and an ultimate tensile strength of 14 tons per sq. in., by GKN Kent Alloys. It was first offered on the XJ 5·3, and lowers unsprung weight usefully, each scaling 17lb 6oz against the previous XJ steel wheel's 23lb 8oz.

Paradoxically, the extra complication of the Lucas/Bosch electronic fuel injection has made the underbonnet appearance of the vee-12 tidier. As suggested by the similar throttle pushrod arrangement with its distinctive capstan, (in mid-vee), the XJ-S has the same superbly progressive throttle control as the XJ12.

The new wheel, first seen on the XJ 5.3, is a low-pressure die-casting (by GKN Kent Alloys) in heat-treated LM 25 aluminium alloy, and provides a 26 per cent weight reduction over the old steel wheel.

In their heyday, Aston Martin were prone to demonstrating that their fastest normal model could accelerate from rest to 100 mph and brake to a halt again in around 20sec. Jaguar claim that the XJ-S will do the same in "just over 20sec", which whatever it says for forward acceleration, speaks well for the brakes, which are essentially the same as on the saloons, Girling power-assisted, with ventilated discs 11·18in. dia. front and 10·38in. rear. Brake circuits are split front and rear, with a pressure difference warning switch to work the brake warning lamp if one circuit fails. The servo is an in-line tandem type on the brake pedal box, with its vacuum reservoir under one wing.

Engine and transmission

A full description of the Bosch eight-cylinder fuel injection, modified by Lucas to suit the V12 Jaguar engine, was given in *Autocar* of 3 May; it is this unit that is used in the XJ-S. Briefly, it is Bosch's electronic system, not their mechanical K-Jetronic one, and its use has raised the power output from 250 to 285 bhp at 5,750 rpm, with 294 lb. ft. maximum torque at 3,500 rpm, and a flatter torque curve overall. The superb 90/× 70mm 5,343 c.c., 60 deg V12 continues mechanically unchanged, except for the adoption (on all European-bound V12s) of air injection behind the exhaust valves, as on North American V12s. The point at which the air pipe enters the head area of each cylinder can surprise the casual onlooker, since it goes in on the inlet port side, being led across the head to the exhaust valve.

The air pump, belt-driven as usual, is an American Manox one; Federal V12s have in addition an engine anti-run-on valve and an evaporative emissions carbon canister (needed only for the fuel tank of course as fuel injection involves no fuel surfaces open to atmosphere), exhaust gas recirculation, and a catalytic reactor for each bank of cylinders. Jaguar say that the power is not seriously down for the US, except for Californian XJ-Ss, which because of more exhaust recirculation, have 244 bhp at 5,250 rpm and 269 lb. ft. maximum torque at 4,500 rpm. A lower axle ratio (3·31) is used to compensate for the reduced output.

Another detail inherited by fuel injected European V12s (and therefore found on the XJ-S) from American dictates is an inertia switch which cuts out the fuel pump in any impact greater than 3g. It can be reset by hand. Lucas breakerless Opus Mk II electronic ignition is fitted.

The exhaust system, from manifolds to the first silencer boxes is double-skinned steel tubing, aluminized for corrosion resistance. As on the saloons from last Spring, the silencers, and all pipes aft are stainless steel, which will be welcomed by prospective owners. Seeing the way the section of the pipe over the rear suspension has to be threaded through the suspension assembly on each side *before* the rear sub-frame is attached to the body, one can see that there are other advantages in fitting exhaust pipes that don't need regular replacement.

Something XJ 5·3 owners are not able to appreciate to the full, because that model is only obtainable with automatic transmission, is the truly extraordinary flexibility of the V12; formerly only owners of manual box E-types could realize the engine's full range. Happily, XJ-S buyers will be allowed the pleasure – and better efficiency – of a normal gearbox, Jaguar's four-speed, with the revised ratios introduced when the XJ 3·4 was announced – first gear being set lower, at 3·328 to 1 instead of 2·933. A 3·07-to-1 final drive is used with a Salisbury Powr-Lok limited slip differential, as on the latest XJ 5·3, giving overall gearing of 24·7 mph per 1,000 rpm. At the engine's peak power speed, this will correspond to 135 mph, so that if Jaguar's claims of a maximum speed "of 150 mph plus" are justified – and driving impressions suggest they will be – then the car is, at the moment, undergeared.

Another Jaguar claim is that their own engineers have returned typical fuel consumptions of between 15 and 18 mpg, which if achieved in practice by most owners, will be a valuable improvement over the XJ 5·3's 11 to 13 mpg. MIRA wind-tunnel drag figures for the current XJ6, E-type coupé 2 + 2 and XJ-S (with frontal areas in brackets) make interesting reading: XJ6 0·48 (19·8 sq. ft.); E-type 0·455 (17·8 sq. ft.); XJ-S 0·39 (19·8 sq. ft.).

Borg Warner's Model 12 automatic transmission (as fitted to the XJ 5·3) will no doubt be favoured by most XJ-S buyers, since it is likely to appeal more to saloon customers than sports-car types. The propeller shaft is a one-piece one.

Equipment

This is comprehensive. The automatically regulated air-conditioning unit developed jointly by Delanair and Jaguar for the Series 2 XJ saloons is standard, and has only two controls, one for temperature and one for automatic or manual working. There are both facia and footwell vents in the usual way, and as mentioned earlier there is a cockpit air extraction in the rear quarters.

The seats are of a new design intended to be both very comfortable and properly locating. Inertia reel seat belts are standard, and are available at the back too, with reels neatly hidden. Seat facings are leather, other side trim is expanded pvc, with nylon pile carpets for the bottom sides of the doors and for the car and boot floor – flame-retardant materials are used throughout the upholstery.

The windscreen is laminated, not toughened, and all glass is tinted to lower heat transmission. Door locks have electromagnetic centralized locking. The release for the bonnet must be positively pushed back to lock – as on a BMW, you cannot slam-lock it.

The dashboard is a one-piece vacuum-forming housing the instruments, adjustable air vents and a locking glove box. The instruments themselves – speedometer, rev counter, fuel, temperature, oil pressure and voltmeter (these auxiliary ones being a new air-cored magnetic type) plus no less than 18 warning lamps – live in a single binnacle wired by printed circuit, with two multi-pin plugs connecting it to the rest of the wiring loom. Jaguar have to some extent followed Citroën with their warning lamp system, in having a red overall "stop at once and investigate" lamp, plus an amber "investigate when convenient lamp", both linked to the other lamps, which deal with heated rear window, main beam, fog lamps, battery overcharge, ignition, fuel level, hazard, water (low coolant level –red), brake failure (low fluid or failure of a circuit – red), oil (low pressure – red), handbrake (amber), seat belts (amber), parking lamp (sidelamp failure – amber), and stop lamp (failure of brake light bulb – amber).

Controls are conventional and thorough, what one might expect on a car of luxurious specification. One little point is worth noting – the horn is worked not by the sometimes dubious business of finding and pushing the end of a stalk, but by hitting the padded centre of the steering wheel.

Another notable feature is the somewhat curious looking headlamps, which for Europe are specially developed Cibié biode units, each with two independent reflectors and halogen bulbs for dip and main beam, plus a system of auxiliary prisms inside the main reflector. America, thanks to inevitable regulations, must make do with GEC tungsten, though Canada, which is sensibly tending away from American thinking towards European in these matters, will allow the Cibiés.

The two-speed and single-stroke wipers have their motor and linkages mounted as an integral piece under the scuttle air intake, and can be removed in one piece after undoing four screws – another aid to easier maintenance.

Production and availability

To give some idea of production scale, when we visited Jaguar to look at the XJ-S in August, XJ saloon production was running at 750 a week, of which 600 were six-cylinder models; the maximum figure is 900, which was achieved just before the fuel flap started. Peak E-type production had been 170 a week, and latterly it was 90. The XJ-S being a more exclusive car, by virtue of its price as much as anything, Jaguar plan to make around 60 a week. Prices were to be announced after this section went to press, and will be found on the News pages in *Autocar*. Sales begin from 10 September; Jaguar intend every *distributor* to have an XJ-S in his showroom by the time you read this.

Driving impressions

Whatever one thinks of the appearance of the XJ-S – for what personal prejudices are worth, not all of us find this Jaguar as immediately beautiful as several of its predecessors – to drive the XJ-S even for an afternoon is to admire it very much.

It is sensationally quiet – the engine was virtually inaudible even at speeds around 60 mph, with a subdued hum when you used the acceleration to the full, very low bump-thump, hardly any rolling noise except when badly provoked by harsh concrete, and very little wind noise except at above an indicated 130 mph when, on the test car, the driver's door vibrated slightly.

It is certainly fast, even allowing for possible speedometer error when showing 145 mph, and in spite of the inevitable torque-converter inefficiencies of the not-terribly-responsive Borg-Warner automatic box. And, perhaps best of all, it handles superbly. We came to it after driving an XJ 3·4, which like all XJ saloons, has beautiful road manners. Therefore it was all the more surprising to find that the XJ-S steers and sticks that little bit better – the extra caster and the bigger rack pinion may not sound much, but combined with the other adjustments to the suspension, the car self centres better, and feels more positive, more responsive, and even more straight-stable over an uneven surface whilst at the same time, in corners, apparently understeering less. It rolls less than the saloon, and the ride is excellent. Although it is everything a luxury saloon tries to be, it is much more of a sporting car than one would have thought possible.

There proved to be just enough leg adjustment for a six-footer. The seats, high-backed and more modern than one is used to in a Jaguar (Sir William Lyons, it is said, always believed that you shouldn't be able to see the seats over the waist from outside) are excellent. We could not quarrel greatly with the controls, except for the usual unnecessary stop between D and 2 on the automatic selector, and none between D and N – only Mercedes seem to understand that it should, for safety and convenience sake, be the other way round. Americans will no doubt find the electric-window lifts slow, and wonder why, unlike a Porsche, you cannot raise or lower both at once. And I wish that rear-quarter view was not so badly blocked by those unnecessary curved-in wings behind the cabin – nevertheless, this is a new world-beater.

MS

Specification

ENGINE	
Cylinders	12, in 60deg vee
Main bearings	7
Cooling	Water
Fan	Viscous and electric
Bore, mm (in.)	90 (3·54)
Stroke, mm (in.)	70 (2·76)
Capacity, c.c. (cu. in.)	5,343 (326)
Valve gear	One ohc per bank
Camshaft drive	Chain
Compression ratio	9 to 1
Octane requirement	97RM
Injection	Lucas/Bosch electronic
Max power	285 bhp (DIN) at 5,500 rpm
Max torque	294 lb. ft. at 3,500 rpm
TRANSMISSION	
Clutch	Single dry plate, 10½in. dia.

Gear	Ratio	(Auto)	mph/1,000 rpm
Top	1·000	(1·00–2·03)	24·7 (24·7)
3rd	1·387	(—)	17·8 (—)
2nd	1·905	(1·45–2·94)	12·96 (17·0)
1st	3·238	(2·39–4·85)	7·63 (10·3)

Final drive gear	Salisbury hypoid bevel, Powr-lok diff.
Ratio	3·07
SUSPENSION	
Front – location	Wishbones
springs	Coil
dampers	Girling Monitube telescopic
anti-roll bar	⅞in. dia.
Rear – location	Modified wishbone (drive shaft and lower links)
springs	Twin coil
dampers	Girling Monitube telescopic
anti-roll bar	9/16in. dia.
STEERING	
Type	Adwest rack and pinion
Power assistance	Yes
Wheel diameter	15½in.
BRAKES	
Front	Girling ventilated, 11·18in. dia. disc
Rear	Girling plain, 10·38in. dia. disc
Servo	
WHEELS	
Type	Die cast aluminium alloy
Rim width	6in.
Tyres – make	Dunlop
– type	SP Super Sport steel braced radial tubed
– size	205–70VR15
EQUIPMENT	
Battery	Lucas CP13 12 volt 60 Ah
Alternator	Lucas 20ACR 60 amp
Headlamps	Cibié biode halogen 110/220 watt
Reversing lamp	Standard
Hazard warning	Standard
Electric fuses	16
Screen wipers	Two-speed, with flick wipe
Screen washer	Electric
Interior heater	Air conditioning
Interior trim	Leather seats, pvc headlining
Floor covering	Carpet
Jack	Screw cantilever
Jacking points	Two each side
Windscreen	Laminated, tinted
Underbody protection	Bitumastic
DIMENSIONS	
Wheelbase	8ft 6in. (259cm)
Front track	4ft 10in. (147cm)
Rear track	4ft 10·6in. (149cm)
Overall length	15ft 11·72in. (487cm)
Overall width	4ft 1·65in. (126cm)
Overall height	5ft 10·6in. (179cm)
Ground clearance	5½in. (14cm)
Unladen weight (no fuel, automatic)	3,710lb (1,683kg)

On the road

The new **XJ-S** costs £8900 and does 153mph. *Paul Skilleter* looks at the latest from Coventry.

Well, here it is, the first new sporting car from Jaguar since 1961 — and one might well be expected to say "at last", because it really has been rather a long time coming. But then, Jaguar's own position within British Leyland has hardly been very assured over the past year or so, and this can't have helped the XJ-S's gestation very much — not to mention Geoffrey Robinson's huge efforts in getting the XJ saloon made in greater numbers during his term of office as MD, the reorganising of the production lines which this entailed no doubt postponing the new car's launch still further.

We have with the XJ-S a vehicle of rather different concept from that of the original E-type, which was purely two seats and more or less the maximum possible performance without much regard for such conveniences as luggage room, ventilation or even completely adequate space for the car's occupants. The XJ-S on the other hand is very definitely a high-speed tourer, not a sports car, giving in return for the loss of an ultimate power-weight ratio a standard of comfort, silence and ride that I am sure has never been offered in a similar car before. And lest it be thought that the XJ-S has sacrificed too much performance, it should be said right away that the new car is faster than any production E-type, including the first 3·8s, with a top speed of well over 150mph.

This brings us to an interesting point — that it is still viable for a production manufacturer to campaign a new car on a 150mph ticket, despite petrol scares and the spread of spoil-sport legislation almost throughout the world. Equally interesting is the fact that for the first time, Jaguar are actively going after the "exotic" section of the market, previously inhabited only by such as Ferrari, Maserati (as was) and Lamborghini. Not that Jaguar's more traditional rivals are by-passed — the Jensen Interceptor and Mercedes 450SLC are two singled out in Jaguar's launch publicity for comparison with the XJ-S.

Although it had long been realised by those with their ears close to the ground that Jaguar's new car would have its engine in the sensible place — the front — there may still have been expectations in some quarters that in the XJ-S we would see the first mid-engined Jaguar road car. However, that it didn't happen should be no cause for disappointment; the mid-engine fashion of the last few years amongst exotic road cars has largely been just that — fashion. In practice, there is virtually nothing to be gained in terms of roadholding/handling with this configuration off the race track, as the Lotus Elite and now the XJ-S prove beyond doubt; on the other hand, sacrifices in space, and in the insulation of occupants from engine noise and heat, can be great and it was this that decided Jaguar to stay with their front engine, rear wheel drive layout. A mid-engined road car was never seriously considered in fact, Jaguar's experiments in this direction being limited to a few styling exercises by the late Malcolm Sayer.

The styling of Jaguars has always been individual and the continued use of a front engine position has meant that the XJ-S isn't just another wedge shape, but still has real Jaguar curves. The basic shape is Sayer, influenced by Sir William Lyons and US safety regulations although probably not in that order. With the XJ-S we do, of course, have the last Jaguar to be designed under the aegis of the immortal trio of Lyons, Sayer and Heynes — the latter deserving inclusion as it was under his direction that the car's suspension was originally laid out, inherited as it is from the XJ and E-type ranges. The body shape itself was wind tunnel tested at MIRA, where for instance it was found that with a spoiler added to the front and an undertray beneath the engine, there was a reduction of 50% in front end lift, and 10% less drag. "We decided from the very first that aerodynamics were the prime concern," says Sir William (now retired of course), "and I exerted my influence in a consultative capacity with Malcolm Sayer ... we originally considered a lower bonnet line, but the international regulations on crush control and lighting made us change and we started afresh."

As for under the skin, it wouldn't be quite fair to say that the XJ-S is basically a short chassis XJ (7in less in wheelbase than the XJ coupé) but the monocoque structure, suspension and mechanics do follow those of the V12-engined saloons. The power unit is the injected 5·3-litre version of Jaguar's all-alloy V12, the new breathing arrangements which displaced the emission Strombergs doing much to hoist the bhp figure back to where it was and more (now quoted at 285 DIN at 5500rpm) in spite of smog equipment — it's enough at least to give the 33½cwt car a top speed of 153mph and a 0–60 time of 6·8 seconds (manufacturer's figures) which surely makes it about the quickest four-seater genuinely in production.

Front suspension is semi-trailing wishbones and coil springs, with inbuilt anti-drive characteristics, but differs from the XJ's set-up in that a stiffer anti-roll bar has been fitted; this is all part of an attempt by Jim Randle, Roger Leng and Jaguar's engineering department to tauten the car's general response without affecting ride quality. Spring/damper relationships and roll stiffness were all examined carefully and resulted in a new rear anti-roll bar plus various changes in damping rates — the rear suspension remains the E-type-inspired independent system with its lower transverse wishbones, radius arms, and drive shafts which form the upper suspension links. The process was also carried on into the steering arrangements, and Jaguar have actually bent their attitude a little and allowed that maybe the XJ steering was a little too light after all — the XJ-S has an 8-tooth rack which results in a reduction of the steering's overall ratio; it's still assisted of course. The actual geometry of the steering was given a lot of thought too, and was "tuned in" to the new Dunlop SP Super Sport tyres. These are fitted on GKN-Kent Alloys aluminium wheels as standard, of the sort offered a short while ago as an option on some XJ saloons.

Brakes are 11·8in ventilated discs on the front, gripped by quadruple piston calipers, and 10·9in discs mounted inboard at the rear in familiar Jaguar fashion; the hydraulic circuit is split front to rear and there's a cut-

off valve which isolates one system should the other lose its pressure.

Coming indoors, perhaps the most striking aspect of the XJ-S is its almost total silence even when most of the performance is being used. Jaguar's engineers have gone to almost incredible lengths to prevent audible vibrations from disturbing the driver or his passengers, and we can begin with the engine bay itself, which was designed from the outset to deflect sound waves away from the passenger compartment. Then to avoid holes in the bulkhead, all wiring passing through it does so via multi-pin plug and socket units; bulkhead and transmission tunnel both have close-fitting heat and noise shields, while under the carpets, moulded rubber-backed undermats follow the contours of the bulkhead and floor pressings. All remaining joints and cracks are sealed with rubber sealer, anti-vibration pads are fitted at strategic spots, while yet another layer of insulating material is sandwiched between the rear seat bulkhead and the 20-gallon tank.

Instrumentation has been extremely well thought out, and there are some novelties too, like the vertical "air-cored" secondary instruments monitoring fuel, water, oil pressure and battery condition; personally I don't think that the rev-counter and speedometer have quite the clarity and authority of Jaguar instruments of previous models, although they do adhere to the white lettering on black background style. At the top of the binnacle which encloses the instruments is a battery of warning lights which indicate a variety of possible malfunctions in the car's equipment, including major items like brakes or minor ones such as sidelights; XJ-type finger-tip stalks control lighting and wipers. Air conditioning is standard incidentally, while transmission can be ordered in automatic Borg Warner Model 12 form, or in the shape of Jaguar's four-speed manual box. Alas no five-speed box yet, though I suppose that with the torque given by the big V12 engine Jaguar thought it unnecessary.

To drive, the XJ-S is without any doubt at all a significant advance on anything that

CONTINUED ON PAGE 55

CAR AND DRIVER • JANUARY 1976

Jaguar XJ-S

The cat comes back.

TEXT BY PATRICK BEDARD
PHOTOGRAPHY BY NOEL WERRETT

• The first thing you should know about the XJ-S is that everybody can't have one. Which, if human nature continues its course, means that it automatically moves to a position near the top of everybody's desirability list.

The main reason why everybody can't have an XJ-S is that Jaguar plans to ship only about 2000 a year (75 percent of total production) to the U.S. So being on a first-name—maybe even golf-playing—basis with your dealer will give you a certain priority. *Maybe.*

Another reason why everybody can't have an XJ-S is that they cost an enormous amount of money: 19,000 U.S. dollars is the number on the window sticker. Even with a liberal trade-in allowance for your old Impala and 48-month paper from First National City, that's still a bit rich. Anyone less than an executive vice-president or maybe a land baron need not apply.

Perhaps that is as it should be. Because the XJ-S is very much a rich man's car. Those who have to worry about where their next tank of gas is coming from won't appreciate the V-12's silken surge of power. All they'll see is that the fuel consumption readily drops below 10 mpg if they indulge themselves. The practical man and the machinery lover, both of whom have visions of doing their own work, will be horrified by the engine compartment. The engine itself is a marvelous device: 326 cubic inches, aluminum block and heads, a single overhead camshaft for each bank enclosed in an exquisite ribbed cover. Except that you can't see any of it. The whole thing is hidden beneath a bewildering tangle of hoses, wires, tubes, cables, pipes, valves, relays, filters, switches and miscellaneous black boxes. It's enough to make even the most determined home tuner snap his toolbox back shut.

But the rich man won't be put off. In fact, this sort of incomprehensible machinery makes the car that much more exclusive. Certainly it's not a source of concern. Does he bother himself with how much trouble it is to dry clean a suit? Of course not. He just waves his hand and it's done for him. Same with the Jaguar. Just send it down to the dealer and let one of the serfs take care of it. That's what they're for, you know.

There is such a shortage of rich man's cars these days. How do you show the world that you've got money to burn? You sure as hell don't do it by buying a Seville: a paltry $13,000 on the window sticker, ironclad warranty, gilt-edge resale value. Nothing chancy about that. It runs like a piece from Tiffany's while you own it and there is always some guy waiting around a used-car lot to take it off your hands for a good price when you grow bored. No extravagance there.

Now we have the XJ-S, a truly expensive car by every measure. It's priced beyond any American-built car, and for the first few years, you could have a new Chevrolet for the cost of its annual depreciation. And if the mechanical and incidental failures we experienced on our test car are any indication, it will require frequent maintenance. So you have to be rich. There is no getting around that. But the beauty of this Jaguar is that it gives you something you can't get elsewhere. It goes to excess.

Even five years ago, when fuel was plentiful and there were few speed limits on the Continent, the XJ-S would have been considered the worldly gentleman's ultimate automobile. Take the well-to-do Londoner of the time; a partner in his firm, a title in the offing, a hunting preserve in Scotland, a villa in the South of France. This Jaguar would have been perfect. Up the Motorway at 140 mph in complete serenity. No laboring. Nary a murmur. Normal cars strain more at 70 mph. Brush aside slow-moving traffic with a flick of the lights. Converse politely with the passengers. Anticipate the weekend while the Jaguar does the work. Off the four-lane and onto the ancient secondary roads, the powerful disc brakes burn off speed as the corners approach. The supple independent suspension lopes gracefully

(Text continued on page 18)

Extravagance is a car
that will happily cruise at 140 mph
when the speed limit is 55

ACCELERATION standing ¼ mile, seconds

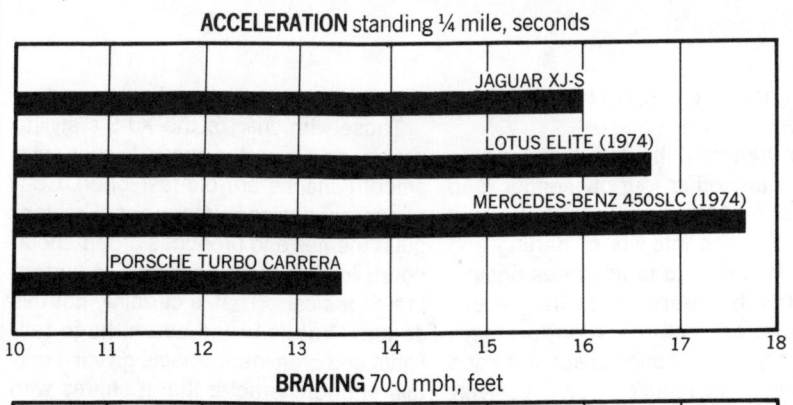

BRAKING 70-0 mph, feet

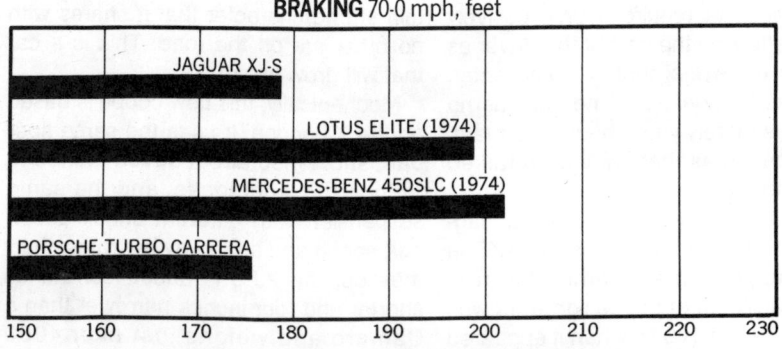

FUEL ECONOMY C/D mileage cycle, mpg

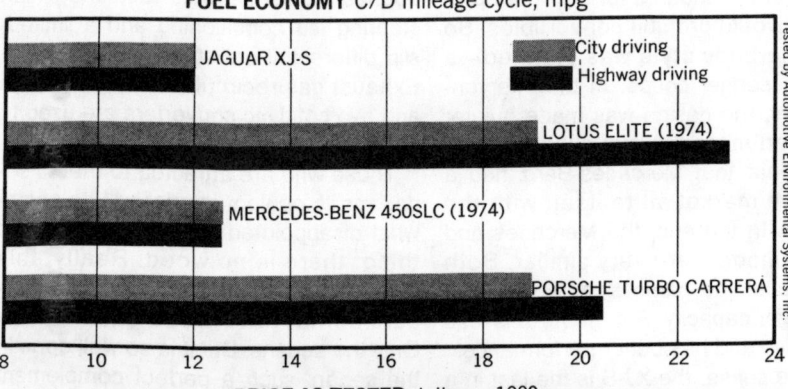

PRICE AS TESTED dollars x 1000

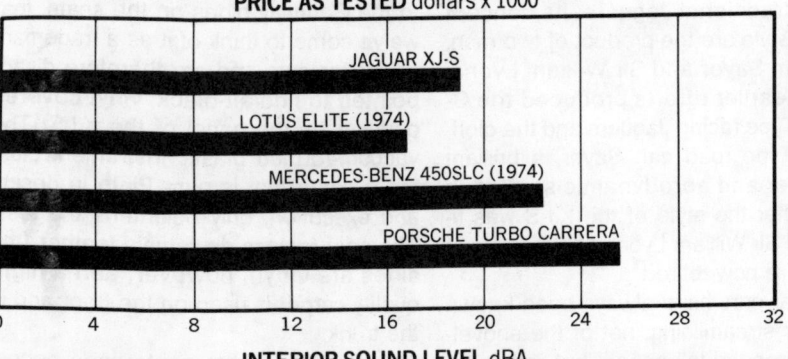

INTERIOR SOUND LEVEL dBA

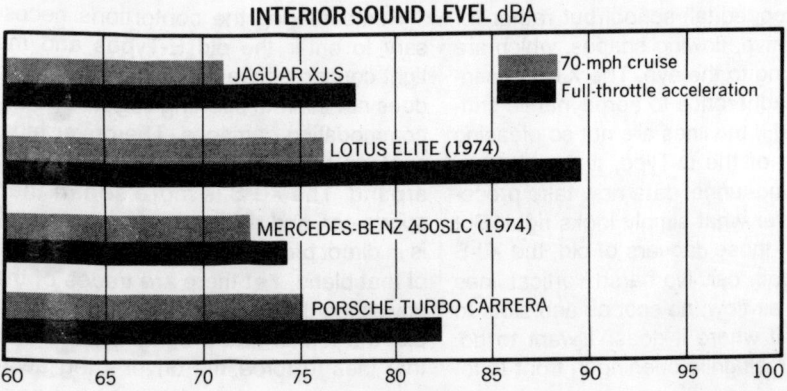

Tested by Automotive Environmental Systems, Inc.

HEINZ MAURER

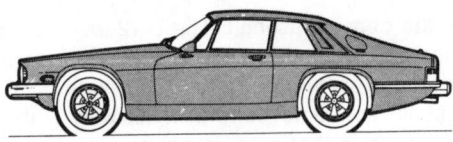

Jaguar XJ-S

Importer: British Leyland Motors, Inc.
600 Willow Tree Road
Leonia, N.J. 07605

Vehicle type: front-engine, rear-wheel-drive, 2+2 passenger coupe

Price as tested: $19,200
(Manufacturer's suggested retail price, including all options listed below, dealer preparation and delivery charges, does not include state and local taxes, license or freight charges)

Options on test car: base Jaguar XJ-S, $19,000; dealer preparation, $200

ENGINE
Type: V-12, water-cooled, aluminum block and heads, 7 main bearings
Bore x stroke .... 3.54x2.76 in, 90.0x70.0 mm
Displacement .... 326 cu in, 5343cc
Compression ratio .... 8.0 to one
Carburetion .... Lucas electronic fuel injection
Valve gear .... chain driven single overhead cam
Power (SAE net) .... 244 bhp @ 5250 rpm
Torque (SAE net) .... 269 lbs-ft @ 4500 rpm
Specific power output .... 0.75 bhp/cu in, 45.7 bhp/liter
Max. recommended engine speed .... 6500 rpm

DRIVE TRAIN
Transmission .... 3-speed, automatic
Max. torque converter .... 2.03 to one
Final drive ratio .... 3.31 to one

Gear	Ratio	Mph/1000 rpm	Max. test speed
I	2.39	9.6	62 mph (6500 rpm)
II	1.45	15.8	103 mph (6500 rpm)
III	1.00	22.9	140 mph (6100 rpm)

DIMENSIONS AND CAPACITIES
Wheelbase .... 102.0 in
Track, F/R .... 58.0/58.6 in
Length .... 191.7 in
Width .... 70.6 in
Height .... 49.6 in
Ground clearance .... 5.5 in
Curb weight .... 4020 lbs
Weight distribution, F/R .... 56.3/43.7 %
Fuel capacity .... 23.0 gal
Oil capacity .... 11.4 qts
Water capacity .... 21.5 qts

SUSPENSION
F: .... ind, unequal length control arms, coil spring, anti-sway bar
R: ...... ind, single lower control arm and fixed-length half-shaft, trailing arm, coil springs, anti-sway bar

STEERING
Type .... rack and pinion, power assist
Turns lock-to-lock 3.0
Turning circle curb-to-curb .... 39.0 ft

BRAKES
F: .... 11.2-in dia vented disc, power assist
R: .... 10.4-in dia inboard disc, power assisted

WHEELS AND TIRES
Wheel size 15x6.0-in
Wheel type .... cast alloy, 5-bolt
Tire make and size ...... Dunlop SP Sport Super, 205/70VR-15
Test inflation pressures, F/R .... 24/26 psi

PERFORMANCE
Zero to   Seconds
    30 mph 3.2
    40 mph 4.4
    50 mph 5.9
    60 mph 7.5
    70 mph 9.6
    80 mph 12.4
    90 mph 15.9
   100 mph 21.5
Standing ¼-mile .... 16.0 sec @ 90.3 mph
Top speed (observed) .... 140 mph
70-0 mph .... 179 ft (0.91 G)
Fuel economy, C/D mileage cycle ...... 12.0 mpg, urban driving
                                     12.0 mpg, highway driving

JAGUAR XJ-S

over the heaving asphalt. The V-12 levels the grades. And it happens so easily, so completely without strain, that you forget all cars do not perform in this way.

The truth is that cars do not perform this way today because they are not allowed to. The speed limit is 55 everywhere in the U.S. Some parts of Europe allow more, but *nowhere* is 140 mph permitted. So the manufacturers make their cars for the speed limits. Most people don't know the difference anyway.

Therein lies the extravagance of the XJ-S. The car has one outstanding quality—poise at extraordinary speeds—and you can't use it. You must understand how special this car really is. Not only will it accelerate to 140 mph on short notice, but once you reach that speed, you can lean over to retune the radio. And then continue your conversation with the passengers. One hundred and forty miles per hour in this car is no big deal. The scenery just rushes past a little faster. Never mind that you don't need the speed and can't use it. What could be more extravagant?

All of this does have some useful rub-off at legal speeds. A car that is silent and surefooted at 140 has to be even better at 55. The engine is quiet—uncannily so. You hear other sounds instead; the expansion of Freon in the air conditioner, a whine from the transmission, even the speedometer cable. It seems uncarlike. Exotic. But when you toe into the accelerator, the engine takes over with authority. Air moans as it enters the twin cleaner boxes. The car deliberately gathers itself up before it moves forward. It's very quick—90.3 mph in the quarter-mile—but it denies you the thrill of speed. It never *seems* to go fast, even when the speedometer needle is pointing out three-digit numbers. It's too composed to seem fast.

The suspension is amazing. It is taut but not hard, and the geometry is accurate. The car seems immune to the laws of physics. It makes you think you could never lose it—that it would compensate automatically for your blunders. If you look hard, you'll find certain areas where the handling misses perfection. If you are hurrying through a fast bend at full power then suddenly lift off the gas, you are aware of the car settling into a new attitude. It needs a steering correction to maintain its path. Not much, but enough so that you know it must have a driver. And you can bottom out the suspension;

we did it only once, in a curve at the bottom of a steep hill.

Other than that, the Jaguar is leagues ahead of any other four-passenger road car. Pitching blacktop roads that would send a Porsche into fits of darting and bring a Thunderbird to its knees against the jounce bumpers cause the Jaguar no discomfort. In turns, it understeers predictably. The steering is accurate and the power assist requires a progressively greater effort on the part of the driver as he turns the wheel farther from center. This is as it should be. The car has no tricks. It is merely your obedient servant, with capabilities that should be talked about rather than used.

The XJ-S is a new type of car for Jaguar. Even though it replaces the XKE in the lineup, it's an altogether different sort of car. Most of the design work was done in the late 1960s, when it appeared that federal standards for occupant protection would prohibit convertibles. So only one body style was planned—a 2+2-passenger coupe. In the intervening years, the design was made quieter and even more luxurious when it became clear that Mercedes-Benz had a lucrative market all to itself with the 450SLC. In the end, the Mercedes and Jaguar models are very similar. Both have about the same size, weight and passenger capacity. But the XJ-S is less expensive and has better performance.

In one sense, the XJ-S is the last in a line of traditional Jaguars. Its concept and its style are the product of two men, Malcolm Sayer and Sir William Lyons, whose earlier efforts produced the C- and D-Type racing Jaguars and the glorious E-Type road car. Sayer, a brilliant engineer and aerodynamicist, died in 1970 after the style of the XJ-S was finalized. Sir William Lyons, the founder of Jaguar, is now retired.

Jaguar cars have always been known for their streamlining, not of the shovelnose/chopped-tail school but rather for their intuitive, flowing shapes, which are so pleasing to the eye. The XJ-S continues this adherence to aerodynamic principle, and if the lines are not so pleasing as those of the D-Type, it may be because wind-tunnel data now take precedence over what simply looks right. But like all of those Jaguars of old, the XJ-S is a smooth car. No harsh vertical lines to upset air flow; no scoops and slots to channel it where it doesn't want to go. Only a small grille opening in front to let

in the right amount of cooling air.

Those who criticize the XJ-S's styling usually do so on the grounds that such smooth shapes are old-fashioned. Contemporary design centers about rectangular themes and produces broad-shouldered images such as those of current Mercedes cars. That is certainly not this Jaguar. Yet its broad beam, huge taillights and prominent wheels give it a certain strong character that it shares with no other car on the road. This is a car that will grow on you.

Mechanically, this new coupe is based on the XJ sedan. It uses the same floor pan, shortened about seven inches to reduce the wheelbase, and the same suspension with different shock absorber, spring and anti-sway bar rates. Fully dressed, the XJ-S is about four inches shorter and four inches narrower than a Camaro and weighs just over 4000 pounds. Power-assisted rack-and-pinion steering, air conditioning and a limited-slip differential are standard equipment. Exhaust gas recirculation, air injection and two catalytic converters are used to control emissions.

Those who are attracted to the XJ sedan for its opulent interior will be somewhat disappointed by the XJ-S. For one thing, there is no wood. Really, this shouldn't be a surprise. Sporting Jaguars do not have wood dashboards. Only the sedans. But it is so well done in the sedan, such a perfect complement to the Connally hides on the seats, that we've come to think of it as a trademark of Jaguar cars and are therefore disappointed to find all-black, vinyl-covered padding on the panel of the XJ-S. The vacuum-formed plastic instrument cluster, a piece that is pure Pinto in design and execution, only heightens the loss. The seat facings do remain leather (the sides are vinyl), however, and a high-quality carpet is used on the floor and in the trunk.

Remembering the contortions necessary to enter the old E-Types and the tight confines once you were inside, one does not think of sporting Jaguars as accommodating carriages. The driver must conform to the car; never the other way around. The XJ-S is more sedan than sports car, and the comfort of its interior is in direct proportion to the components of that blend. Yet there are traces of the unyielding Jaguar sports car. For example, the tunnel has a bump on the side that tries to force the driver's leg away

CONTINUED ON PAGE 69

New from Britain
JAGUAR XJ·S

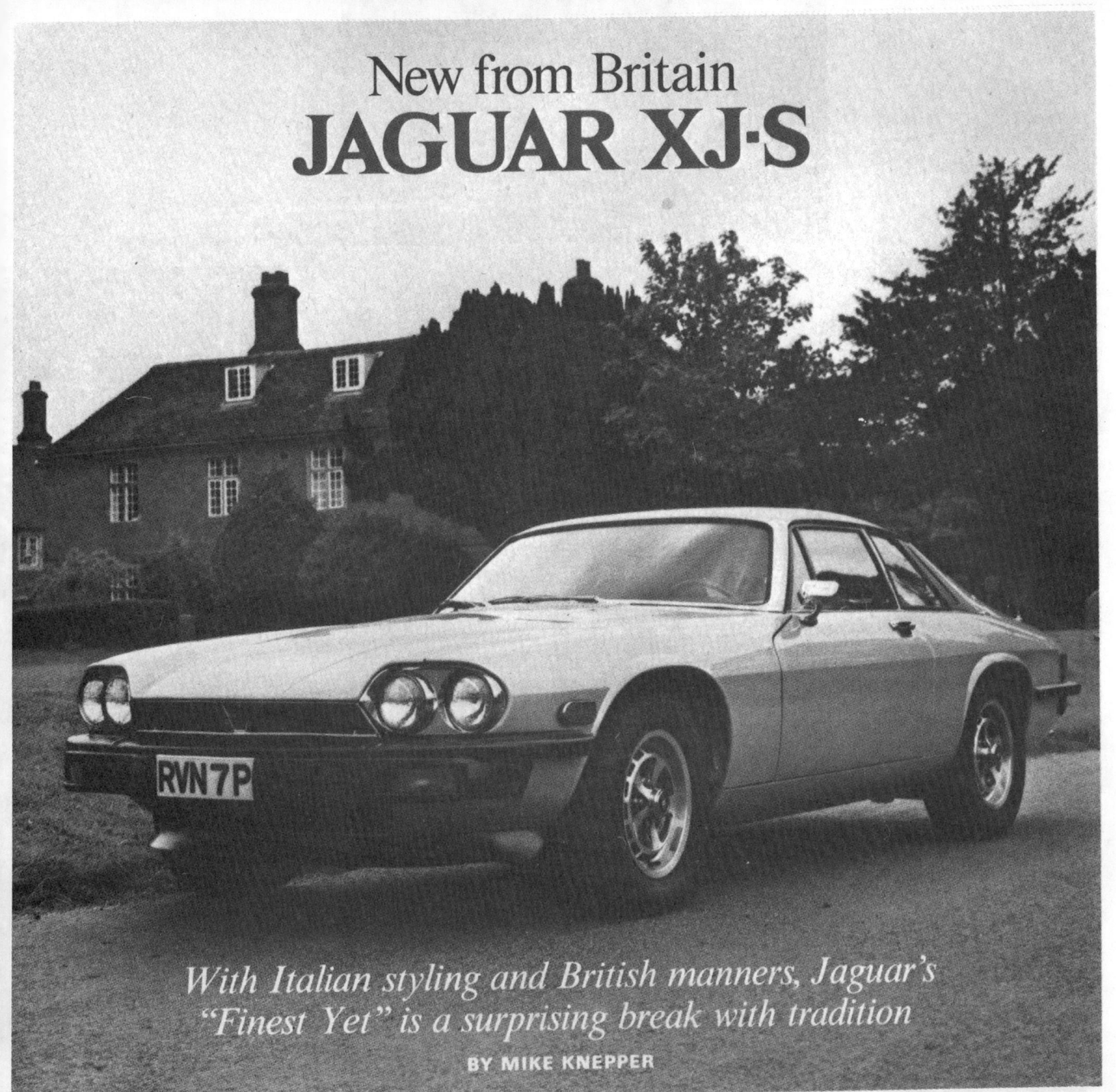

With Italian styling and British manners, Jaguar's "Finest Yet" is a surprising break with tradition

BY MIKE KNEPPER

"WE WANTED TO give you a route that would provide some real challenging bits of road so you could really give the cars a workout," Jaguar's Tony Spaulding smiled over his pint of bitter at the Boar's Head Inn in the little English town of Beeston near the Welsh border.

"Thanks, Tony," I thought to myself. The only thing that had been getting a workout was me. For the past three hours I had been trying to keep an unfamiliar luxury GT car somewhere within the boundaries defined by the imaginary center stripe of the narrow English country lanes and the hedgerows that marked the no man's land on the left. Not only was the car unfamiliar and the road narrow, the steering wheel was on the wrong side of the front passenger compartment and local custom dictated the car be kept on the left side of the road. All very confusing to a boy from Missouri.

The venerable and much-loved XK-E is gone. Last year Jaguar decided 14 years was enough for the E-Type. And, Jaguar people swear, the oft-rumored F-Type was never more than a rumor. Rather, since the late 1960s the XJ-S has been dominating the designers' and engineers' drawing boards at Coventry. The result, quoting Jaguar's own press release, is "the quietest, most comfortable and luxurious high performance car on the market." And, "the finest Jaguar yet." Heavy. And quite an advance build-up to live up to.

Although the XJ-S is a new model and although British Leyland—Jaguar's parent company—spent something like $100 million tooling up for the new car, it isn't intended to be the great profit leader that will lead the company into fiscal Nirvana. The bread and butter of the Jag line will continue to be the XJ-6/XJ-12 sedans. Last year Jaguar produced 29.000 XJ-6/12s and expects to do more this year. Projected production of the new XJ-S is only 2000. Rather, the new car will be a top-of-the-line prestige piece. It will, according to a Jaguar spokesman, "carry the banner for all British Leyland cars."

This particular banner will also carry a rather healthy five-figure price tag. According to Graham Whitehead, who is in charge of all British Leyland North American operations, the XJ-S will be priced "substantially under $20,000." When we talked he could hardly have been more specific. It was still three months from intro time and with the international monetary situation in such a state of flux, zeroing in on a more precise figure would have been foolhardy. My own best guess

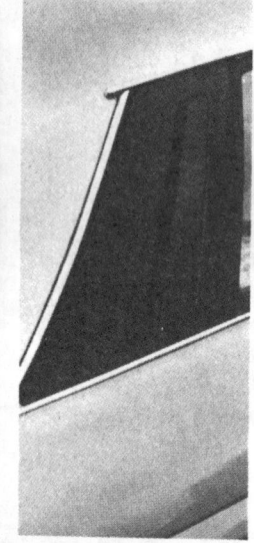

PHOTOS BY GEOFFREY GODDARD

is that the top-of-the-line British Leyland banner-carrier will hit U.S. showrooms this fall for something between $18,500 and $19,500. (The current XJ-12L is POE listed at $14,900.)

So what will one get for nearly 20 grand? It is, for better or for worse, an XJ-12 with a new body of decidedly questionable style. For the XJ-S to be a mechanical twin of the XJ-12 is not a detriment. The XJ-12 enjoys a large group of proponents who maintain the car is the best sedan in the world. (Mercedes 450SE fans will take issue, but then they have to pay considerably more for their favorite. There is something to be said for value-per-dollar.) And Jaguar has come up with some technical refinements to the XJ-12 mechanicals that seem to have made a good thing even better.

Chief of the refinements is the new fuel injection system. Jaguar has given its 5.3-liter, 326-cu-in. V-12 the Lucas D-Jetronic electronic fuel injection system. The result is a claimed increase in both horsepower and fuel economy over the carburetor version in the XJ-12. The 49-state XJ-S, Jaguar says, will produce 250 SAE net horsepower, up from the 244 claimed for the 1975 XJ-12. Fuel consumption for the 49-state XJ-S, as determined by the EPA test, is 10.5 mpg in the city cycle and 17.4 mpg in the highway cycle. The XJ-12 with carburetors did 8.2/12.0. The California XJ-S delivers 9.5/13.0; it differs only in the rate of exhaust gas recirculation as both versions have a catalyst for each bank of cylinders. In European trim the engine is rated by the factory at 285 DIN net horsepower. That, Jaguar says, is good for a 0–60 mph time of 6.8 sec, 0–100–0 mph in "slightly more than 20 sec" and a top speed in excess of 150 mph. The XJ-12L we tested in October 1974 turned in a 0–60 mph time of 9.1 sec, so it's safe to assume the U.S. version of the XJ-S, which is lighter and more powerful, should be in the 8-sec bracket. Suiting the sporting image Jaguar feels its new car portrays (and the claimed performance seems to verify), one can order up Jaguar's own 4-speed manual transmission if the Borg-Warner Model 12 3-speed automatic doesn't seem quite the thing.

So suffice it to say the XJ-S has the mechanical/technical/engineering pedigree a luxury 2+2 GT car should have, and pending verification by a complete road test, we'll also assume Jaguar's performance figures to be accurate. But what about the car as a whole, as a complete package? With the XJ-S Jaguar has ventured into unfamiliar territory and will be courting an

unfamiliar segment of the market. Although Jag marketing types talk of their new creation going head-to-head with the Mercedes 450SLC with a "style and flair the 450SLC doesn't have," it will also attract the attention of Merak, Urraco and Dino buyers. Jaguar apparently doesn't intend for their new car to keep such heady company, but if you put an exotic V-12 engine in an exotic 2+2 GT body, what can you expect? I think there's no question the front-engine XJ-S will be able to more than hold its own in performance with those three mid-engine exotic cars. Its refinement really can't be matched by the Dino, Merak or Urraco and it will cost less than any of that trio. On paper, it's a strong threat. Maybe Jaguar has a double-threat in the XJ-S. But maybe not. The 450SLC and the aforementioned Italian trio represent two ends of the supra-expensive "sporty" car market: refinement and manners vs sensuality and performance. The Jaguar is a strange combination of both which just may not appeal to either segment of the market.

Driving Impressions

I REALLY didn't know what to expect. In our May 1975 issue we ran a couple of semi-in-focus shots of the new XJ-S smuggled out of the secret dealer introduction in a Florida swamp early last spring, along with an artist's idea of what the car probably looked like in detail. I knew it was a new Jaguar, but what did "new" mean. It reportedly had the familiar V-12 engine and probably had the XJ-12's underpinnings. Yet our sneak preview revealed something very different and very un-Jaguar like. Certainly it wasn't a sports car, a new version of the E-Type. And it definitely wasn't a sedan. And when, after tea and introductions at the Jaguar factory in Coventry, a drape was pulled back to reveal a sparkling red, in-the-flesh XJ-S I still wasn't sure.

Visually, the car is Italian, and a little concentration on the styling features reveals some remarkably familiar lines. The rear window and the flying buttresses are pure Dino 246GT. Even the tail, despite the huge lights in each corner, is reminiscent of the Dino. The sloping nose that falls away from the headlight nacelles easily conjures up the Lamborghini 400GT. A little German (450SLC) influence sneaks in via the flat-back vent panels behind each rear side window. Only when looking down the car's flank from the rear does a little familiar Jaguar creep in: there is just a taste of XJ-6/XJ-12 in the fender flares. Taken as a whole the styling is unique, but at that first impression not altogether pleasing; not a design of the 1970s.

The interior is neither opulent nor spartan. Immaculate is applicable, as is refined. Definitely refined. The seats are fully adjustable and comfortable although they don't quite live up to Jaguar's stated goal of ultimate luxury. Sadly, any vestige of a traditional Jaguar cockpit is gone. Burnished chrome and black metal have replaced the familiar wood. Round tach and speedometer dials are set in square recesses and flank a rectangular recess housing four vertical gauges for fuel, water temperature, oil pressure and battery. The instrument cluster is topped off by a row of 18 warning lights which flash amber if a casual check is required or red if immediate attention is called for.

On the road the car's performance is an extension of its visual refinement with a dash of raw-bone to make things interesting. Although acceleration, with the 3-speed automatic, was not neck-snapping from a full stop, the big V-12 just winds and winds and winds, turbine-like in its smoothness. Jaguar claims a top speed of 150+ mph and I have no reason to doubt the figure. On the M1 Motorway—where the speed limit is 70 mph but the fast lane moves at more like 90 mph—I had the Jag up to an indicated 110. No fuss, no bother and the car is so quiet it is almost eerie. Jaguar engineers spent a lot of time and effort making the XJ-S quiet. They get a gold star.

I quickly decided the narrow country lanes where most of the driving was done were no place to play Ronnie Racer, so I can't give you a first-hand impression of the car's handling and braking when pushed hard. I do have the definite impression handling will be commensurate with its sporting flavor.

So what we have is an admittedly subjective editor's-eye view of a significant new car. Not significant technically. Indeed, there is little new under the new skin. But significant because it is a radical departure from things traditionally Jaguar, and significant because it falls, I think, into a heretofore unoccupied, indeed unexplored, niche between the sedan-like Mercedes 450SLC and the exotic Italian machinery.

I like the XJ-S, but my enthusiasm doesn't go much beyond that. In their attempt to create an ultra-refined, luxurious GT machine, Jaguar engineers have engineered out any personality the car might have had. And the styling—which is basically as originally conceived in 1970—is, well, not good.

The Jaguar XK-120 set a standard for sports cars that was extended by the 140 and 150. The XK-E set the motoring world on its ear when it was introduced in 1961. The XJ-6 stirred the juices when it was brought out and the maturation of the XJ-6 into the XJ-12 was a significant occurrence. I guess I expected more of the same with the XJ-S.

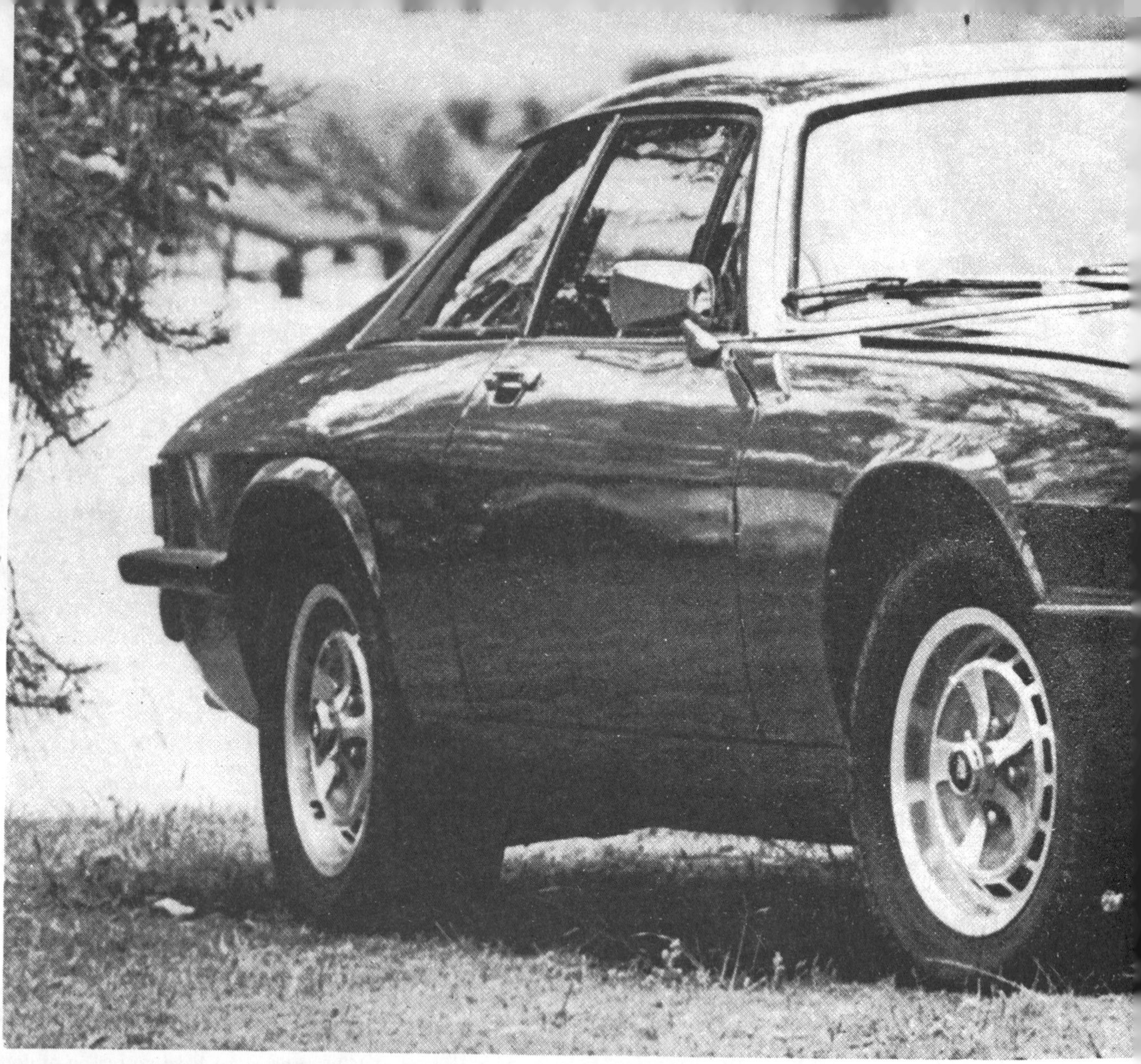

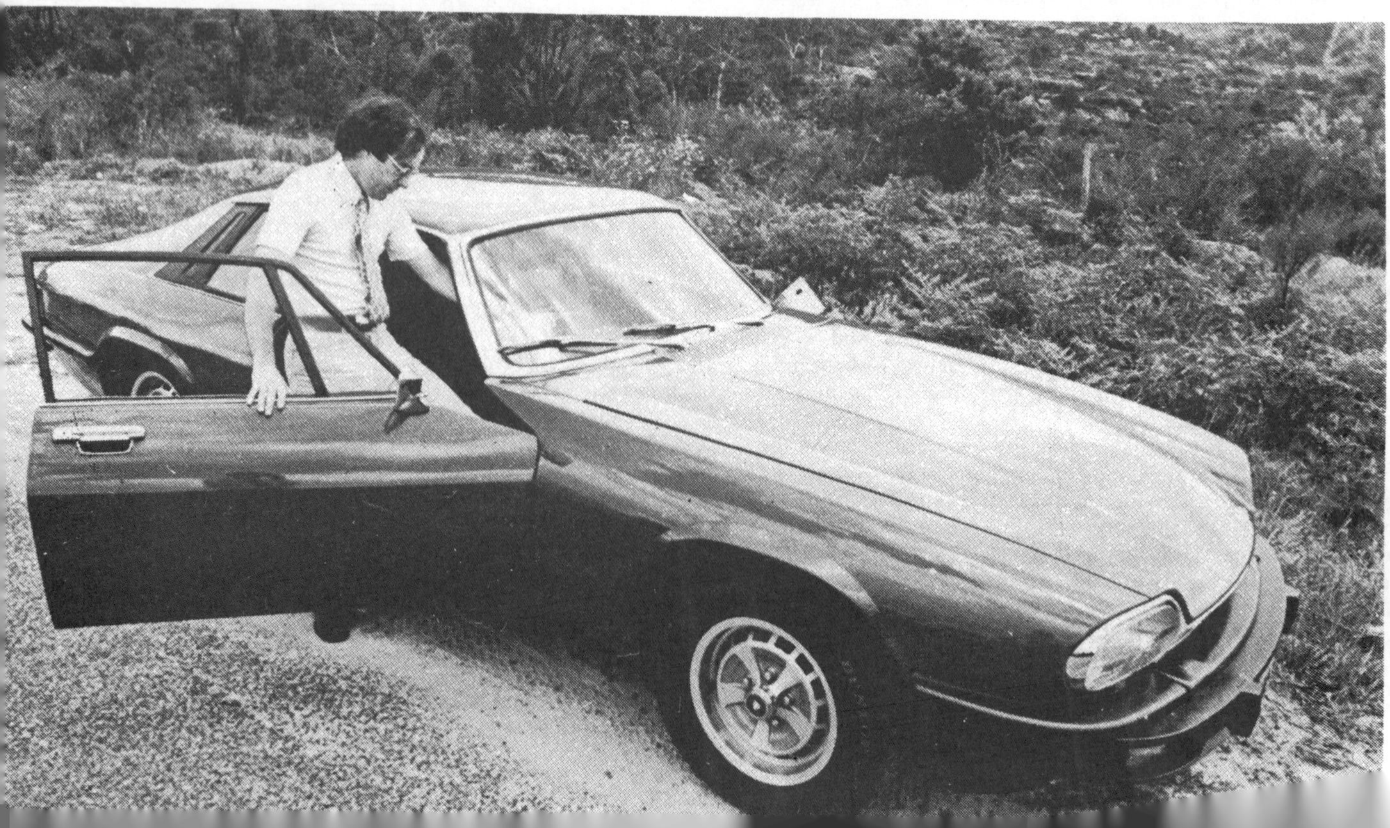

Driving Australia's First XJ-S...

QUIETLY, CAT, QUIETLY

THE JAGUAR XJ-S was released in Britain in mid-September and already Australia is seeing its first car, an experimental RHD unit for local evaluation. MODERN MOTOR editor, John Crawford, was one of only two leading motoring writers to be offered a first test drive...

MUST I BE denied the continuing pleasure of driving the XJ-S simply because of my inability to rake up *at least* $28,000 with which to place my order? In a word, YES. My accountant and my bank manager (not to mention my wife) are all reasonable people — but not when I confront them with what seems at first to be a rash scheme to sell my house, car, gold watch, stereo, movie camera and even my gold Jaguar cufflinks — just so that I may warm the Supercat's seat with my posterior.

You see the XJ-S isn't just a *new* Jaguar, it's a *better* Jaguar. If you've ever had the pleasure of a few kilometres behind the wheel of either the XJ-6 or XJ-12 sedans then you're already acquainted with the ride and handling finesse these big British luxury saloons boast. Not only are they *right* in the suspension department, but the efforts of the Jaguar engineers have resulted in almost unequalled quietness and smoothness plus a powertrain more

JAGUAR XJ-S

sophisticated than just about any other luxury saloon on the road today — yes I'll be game and say better than Rolls Royce and Mercedes.

Now, here is the XJ-S and not only have the designers made it handle better than its sedan counterparts, but it's almost indecently quieter and smoother.

As pointed out in our first report on the XJ-S (by correspondent Karl Ludvigsen, MODERN MOTOR, November '75) the coupe is similar in style of design to the sedans, but it's on a shorter wheelbase, it's got an even stiffer body and its adhesion levels are incredibly *higher!*

My brief acclimatisation drive could by no means be called a road test in the strict sense, but that short time behind the wheel told me a lot about the car we hope to test in detail after release in Australia in the middle of next year.

I was impressed by the fuel-injected engine, the precise steering, the ride, the quietness and its surprising agility. The power steering is very well-engineered. It gives beautiful feedback and seems to have just the right loading at all times. It's surprisingly direct. Before jumping into the coupe Leyland's engineering director, Kell Eriksen, warned me not be too enthusiastic first-up because the steering is so direct the big cat acts a bit like a Go-Kart if you apply too much lock in the wrong spot. I heeded the warning and got used to the steering action slowly and carefully. Of course there's absolutely nothing Go-Kart-ish about the steering at all — it's just so positive, but then everybody who knows Kell Eriksen knows he has a tendency to exaggerate a little.

You're probably surprised to hear that Leyland Australia has been able to procure an XJ-S so soon after the car's release — all initial production is headed to the US — but the secret is that the car which was sent Down Under is a pre-production model specifically ordered for Kell Ericksen's engineers to begin arranging compliance with Australia's difficult Design Rules.

As you read this the XJ-S will be stripped in a million pieces, then painstakingly nailed back together again, tested in various temperature extremes, driven over thousands of kilometres of varied terrain and finally stripped down again and studied for flaws and wear points. Although Australia will only see a limited number of the XJ-S coupes, Leyland Australia intends to see that the paying customers get a proven product for their money. The price hasn't been decided and what with inflation and all the usual cost pressures which we'll be subjected to between now and release time I believe the price will be $30,000-plus! It's unfortunate, but I guess for the people who buy this car it will be well worth it.

DRIVING IMPRESSIONS

On the road it's so incredibly quiet. You have to throw away dependence on hearing and seat-of-the-pants stuff and open your eyes to the speed you're travelling at. The coupe runs in excess of 205 km/h and I find difficulty in getting used to the idea of travelling at that speed. Even bursts to 160 km/h (100mph) from normal cruising speeds are achieved so effortlessly that it just feels like you're doing 'a bit over the limit'. It's all eye-power that keeps you honest — a sideways glance out the side window and a second quick look at the speedo is sufficient to tell you that you're way over it. At the risk of sounding like an unfulfilled hooligan can I impress that I drove the XJ-S solely within my own pre-set limits and that attaining a maximum well over the legal limits merely indicates the car's exceptional mid-range performance — a factor which makes it incredibly safe at normal touring speeds.

Indeed the XJ-S is so superbly responsive it slows decision-making down to an almost slow-motion process. Apart from that I've got to admit that being able to hunt the 240 km/h coupe along very quickly over a long and winding stretch of road is totally exhilerating. It's a pity this sort of driving is becoming so anti-social, because to a driver who *really enjoys* it, this is the car to do it in.

I mentioned agility before — the XJ-S feels much more compact than it really is. All the feedback sources relay their information so quickly and accurately, and the car responds so equally fast, you don't feel like you're pushing a GT car of 5.3 litres, weighing 1683 kg, which is 5843 mm long.

The doors open wide and the car is very low, but access is relatively easy. The British-made inertia-reel belts are hopeless, but the seat/steering wheel adjustments make it easy to tailor the driving position. The wheel is leather-bound (about time too) and the new instrument panel is quite a change for Jaguar. There is absolutely no trace of walnut veneer, it's all trimmed in aluminium and vinyl. The XJ-S is one of *those* two-plus-twos — really, it's restricted to kids under ten, legless midgets and masochists. There *is* room for four people, but use your imagination when you set up the seating positions.

The really surprising thing about the XJ-S is that it turns out to be incredibly practical. Given that it's a restricted four-seater, but a generous two-seater the boot offers quite an incredible amount of space. A quick look at the tail end of the 'S' would perhaps lead you to believe the boot was tiny, and taken up with things like spare, tools and battery. In fact it *is* taken up with spare, tools and battery, but the designers have given the owner a useful, deep, fully carpeted well underneath the relatively small boot lid.

Controls are similar to those in the XJ sedans and so is the superb air conditioning. The seats have leather facings and suited me, but then I'm not your two metre tall person. Long people tend to get a little bent when entering and exiting, and the seats look to small for big 'uns to be really comfortable.

Of course it's hard for us, without the necessary money, to visualise spending thirty-grand on a 'mean man's car', but for those who can afford it and enjoy exceptional smoothness, stability and safety in a grand touring car then Jaguar's XJ-S is just their cup of tea.

What's that you say? Did I like it? Well, I suppose I've got to say yes — it was bloody *triffic*.

THE DASHBOARD has been redesigned and is very efficient. Note the handbrake alongside the driver's seat. Entry to the front compartment is easy, but the seat backs restrict access to the rear of the cabin. The boot is truly enormous, considering the exterior shape of the coupe — the battery rides next to the spare.

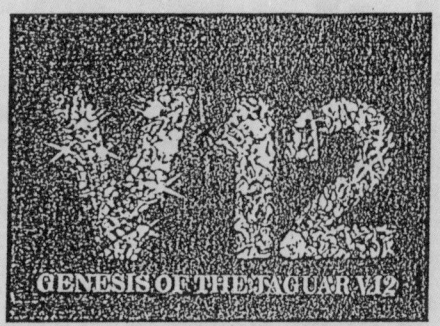

In 1971 to commemorate the introduction of their V12 power unit Jaguar published a prestigious brochure - Genesis of the Jaguar V12 - that was issued as part of the press kit to launch the E-Type Series 3.

Within the 12 pages of text are an introduction to the factory personalities of the period who were responsible for the engines evolution, a full illustrated technical discription, a question and answer segment on reasearch and design, full specifications, and a photographic section on production techniques.

The most outstanding feature of the brochure is however the four page Transart 'lift and reveal' engine cutaway especially drawn for Jaguar by Alex Sinclair and printed in full colour on clear acetate film.

The availability of this historical piece is stricly limited. This is an original works publication and not a reprint.

Please write or phone regarding price and availability to:

BROOKLANDS BOOK DISTRIBUTION, 'HOLMERISE', SEVEN HILLS ROAD, COBHAM, SURREY KT11 1ES, ENGLAND. Telephone: Cobham (09326) 5051
MOTORBOOKS INTERNATIONAL, OSCEOLA, WISCONSIN 54020, USA.
Telephone: 715 294 3345 & 800 826 6600

ROAD TEST
JAGUAR XJ-S

Tank or Supercar? A bit of both. It's large, heavy, thirsty and cramped in the back. It's also superbly engineered, sensationally quick, very refined and magnificent to drive—a combination of qualities that no other car we've driven can match at the price

Despite what it is—an outrageously large and heavy two-plus-two—the Jaguar XJ-S is a magnificent car, not just for what it does, but for the way it does it. The XJ-S combines a startling performance with exceptional smoothness and tractability and a standard of refinement that few cars can match. Others may be quicker; some handle better; a Rolls-Royce has more prestige; many have more room; quite a few are prettier. But none (other Jaguars apart) are smoother, quieter or more flexible.

In concept, the latest Jaguar must be regarded as a splendid anachronism. For a start, it was conceived when there were fewer speed limits, and petrol was relatively cheap and apparently abundant.

Much the same result could be achieved with less weight, capacity or complication, and thus better fuel consumption—the Lotus Elite shows the way. Yet we don't doubt that throughout the world there are many people who have both the taste and finance to appreciate the skill and technology that has gone into this superb machine, even if it does belong to a bygone era.

The XJ-S is a classic Gran Turismo, providing not just effortless performance but luxury, comfort, tractability and refinement as well, a species exemplified up to now by such cars as the Mercedes 450SLC, the Aston Martin V8 and the Jensen Interceptor III. At £9500 it is not cheap but you have to pay significantly more for any other car with anything approaching the same specification and qualities. Most of the mechanical components are lifted from the XJ12, and the floor pan is identical to that employed on the 109 in wheelbase saloons. The reduction in wheelbase to 102 in has been accomplished by moving the rear axle forward under the rear seat pan which has itself been shortened. The shell is shorter and lighter than the equivalent saloon, weighing 720 lb compared to 820 lb, as well as torsionally stiffer.

The heart of the XJ-S is, of course, Jaguar's superb 5.3 litre V12. In fuel injected form it produces 285 bhp (DIN) at 5500 rpm as against 250 bhp (DIN) for the carburetted version, while the torque curve is flatter, peaking at 294 lb ft (DIN) at 3500 rpm.

The engine is of all-alloy construction, with single overhead camshafts per bank, Lucas Opus electronic ignition, Lucas fuel injection—and splendid ribbed covers over the camshafts.

The manual four-speed transmission of the test car (the others we have driven were automatics) is Jaguar's own, and the suspension follows XJ practice. The twin wishbones at the front are angled to give anti-dive characteristics, springing is by coils with concentric shock absorbers. The geometry and roll-bar stiffness have been altered to suit the new car and the special Dunlop SP Sport Super 205/70VR15 tyres.

At the back the familiar Jaguar independent system is employed, with lower tubular transverse wishbones, fixed length driveshafts acting as upper links and two radius arms running forward to mounting points on the body. Twin coil spring/damper units each side are located fore and aft of the driveshaft, and for the first time an anti-roll bar is fitted. The powerful braking system is by 11 in ventilated discs at the front and smaller discs inboard at the rear. The rack and pinion steering has Adwest power assistance.

The XJ-S comes comprehensively equipped with such luxuries as electric operation for the windows and the door locks, air conditioning and leather upholstery. But it also breaks new ground in a number of other respects: the special Dunlop tyres, new instruments (by Smiths), unique Cibie biode headlamps, and new bumpers and mountings.

PERFORMANCE

★★★★ The recent spell of continuous bad weather kept our road test staff in the UK, waiting for it to break, so we didn't make the planned trip to the Continent to do a maximum speed run. We have therefore quoted the manufacturer's claimed top speed (155 mph). The ease with which the car reached an indicated 149 mph on one occasion would suggest that the maximum is certainly well above 150 mph.

Thus it would seem that the XJ-S, with the 285 bhp version of the V12 engine and weighing over 3 cwt more, is faster all out than the 30 cwt E-type V12 (146 mph) with 250 bhp.

Accelerating from rest, though, the E-type is fractionally quicker to 60 mph, taking 6.4 s compared to the XJ-S's 6.7 s—but there's not much in it. And 6.7 s to 60 mph and 16.2 s to 100 mph are electrifying times; only the Aston Martin V8, various Porsches and a handful of exotics are any quicker.

But pure performance, impressive though the figures may be, is just one aspect of the XJ-S. What makes it so special is the way the performance is achieved. From the moment the pre-engaged starter whirrs the engine into life, aided by automatic enrichment from the fuel injection system, followed by an undramatic warm-up period, the V12 provides instant and uncannily smooth power—in any gear, at any speed and revs, on up to the maximum in top.

There is no temperament, no fuss or bother, no abrupt change, no "coming on the cam," in fact none of the characteristics which are often the hallmarks of cars with equal or better performance but with more highly tuned and stressed engines. The engine is so flexible that it is possible to pull away cleanly and vigorously from a walking pace in top.

This potency and flexibility is allied to a smoothness and quietness that, to use the words of one of our testers, is "miraculous." We would expect a V12 to be smooth, but that of the XJ-S is exceptional, while it is so well muffled that passengers are often unaware of the high speed at which they are travelling. As revs rise there is just a slight change in engine note, but it only becomes really noticeable above 5500 rpm—and we can't see many drivers using such revs very often. With such massive torque low down you don't need to extend the engine, 70 mph in top is a very restful 2800 rpm, at which point the engine just emits a muted hum. Throttle pedal travel, though rather long, is pleasantly progressive.

ECONOMY

★★ As expected the XJ-S is very thirsty, as our overall consumption of 12.8 mpg confirms, though it's no worse than that of other cars of similar performance and weight. Note, however, that the lighter Porsche 911 lux Targa, with a smaller, more efficient engine returned 19.8 mpg overall, and the (slower) Lotus 21.7 mpg.

We have recently taken delivery of a new fuel-flow measuring device that allows us to take figures on recirculating high-pressure and return-flow fuel injection systems. The equipment could not be calibrated in time, however, to be used on the XJ-S so we have broken with usual procedure and quoted manufacturer's figures for steady-speed consumptions on which our touring figure of 14.5 mpg is based.

This gives a range of about 290 miles from the 20-gallon tank, which in practice seems to be possible, although the low-fuel warning light that comes on when there's 4 or 5 gallons left gives a peace-of-mind limit of only 250 miles.

TRANSMISSION

★★★ One of the less endearing aspects of the XJ-S is the transmission. The clutch is a little heavy and unprogressive, and initially there was some stickiness, though this later seemed to clear up.

The change is stiff and notchy if you rush it so it pays to ease the lever around rather than snatch at it. However you treat it, the change is far removed from the flickswitch shifts we have come to expect of most ordinary family saloons.

With so much torque available, the XJ-S romped up the 1-in-3 test hills with no problem. The gear ratios, giving 50, 85 and 115 mph approximately in each gear at 6500 rpm, are well spaced—although with the torque available that would only be noticeable under full-bore acceleration anyway. A distinctly audible whine in first and second gear detracts from the overall refinement of the car.

HANDLING

★★★★ One of our constant criticisms of Jaguars is that their power steering has, up to now, been too light. That fitted to the XJ-S has more resistance but no more feel than that of other models—but the extra weight is a move in the right direction. It is still not up to the standards set by Mercedes or Aston Martin. Otherwise we have no complaints, for it is direct and responsive with no slop or lost motion, and parking is easy.

With fat, 70 series tyres and a sophisticated and developed suspension the roadholding is all you would expect—the grip in the wet from the Dunlop tyres is remarkable, though it is possible to break traction and spin the wheels, even after a leisurely getaway. The supple suspension absorbs irregularities extremely well so that the XJ-S is not put off line by bumps in mid-corner.

The handling is exemplary and the car always feels well balanced and safe. You have to be cornering very hard before the normal mild understeer changes to predictable and controllable oversteer under power. Lifting off or braking in a corner simply tightens the line without any drama. When the tail does break away—and that requires a fair amount of throttle—the XJ-S can be untidy in that, possibly due to the limited slip differential, poor steering feel or a slightly too stiff anti-roll bar, the rear end tends to lurch around.

The XJ-S is fun to drive fast along country roads, for its taut, balanced handling, excellent traction, lack of roll and precise (if feel-less) steering allow you to hustle the car along without drama. It is possible that the Mercedes and some of the more expensive exoticars can and do behave even better, but this is only noticeable at limits which few owners are ever going to reach.

The power steering naturally takes all the effort out of parking, and the turning circle is a commendable 34 ft or so, comparatively small for what is a long and bulky car. The front spoiler and general shape not only prevent the nose lifting at speed but also seem to keep the XJ-S tracking straight in high cross-winds.

BRAKES

★★★★ The XJ-S has the brakes to cope with its weight and performance. There are 11.2 in ventilated discs at the front, 10.38 in inboard discs at the rear, twin hydraulic circuits, one for each end, a pressure differential warning actuator which shuts off a faulty circuit in case of failure, and naturally a servo.

In use the brakes live up to their specification. They are powerful and progressive, but possibly the most impressive aspect of their performance was that, in conjunction with the excellent grip from the tyres, they pulled this heavy car to a standstill with a deceleration of over 1g—in the wet. Some cars cannot achieve that in the dry. Nor did they show any signs of fade during our test, which involved 20 stops, one minute apart, from a speed of 100 mph. In addition there is very little dive, due to the front suspension geometry.

In contrast the handbrake could only manage a disappointing 0.22g—below the legal limit.

ACCOMMODATION

★★ There is plenty of space for those in the front, but the back is cramped. Not only is leg and footroom severely restricted with the front seats anywhere near their rearmost position (a situation rarely likely to occur) but even for short people headroom is limited and width is just about adequate due to intrusive armrests—but shoulder width is reasonable. The bench seat itself is well padded and reasonably comfortable but you sit noticeably higher than those in the front, the cushion is short, and you have to adopt an upright position.

This lack of rear seat room is not unique to the Jaguar, of course: the Jensen Interceptor and to a certain extent the Aston V8 (plus many others) suffer in the same way. The Lotus Elite again shows how it can and should be done.

With a low roof-line and minimal headroom getting into or out of the back is not very easy.

The boot is deep and square and looks quite big; it swallowed 10.9 cu ft of luggage. But the looks

Plenty of room and shapely seats in the front but a cramped, rear seat.
Left: attractive, ergonomic facia. Both major and minor controls are well sited.
Right: grouped around the gearlever are the electric window switches, central door lock, cigar lighter and ashtrays
Below: modern, clear instrumentation topped by a battery of warning lights

More than enough for a long weekend. The boot held 10.9 cu ft of our best suitcases

are rather deceptive for much of the space is in the form of pockets in the wings and a ledge beside the vertically stowed spare wheel and above the battery, so that there is quite a lot of room for squashy bags but not all that much for ordinary suitcases.

Inside there is an averagely-sized compartment, a usefully large bin beneath the central armrest and little bins on the doors to take small items—cigarettes and pens, for example, plus pockets each side of the rear seat.

RIDE COMFORT

★★
★★
There is some slight trembling at most speeds which can turn into small-bump harshness at low speeds, but these are the sort of characteristics that indicate to us firm damping and a taut suspension. There is also some float at speed over undulations—waves on a motorway for example. Otherwise, the ride is up to the usual very high XJ standards.

The XJ-S will cope with quite rough surfaces and severe irregularities with aplomb and no trace of sogginess. Add to this the lack of bump-thump and tyre noises and the ride becomes remarkable: not as soft and resilient as the saloon's perhaps, but still very good. The lack of roll, dive, squat, pitch or bounce and the comfortably padded and supportive seats increase even more a passenger's peace of mind —in the front at least.

AT THE WHEEL

★★
★★
The XJ-S is designed to provide maximum comfort for two in the front, and in this on the whole it succeeds. The shapely seats are comfortable and provide the right sort of support in the right places, while the extensive range of fore-and-aft adjustment, the reclining backrest and the adjustable (for reach) steering column meant that most drivers could find a good position (although one or two would have liked a cushion tilt control or some form of seat height control). In the XJ-S you can cover long distances or drive for long periods without feeling stiff and tired at the end of a journey.

The major controls are well sited, the pedals allowing heel-and-toeing and the gearlever being within easy reach—but reverse (across to the left, down and back) can be a little awkward if you sit well back. The handbrake is of the type that flops to the floor when applied so it doesn't get in the way when stepping aboard.

Most of the minor control functions are operated by two stalks on the steering column, one for the washers and two-speed-plus-flick-wipers, the other for the indicators and dip/flash. The main rotary light switch is buried down on the facia beside the steering column on the left, matching the ignition lock on the right.

The air conditioning controls are located each side of the radio in the centre of the facia, and above them is a row of four push-button switches, for the hazard warning lights, the heated rear window, a map light on the passenger's side, and all the interior lights. The three toggle switches on the centre console operate the electric windows and the central door-locking system— no bad thing since both the XJ-S and the centre console are wide. The four push-button and the rotary light and ignition switches are illuminated at night, the latter two by fibre optics.

The inertia seat belts fitted for the front seats are awkward to reach but comfortable to wear; the test car was also fitted with (optional) rear seat belts.

INSTRUMENTS

★★
★★
Some of our testers thought that Jaguar had spoilt the styling of the instruments by surrounding them with a thin, painted silver line which made them look cheap. They are otherwise attractive to look at, well lit via a rheostat, and reasonably comprehensive.

Flanking the four smaller instruments in the middle are the large speedometer plus odometer and trip meter on the left (reading to 160 mph) and the matching tachometer, red-lined at 6500 rpm, on the right. With so many figures on them the calibrations look a little fussy but they are clear and readable.

The four smaller instruments are clever and unusual in that they are arranged so that the needles move vertically, and with everything working correctly (and the tank half full) they form a single yellow line across the panel, so that only the briefest glance is needed to check that all is well. This is in accordance with aircraft practice, and the instruments themselves are completely new air-cored devices developed by Smiths Intruments.

Claims made for them are that they are almost instantaneous in action and more robust than previous types. Reading from left to right the individual instruments are the water temperature gauge, the oil pressure gauge, the fuel gauge and the voltmeter. Above them is a battery of no fewer than 18 warning lights.

HEATING VENTILATION

★★★ The XJ-S comes as standard with a fully automatic air conditioning unit made by Delanair. This is controlled by two rotary switches, one to select temperature (from 65° to 85°), the other to control the function from "off," through "lo," "auto," "hi" to "def" (rost).

In practice the system does not seem to give all that it promises. In the cold weather of the past few weeks we had no real chance properly to assess the cooling mode (although a quick test gave a really cold blast from the centre vent) but on the "auto" setting (or the "hi" or "lo" setting for that matter) the flow from the vents at each end of the facia is poor so that it is difficult to get warm feet and a cool face, which leads to stuffiness, in spite of a temperature differential between upper and lower circuits.

The controls themselves are simple to understand and operate, while the "defrost" mode is quick and powerful—fortunately, since the inside of the car mists up instantly if it is started up when hot but after sitting for some time.

NOISE

★★★★ Jaguar are masters at suppressing noise, vibration and harshness and the XJ-S is a brilliant example of their technological achievement. Even so, individual noises can be identified from the general light murmer: there is the gearbox whine in first and second, road noise is sometimes noticeable at speed on some surfaces, the power steering hisses when parking or manoeuvring, and the engine note rises at high revs, of course. But many other manufacturers would be glad to get their noise levels down even to these standards, and they must envy the lack of wind noise (no louder, it seems, at 120 mph than it is at 70 mph), the well-muffled bump-thump and the suppression of tyre roar at low speeds in town. There was no noticeable drumming from the wheelarches from spray or "singing" from the tyres.

VISIBILITY

We expected the rear three-quarter pillars and curved Dino-style fins to create blind spots, but they are

MOTOR ROAD TEST No. 8/76 • Jaguar XJ-S

PERFORMANCE

CONDITIONS
Weather	Dry: Wind 10-20 mph
Temperature	40°-45°F
Barometer	29.7 in Hg
Surface	Tarmac

MAXIMUM SPEEDS
	mph	kph
Banked Circuit	155 (see text)	250
Terminal speeds		
at ¼ mile	96	154
at kilometer	122	196
Speed in gears (at 6500 rpm):		
1st	50	80
2nd	84	135
3rd	116	187

ACCELERATION FROM REST
mph	sec	kph	sec
0-30	2.8	0-40	2.3
0-40	3.8	0-60	3.5
0-50	5.1	0-80	5.0
0-60	6.7	0-100	7.0
0-70	8.4	0-120	9.4
0-80	10.5	0-140	12.4
0-90	13.4	0-160	16.1
0-100	16.2	0-180	21.5
0-110	20.2		
0-120	25.8		
Stand'g ¼	15.0	Stand'g km	27.2

ACCELERATION IN TOP
mph	sec	kph	sec
10-30	7.5	20-40	4.4
20-40	6.8	40-60	4.5
30-50	6.6	60-80	4.0
40-60	6.8	80-100	4.3
50-70	6.9	100-120	4.3
60-80	7.1	120-140	4.4
70-90	7.2	140-160	5.0
80-100	8.0	160-180	6.0
90-110	8.6		
100-120	10.3		

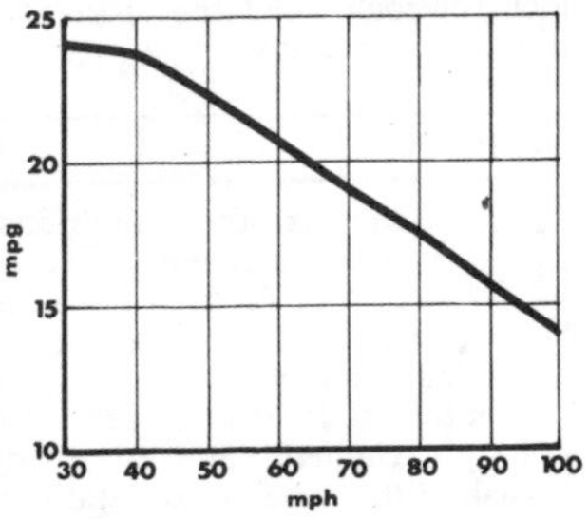

FUEL CONSUMPTION
Touring	14.4 mpg
	19.6 litres/100 km
Overall	12.8 mpg
	22.1 litres/100 km
Fuel grade	97 octane
	4 star rating
Tank capacity	20.0 galls
	90.9 litres
Max range	288 miles
	463 km
Test distance	1502 miles
	2417 km

*Consumption midway between 30 mph and maximum less 5 per cent for acceleration.

BRAKES
Pedal pressure deceleration and stopping distance from 30 mph (48 kph)

lb	kg	g	ft	m
25	11	0.32	94	92
50	23	0.64	47	14
75	34	0.96	31	9
85	39	1.00+	30	9
Handbrake	0.22		136	41

FADE
20½ stops at 1 min intervals from speed midway between 40 mph (64 kph) and maximum (97.5 mph, 157 kph)

	lb	kg
Pedal force at start	45	20
Pedal force at 10th stop	40	18
Pedal force at 20th stop	44	20

STEERING
Turning circle between kerbs:
	ft	m
left	34.0	10.4
right	34.4	10.5
Lock to lock	3.0 turns	
50ft diam circle	1.1 turns	

CLUTCH
	in	in
Free pedal movement	0.5	1.3
Additional to disengage	3.5	8.9
Maximum pedal load	38 lb	

SPEEDOMETER (mph)
Speedo	30 40 50 60 70 80 90 100 110 120
True mph	30 40 50 59 68 77 86 95 105 115

Distance recorder: accurate

WEIGHT
	cwt	kg
Unladen weight*	33.4	1696.8
Weight as tested	37.1	1884.8

*with fuel for approx 50 miles

Performance tests carried out by Motor's staff at the Motor Industry Research Association proving ground, Lindley.

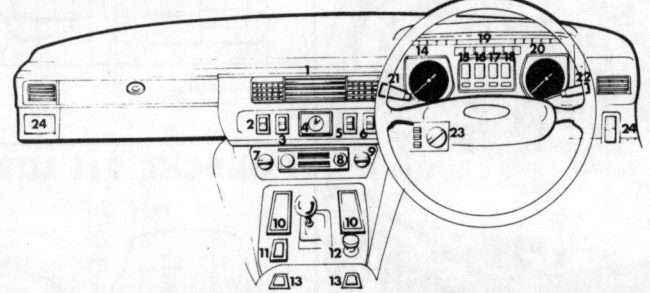

GENERAL SPECIFICATIONS

ENGINE
Cylinders	V12
Capacity	5343 cc (326 cu in)
Bore/stroke	90 x 70 mm (3.54 x 2.76 in)
Cooling	Water
Block	LM 25 aluminium alloy
Head	LM 25 WP aluminium alloy
Valves	Sohc per bank
Valve timing	
inlet opens	17° btdc
inlet closes	59° abdc
ex opens	59° bbdc
ex closes	17° atdc
Compression	9.0:1
Carburetter	Lucas fuel injection
Bearings	7 main
Fuel pump	Electrical
Max power	285 bhp (DIN) at 5500 rpm
Max torque	294 lb ft (DIN) at 3500 rpm

TRANSMISSION
Type	4-speed, all synchromesh
Clutch	Sdp
Internal ratios and mph/1000 rpm	
Top	1.000:1/24.7
3rd	1.389:1/17.8
2nd	1.905:1/13.0
1st	3.238:1/7.6
Rev	3.428:1/
Final drive	3.07:1

BODY/CHASSIS
Construction	Unitary, all steel
Protection	Bare shell sprayed with 0.020 in bitumen. Electro-dip. Entire underside sprayed with water-repellant wax after manufacture.

SUSPENSION
Front	Independent by wishbones, coil springs, telescopic dampers, anti-roll bar
Rear	Independent by lower wishbones, driveshafts as upper links, radius arms, twin coil spring/damper units per side

STEERING
Type	Rack and pinion
Assistance	Yes
Toe in	1/16"–1"
Camber	+¼ ± ⅓°
Castor	3½ ± ⅓°
King pin	1½ ± ⅓°
Rear toe in	Parallel ± 1/32"
Adjustment	Yes

BRAKES
Type	Discs all round
Servo	Yes
Circuit	Split front/rear
Rear valve	Yes
Adjustment	Yes

WHEELS
Type	Light alloy, 6JK
Tyres	Dunlop SP Sport Super, 205 70 VR 15
Pressures	Normal use, 26F, 24R. High speed (120 mph +) 32F, 30R

ELECTRICAL
Battery	12V, 68 Ah
Polarity	Negative
Generator	Alternator
Fuses	12
Headlights	2 x Cibie, 2 quartz halogen filaments each: 55/55W

IN SERVICE

GUARANTEE
Duration 1 year, unlimited mileage (2 year, unlimited mileage extra cost option)

MAINTENANCE
Schedule	every 6000 miles
Free service	at 1000 miles
Labour for year	17.5 hours

DO-IT-YOURSELF
Sump	20 pints, SAE 20W40, 20W50
Gearbox	4.5 pints, SAE EP 90
Rear axle	2.75 pints, SAE EP 90
Coolant	37 pints
Chassis lubrication	6000 miles
Contact breaker gap	Zero-electronic contactless ignition
Spark plug type	Champion N10Y
Spark plug gap	0.625 mm
Tappets (hot/cold) inlet	0.3-0.35 mm
exhaust	0.3-0.35 mm

1 air vents
2 hazard
3 heated rear window
4 clock
5 map light
6 interior lights
7 temperature control
8 radio
9 distribution control
10 ashtrays
11 central lock
12 lighter
13 electric windows
14 speedometer
15 Water temp
16 oil pressure
17 fuel
18 volts
19 warning lights
20 tachometer
21 washer/wipers
22 dip/flash/indicators
23 lights
24 map lights

Make: Jaguar **Model:** XJ-S

Makers: Leyland Cars, Longbridge, Birmingham B31 2TB

Price: £8143.00 plus £678.58 car tax plus £705.73 VAT equals £9527.31

STAR GRADE KEY

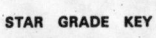 excellent 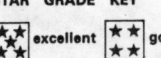 good average poor bad

far enough back not to be a problem. Yet the windscreen pillars get in the way (there are the slender quarter window pillars and door pillars as well to clutter the side view) and the windscreen header rail is low and far back, giving a beetle-browed impression and a slight feeling of claustrophobia. And the side windows get very dirty in foul weather.

The edges of the headlights and the bootlid are visible from the driver's seat, in spite of a lowish seating position and a high waistline, but when parking you have to remember that there are a few more inches of bumper front and rear, and that the sides of the bodywork curve outwards.

The two-speed wipers clear an effectively large area of the windscreen, and unlike those of other XJs do not perform a little dance on the screen when switched on or off. Instead they pause before setting to work, and again when parking—irritating when travelling quickly. We like the flick-wipe facility, but feel that an intermittent wipe, especially one with time control, would be far more useful. Cheaper cars have it—why not the XJ-S?

We expected great things from the headlights but those on the test car seemed to be poorly set, and are nothing like as powerful as those (optional QI ones) of our long-term XJ 3.4.

EQUIPMENT

★★ ★★ As befits the price and image, the XJ-S is well equipped with such still comparatively rare items as air conditioning, leather upholstery, electric window operation, central locking, a multitude of interior lights (one for everyone), a radio, alloy wheels, low-profile tyres, adjustable steering, fuel injection, electronic ignition, an impact switch to isolate the fuel pump in the event of a collision, an adaptor to turn the cigar lighter socket into a trickle charge point for the battery or an inspection light point, plus power steering, a brake servo, carpeting and so on. Omissions include a petrol filler lock (optional) and the aforementioned wiper delay.

FINISH

★★ ★★ Gone is the traditional wooden instrument panel. The new facia, trimmed in black and matt silver, is attractive and functional, while the rest of the interior is subtly and discreetly trimmed and finished: stylish without being ostentatious. The seats look particularly sumptuous, and there is a lovely smell of real leather, while the reels for the inertia belts are hidden within the side trim.

The finish in general is to a high standard: all carpets and other items fit well, the seams on the seats and elsewhere are straight, the doors shut with a satisfying clonk and shut lines are even.

IN SERVICE

Jaguar now offer the Leyland Supercover warranty, so the XJ-S is covered for the first year with unlimited mileage and, at extra cost, for a second year as well, once again with unlimited mileage: owners should have at least two years of relatively worry-free motoring. Naturally there is a comprehensive service network throughout the UK so routine servicing should be no problem.

We can offer only one word of advice for would-be do-it-yourself enthusiasts: don't. The dipstick, radiator cap and one or two other items are reasonably accessible, but (for example) the distributor is under the fuel injection pipes which would thus have to be removed to get at it (fortunately with a contactless electronic ignition system this shouldn't be necessary), and the engine bay is filled with engine, plumbing, pipework and wires. The battery is a little inaccessible, up against the bulkhead next to the spare wheel in the boot.

Under this there is an engine! From 5.3 litres, Jaguar produce 280 fabulously smooth horsepower

THE RIVALS

JAGUAR XJ-S £9,527
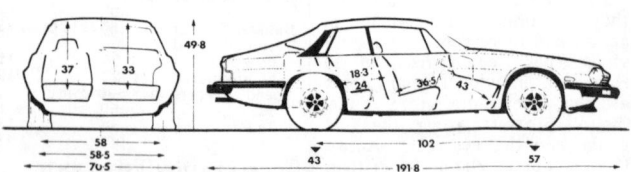

ASTON MARTIN V8 £13,631
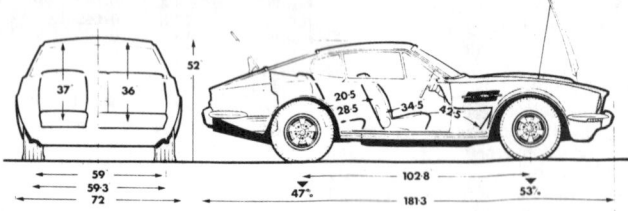

JENSEN INTERCEPTOR III £10,764
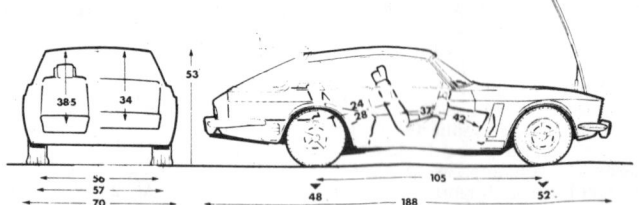

LOTUS ELITE 503 £8,153
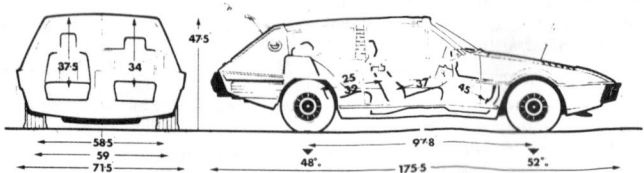

MERCEDES-BENZ 350SL £8,333
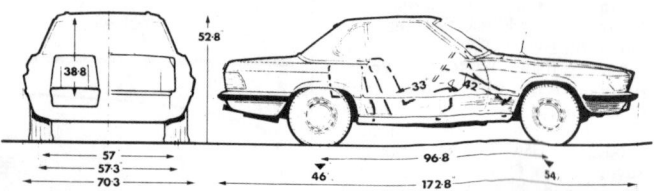

PORSCHE 911 LUX TARGA £8,898
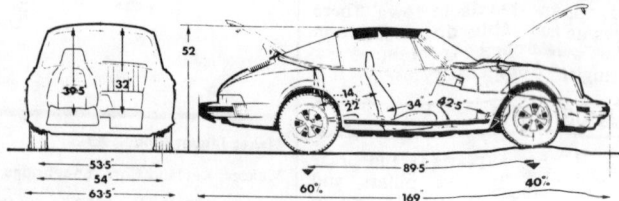

We have not tested the Jaguar's obvious rival, the Mercedes-Benz 450 SLC Coupe. Other possible competitors include the Ferrari 365 GT4 2+2 (£15,720) or the four-seater Lamborghinis

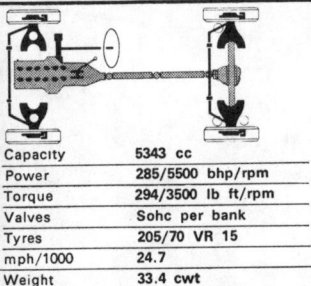

Capacity	5343 cc
Power	285/5500 bhp/rpm
Torque	294/3500 lb ft/rpm
Valves	Sohc per bank
Tyres	205/70 VR 15
mph/1000	24.7
Weight	33.4 cwt

In true Jaguar tradition, the XJ-S combines exceptional performance and refinement at a very competitive price. The styling may not be to everyone's taste and the rear seat accommodation is cramped for so large a car. But if you can afford the fuel bills, the XJ-S is without doubt one of the world's most desirable vehicles. Although faster than the E-type, it's no replacement for Jaguar's classic two-seater.

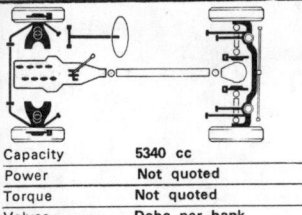

Capacity	5340 cc
Power	Not quoted
Torque	Not quoted
Valves	Dohc per bank
Tyres	215/70 VR 15
mph/1000	26.9
Weight	34.7

An early record of poor reliability and, more recently, of Aston Martin's financial troubles have tended to overshadow the many virtues of the exciting V8. Even faster than the Jaguar it has superb handling, excellent power steering, impressive brakes and more room for rear seat passengers. Lacks the refinement of the XJ-S but is perhaps even more exhilarating to drive.

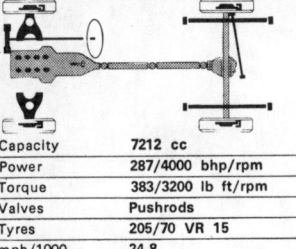

Capacity	7212 cc
Power	287/4000 bhp/rpm
Torque	383/3200 lb ft/rpm
Valves	Pushrods
Tyres	205/70 VR 15
mph/1000	24.8
Weight	34.8

The striking shape doesn't seem to date but the Jensen is beginning to show its age in other departments. The ride and handling of this live-axled Anglo-American-Italian car are bettered by most competitors and it is very heavy on petrol. Though still very impressive the performance is bettered by many smaller capacity machines. The smoothness of the automatic gearbox is one of the best features.

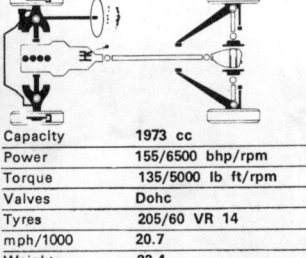

Capacity	1973 cc
Power	155/6500 bhp/rpm
Torque	135/5000 lb ft/rpm
Valves	Dohc
Tyres	205/60 VR 14
mph/1000	20.7
Weight	22.4

A trend setter for exotic cars of the future. Exceptional handling and roadholding and good performance and refinement despite an engine of only 2 litres. Economical for such a fast car with four proper seats and a useful tailgate. Function with fashion. Very comfortable and lavishly equipped. Enormously improved since the early rather noisy cars that lacked low-speed torque.

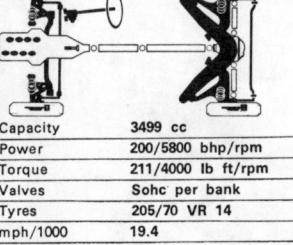

Capacity	3499 cc
Power	200/5800 bhp/rpm
Torque	211/4000 lb ft/rpm
Valves	Sohc per bank
Tyres	205/70 VR 14
mph/1000	19.4
Weight	30.4

A luxurious two-seater, the 350SL offers all the usual Mercedes attributes of sound, safe engineering, predictable handling and adequate rather than exciting performance. No attempt to provide occasional rear seats so the boot is exceptionally large. The heavy hardtop can be removed and there's also a tight-fitting hood. Road noise mars this otherwise refined and beautifully made car.

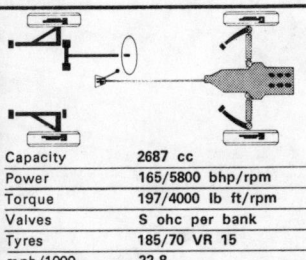

Capacity	2687 cc
Power	165/5800 bhp/rpm
Torque	197/4000 lb ft/rpm
Valves	S ohc per bank
Tyres	185/70 VR 15
mph/1000	22.8
Weight	21.3

Like Lotus, Porsche manage on under 3 litres what others barely achieve with five or more. Their exceptionally smooth flat six engine gives exhilarating performance and runs on two-star fuel. Few cars "come alive" like a Porsche to those who can exploit its handling. Poor face-level ventilation but otherwise well equipped and superbly made and finished. A great driver's car.

PERFORMANCE

	Jaguar	Aston V8	Jensen**	Lotus	Mercedes**	Porsche
Max speed, mph	155†	154.8	129	120.1	127.8	140†
Max in 4th	—	136	—	116	—	113
3rd	116	112	—	85	87	83
2nd	84	77	89	58	53	57
1st	50	47	53	36	28	33
0-60 mph, sec	6.7	5.7	7.7	7.8	8.1	6.5
30-50 mph in 4th, sec	6.6	5.5	2.7	8.0	3.8	6.2
50-70 mph in top, sec	6.9	6.7	4.1	13.5	5.1	9.6
Weight, cwt	33.4	34.7	34.8	22.4	30.4	21.3
Turning circle, ft*	34.2	38.6	35.5	34.7	30.4	32
50ft circle, turns	1.1	1.15	1.2	1.15	0.9	1.0
Boot capacity, cu ft	8.4	8.9	8.5	6.6	7.6	9.8

*mean of left and right.   **Automatic.   †claimed max

COSTS

	Jaguar	Aston	Jensen	Lotus	Mercedes	Porsche
Price, inc. VAT/tax, £	9527	13631	10764	8153	8333	8898
Insurance group	7	7†	7	7	7	7
Overall mpg	12.8	13.2	10.0	21.7	15.4	19.8
Touring mpg	14.4	14.7	15.6	26.4	19.5	††
Fuel grade (stars)	4	5	2	4	4	2
Tank capacity, galls	20	21.0	20	14.75	19.8	17.6
Service intervals, miles	6000	2500	5000	5000	5000	12,000
Set brake pads (front), £*	14.40	21.67	22.46	8.92	12.16	23.22
Oil filter, £*	3.00	1.75	1.10	3.36	2.32	3.92
Starter Motor, £*	82.25	79.31	96.77	72.94	116.14	76.00
Windscreen, £*	26.75	102.60	84.74	74.13	108.99	170.45

*inc VAT   *Laminated   ††to be decided   †††fuel injected

EQUIPMENT

	Jaguar	Aston	Jensen	Lotus	Mercedes	Porsche
Adjustable steering				●	●	
Carpets	●	●	●	●	●	●
Central locking	●		●		●	
Cigar lighter	●	●	●	●	●	●
Clock	●	●	●	●	●	●
Cloth trim				●		
Dipping mirror	●	●	●	●	●	●
Dual circuit brakes	●	●	●	●	●	●
Electric windows	●	●	●		●	
Fresh air vents	●	●	●	●	●	●
Hazard flashers	●	●	●	●	●	●
Headlamp washers					●	
Head restraints	●		●		●	●
Heated rear window	●	●	●	●	●	●
Laminated screen	●	●	●		●	
Locker	●	●	●	●	●	●
Outside mirror	●	●	●	●	●	●
Petrol filler lock				●		
Radio	●					
Rear central armrest			●			
Rear wash/wipe				●		
Rev counter	●	●	●	●		●
Seat belts Front	●	●	●	●	●	●
Rear						
Seat recline	●	●	●	●	●	●
Sliding roof						
Tinted glass	●	●	●	●	●	●
Windscreen wash/wipe	●	●	●	●	●	●
Wiper delay					●	●

CONCLUSION

The only real criticism that could be levelled at the XJ-S is that it's dated in concept. Like Concorde it is a superb technological achievement with perhaps a questionable future, though from what we hear the XJ-S has been well received abroad and sales are high, especially in America.

Having said that, we must emphasise just what a magnificent car the XJ-S is. It is not a successor to the original E-type—but many will consider it so for the later V12 2 plus 2, itself no mean performer, set very high standards. But the XJ-S is a significant advance in almost every respect. It has a comparable performance but is quieter, smoother, roomier, handles better and is more refined.

Each of the rivals may be better than the Jaguar in one or two ways, but none have the unique combination of qualities that makes the XJ-S, for us, the best of the bunch. The Aston Martin is quicker but more expensive and not so refined; the Jensen is slower, more thirsty and not quite as wieldy; the Lotus is of course slower but much more economical and roomier inside, even though it is a smaller car; the Mercedes (in 350SL form) is also slower, while none of the last three are as effortless or refined as the XJ-S. At £9527 the Jaguar is expensive—but still remarkable value for money.

JAGUAR XJ-S

Speed and style: all you need
is a winding road and the right point of view

by Don Fuller

ROAD TEST

This must be the mother and father of all cloudbursts. The wipers on the XJ-S are frantically trying to keep the windshield clear, but at best the winding mountain highway with its attendant 45 mph warning signs at the corners is a vague outline. Inside, I'm fiddling with the radio because we are continually going behind mountains which block the signal and interrupt the soft muzak, while the passenger is idly bemused by the pattern of the water making its way up and across the side window, a pattern we finally decide would be the subject of a four-hour Andy Warhol movie called "Essence of Flotation" (or something like that) and climaxing with the face of a screaming woman (or something like that).

And all the while, the speedometer needle stays between 85 and 100.

A few days later, on a clear morning, I go back for a solo. This time the road is dry, the day is bright and things are clear.

And this time, the needle stays between 120 and 130.

This is one damned competent automobile, this new Jaguar XJ-S.

Jaguar's new car has been rumored for at least a decade, and there was almost a different rumor for every one of those years. Mid-engine roadburners to blow the doors off the Italian rockets and make the same mark left by the first XK-120s and E-Types. Super sedans, able to humiliate Mercedes-Benz. Refined touring cars, built of existing components. Racers, designed for the Mulsanne and adapted to satisfy an insatiably hungry market. The XJ-S emerges finally as the car with a little bit of all those.

The layout is conventional, with the proven V-12 in front, followed by the transmission then on back to the familiar Jaguar independent rear suspension. For all practical purposes the drivetrain is shared with the XJ12 sedan. So even if Jaguar components don't have the greatest reputation for dogged reliability with minimum maintenance, at least there is nothing in there to scare off an experienced Jaguar repair man. In fact, there are some significant changes which should help that long-suffering mechanic.

Under the heading of "Worthwhile Improvements" the most important is the electronic fuel injection which replaces the previous four carburetors. It's the popular Bosch D-Jetronic, with electronically pulse-timed injector openings controlled principally by intake manifold pressure. For the time, all the parts are Bosch with the exception of the Lucas-built power supply, but eventually Lucas is going to manufacture more and more of the components under Bosch license. In operation it works as well as we have come to expect from the Bosch injection on other cars, which is well indeed. Cold starts, hot starts and throttle response are efficiently unobtrusive; the system works so well you're not even aware of it. The injection is also used on the XJ12 and has been in production on the 12-cylinder since the early spring of '75. It all works so well that the only change required to meet California emissions is a simple alteration to the EGR.

Our only complaint with the driveline is the transmission, a Borg-Warner automatic. Left to itself, it shifts lazily, particularly from 2nd to 3rd, and the kickdown performance is only mediocre, requiring full throttle and a lot of time before it finally goes down into second for passing or the occasional quick shot past Porsches and Corvettes. In comparison to the excellent automatic transmissions from Mercedes or the American manufacturers, the Borg-Warner isn't up to the fight. Still, even with the shortcomings of the automatic, we preferred the car that way, as the automatic seems more suited to the car's refined character. The Jag isn't really

a racer, even though it has performance in *big* bunches, and the four-speed manual unit (which will be available perhaps sooner than originally planned due to demand) would almost be a case of overkill. The car is so quiet and refined, that a clutch and shift lever seems incongruous.

Nothing need be said of the rest of the driveline and chassis except that it is superb. Supremely competent. Suspension travel is a little limited, (it's a low car, remember), so the XJ-S doesn't have the rough-road capabilities of a Mercedes, but other than that it handles most surfaces with grace and aplomb. The Jaguar engineers have obviously spent some time on shock absorber valving, because the car just *never* gets out of shape or upset. The usual practice with large, high-performance European cars is to have the damping on the soft side to provide a good rough-road ride. At Jaguar, they have accurately perceived the usual habitat of the XJ-S to be smoother roads, and for those situations the damping on the XJ-S is near-*perfect*.

With weight distribution of 57/43 understeer is expected. This is a car meant for the fast sweepers, not winning the local gymkhana. And on those fast sweepers, the high-speed bends, the XJ-S is right in its element. The factory claims it will corner at 0.9g. We never approached that, but we did record some impressive cornering numbers. With our standard procedure the Jag would circulate around the skidpad at 0.76g, which puts it right in the very upper level of cornering performance. A little fiddling with tire pressures to reduce the understeer and we could wail around consistently at 0.82g, and with more fiddling, a careful touch and a little encouragement the final number was 0.85g, far and away better than anything we have ever tested.

And—once again—the car's attitude during this entire exercise was quiet competence. It never ruffled its feathers, never tried to turn around and snap. It just flexed its big broad shoulders and *worked*.

It is more than a little eerie. The way it's so capable and confidence-inspiring. You can *trust* it. Not like a big domestic car that takes all the command functions away from you, leaving you sitting in a mobile living room. The XJ-S lets you command, it just relieves you of the dirty work.

And riding along, inside, is where you really begin to notice. Suitcase packed? Then on to Geneva for the weekend. You feel transported into the heretofore inaccessible world of the megarich, as it whistles almost silently along the freeway at 90 or so, the barely audible V-12 whirring along at a comfortable canter and the wind just whispering past the windows.

Our test car had been used pretty well before we got it, and there were some things not quite right, little rattles or nuisances here and there. Still, interior noise at 70 mph was a quiet 68 dBA. It seemed, subjectively, even quieter.

The eerie silence, the total dignity of the car at speeds well over twice the national limit could get the careless driver a lot of long conversations with the law. Not that you necessarily try to ignore speed limits. It's just that the XJ-S is, to use an incredibly worn cliché, as smooth and effortless at 90 and more as most other cars *try* to be at 55, and so by accident rather than intent you find yourself continually above legal speeds.

And if you have the *intent* to play havoc with speed limits, the XJ-S is willing and able. At nearly two tons and with a high-speed engine coupled to an automatic transmission it's not especially fast off the mark; wheelspin was almost impossible to provoke with the superb traction afforded by the large tires and independent suspension. But it's deceptive; with negligible wheelspin and a sluggish start it still gets to 60 in 7.8 seconds, and above that it really *hurries*. The acceleration curve from 60 to 100 is nearly straight, and three digits come up in substantially under twenty seconds. The tractability of the engine—its flat torque curve—is illustrated in no better way than by comparing the various dragstrip runs we made, first by letting the transmission shift itself at 5500 rpm and then by holding it in gear until 6000 and then 6500. All of the runs were within a tenth of a second and less than a mile per hour of each other. Any way you try it, the XJ-S will do a quarter in a couple of ticks over 16 seconds with the speed on the high side of 90. It has the kind of high-speed performance that will absolutely blow most other quick cars right into the weeds, and you find yourself lurking the freeways, ready to pounce on the unwary Porsche driver. (Unless, of course, it says T U R B O across the back.)

Good as this car performs, most owners should be able to do better, as our test car was not in the best tune. Giveaways were the slight low-speed unevenness and mediocre fuel economy. Jaguar says it should be good for 15 or so, but our numbers came out to 12.7 mpg, and we were frequently below 10. On the one hand, we believe it should get closer to the Jaguar claim, but on the other hand, chances are good that the XJ-S buyer won't care about fuel economy anyway . . . only total performance.

Total performance includes being able to stop the thing if the need arises, and the four-wheel discs, large Dunlops and excellent chassis all play their part in producing 60 mph emergency stops in consistently less than 150', with a best of 147. Controllability, or lack of it, is never a problem, and fade is simply not in the discussion. Braking performance, like just about everything else about this car, is at the head of the class.

Can it be that good? Is it possible the English have finally figured out how to build complete automobiles? They've always been acknowledged chassis wizards, but other areas, like execution and reliability, have often been politely omitted from conversation. Observing puddles of oil on garage floors, more than one wag has suggested that Jaguar has cleverly built engines with porous cylinder blocks to aid in oil flow throughout the engine. Our test car annointed the ground wherever we parked it. Another bit of dark humor is that the Jaguar is cunningly designed to disassemble itself, thereby saving the mechanic the trouble and leaving him only to put it back together again. The exhaust system on the test car fell off twice. Is the International Brotherhood of Labor, (or whatever it's called) so strong and close-knit that the good folk who assemble Jaguars are only looking out for their fellows when they craftily put the cars together on a temporary basis only? The left front shock on our car was already making noises, and the climate control was on strike about half the time.

Well, all that is only partially true. The problems with our test car were traceable; it had a lot of hard miles on it when we received it, and it had been subject to only casual inspection and maintenance by the local distributor, who, among other things, neglected to note that of the two bolts securing the steering rack to the chassis only one of them was fitted with a complementary nut. So our feelings about the quality control of the XJ-S are mixed. We know, in our hearts, that this car is a quality item, probably the best car Jaguar has built. But our only experience with it in this country has revealed shortcomings. Certainly, the individual owner starting with a new car and giving it proper maintenance should find it a vast improvement over previous Jaguars.

The people at Jaguar have also spent some time with ergonomics, although the back seat passengers don't fare too well. After all, it's really a two-passenger car, although we did put four adults in it and nobody threatened a civil action. But the two in front go first class. The driver's seat has only two adjustments, for seat travel and backrest rake, but the seat and controls are placed so well that those two adjustments, combined with the telescoping steering wheel, are all that's needed. The seatback rake, like the travel, is adjustable via a lever and notched, so it's not infinitely adjustable like some other seats with knobs to turn. But the more we used it the better we liked it, because the notches are close together, which allows enough positions to suit any reasonable person, and also makes it easier to find your particular favorite position after someone else has driven the car and readjusted everything. The controls are, for the most part, the usual Jaguar XJ-series fare, with the same knobs and levers for lights, air conditioning, directionals, wipers and gearshift as used in the sedans. The steering wheel is the same except for a thicker, padded rim which is a simple, but important, improvement. Even with the same controls and knobs, however, the

ROAD TEST DATA — JAGUAR XJ-S

SPECIFICATIONS

ENGINE
Type	OHC V-12
Displacement, cu in	326
Displacement, cc	5343
Bore x stroke, in	3.54 x 2.76
Bore x stroke, mm	90.0 x 70.0
Compression ratio	8.0:1
Hp at rpm, net	244 @ 5250
Torque at rpm, lb/ft, net	269 @ 4500
Carburetion	fuel injection

DRIVELINE
Transmission	3-spd auto
Gear ratios:	
1st	2.39:1
2nd	1.45:1
3rd	1.00:1
Final drive ratio	3.31:1
Driving wheels	rear

GENERAL
Wheelbase, ins	102.0
Overall length, ins	191.7
Width, ins	70.6
Height, ins	49.7
Front track, ins	58.6
Rear track, ins	58.6
Trunk capacity, cu ft	8.5
Curb weight, lbs	3845
Distribution, % front/rear	57.1/42.9
Power-to-weight ratio, lbs/hp	15.8

BODY AND CHASSIS
Body/frame construction	unit
Brakes, front/rear	vented disc/disc
Swept area, sq in	400.0
Swept area, sq in/1000 lb	104.0
Steering	rack & pinion
Ratio	17.5:1
Turns, lock-to-lock	3.0
Turning circle, ft	36.3

Front suspension: independent, upper and lower control arms, coil springs, tubular shocks, anti-roll bar
Rear suspension: independent, halfshafts as upper links, lower lateral links, trailing arms, two coil springs and two tubular shocks per wheel, anti-roll bar

WHEELS AND TIRES
Wheels	15 x 6
Tires	205/70 VR 15 Dunlop SP Sport Super Steel Radial
Reserve load, front/rear, lb	421/1014

INSTRUMENTATION
Instruments: 0-160 mph speedo, trip odo, 0-7000 rpm tach, water temp, oil press, fuel level, volts
Warning lights: directionals, high beam, rear defog, fog lamps, volts, ignition, fuel level, hazards, water temp, brake failure, oil press, parking brake, seat belts, parking lamps, stop lamps, EGR service/catalyst service

PRICE
Factory list, as tested: $19,000
Options included in price: std. equip includes AM/FM stereo tape, climate control, auto trans, and other accessories

TEST RESULTS

ACCELERATION, SEC.
0-30 mph	3.7
0-40 mph	4.9
0-50 mph	6.2
0-60 mph	7.8
0-70 mph	9.9
0-80 mph	12.3
Standing start, ¼ mile	91.46
Speed at end ¼ mile, mph	16.22
Avg accel over ¼ mile, g	0.26

SPEEDS IN GEARS, MPH
1st (6500 rpm)	65
2nd (6500 rpm)	108
3rd (6000 rpm) (observed)	141
Engine revs at 70 mph	2950

SPEEDOMETER ERROR
Indicated speed	True speed
40 mph	40 mph
50 mph	50 mph
60 mph	60 mph
70 mph	70 mph
80 mph	80 mph

INTERIOR NOISE, dBA
Idle	53
Max 1st gear	77
Steady 40 mph	64
50 mph	66
60 mph	67
70 mph	68

HANDLING
Max speed on 100-ft rad, mph	33.7
Lateral acceleration, g	0.76
Transient response, avg spd, mph	22.7

BRAKES
Min stopping distance from 60 mph, ft	147
Avg deceleration rate, g	0.82

FUEL ECONOMY
Overall avg, RT cycle	12.7 mpg
Range on 24.0 gal tank	304 miles
Fuel required	unleaded

Graph Of Recorded Data Expressed in Percentage of 100 (100 = best possible rating)*

- Acceleration
- Brakes
- Handling
- Interior Noise
- Tire Reserve
- Fuel Economy
- Overall Rating
- Highest Rated To Date — Carrera, Jensen-Healey

*Acceleration (0-60 mph): 0% = 34.0 secs., 100% = 4.0 sec.; Brakes (60-0) mph: 0% = 220.0 ft., 100% = 140.0 ft.; Handling: skidpad lateral accel., 0% = 0.3 g, 100% = 0.9 g, transient response, 0% = 20 mph, 100% = 25 mph (average skid pad and transient response for overall handling percentage); Interior Noise (70 mph): 0% = 90.0 dBA, 100% = 65.0 dBA; Tire Reserve (with passengers): 0% = 0.0 lbs., 100% = 1500 lbs. or more; Fuel Economy: 0% = 5 mpg, 100% = 45 mpg or more. Test Equipment Used: Testron Fifth Wheel and Pulse Totalizer, Lamar Data Recording System, Esterline-Angus Recorder, Sun Tachometer, EDL Pocket-Probe Pyrometer, General Radio Sound Level Meter.

Jaguar's fuel-injected, single overhead camshaft, twelve-cylinder engine: a model of smooth, effortless power but no place for the unwary or inexperienced at service time.

The backseat is really fit for adults for short duration only.

Clever positioning of the spare has resulted in a deep trunk; even large suitcases can be stored vertically.

dash is different in character. Gone is the real wood dash, replaced by metal and leather and plastic, and the whole thing is modern and clean looking, with none of the paneled, English-gentleman's-library look to it.

But let's be honest about it. How much of this matters? With a limited production car carrying a revered and legendary name and a price tag near twenty grand, there are two main things that really count, and if those two work, everything else is just icing. The car must have image. Exclusivity, impact, be able to Wilt the Knees of Sweet Young Things, whatever you may call it, it must have that status that *immediately* singles its owner out from the crowd. And, it must have performance that delivers on the pledge of the looks. In those two most important categories, the Jaguar XJ-S is an unqualified success.

Just *look* at it. Especially from the front. It *squats* there, looking for all the world like a big cat waiting to spring. Take it to the overpriced restaurants, where the stuffed-shirt parking lot attendants try to be cool while falling all over themselves and the guy who just drove in ahead of you in his Carrera with the ski rack is *boiling* because his date is looking at you in a way she never looked at him. Cruise the freeways, and traffic parts when you become visible in the mirrors. They don't know exactly what it is, but *whatever* it is, it is definitely a heavy hitter. The *nouveau riche* in their Cadillacs nearly fall out the windows while the Mercedes types try to be less obvious about it, but are betrayed by the constant sideways glances. Leave it in a parking lot, and when you return the windows will be covered with nose prints. A word to the wise, though. If you're going to be seen in and with this car, you had better be able to handle yourself, because the rabble can be merciless when they spot a fraud.

The Jaguar XJ-S has the performance to go with the image; a combination of image and performance perhaps never before realized in an English car. The quiet, reserved, capable, total performance you would expect from some extraordinarily experienced English engineer, who would finally sit back, take a slow draw from his pipe, and then, holding his pipe familiarly in one hand and stroking his chin with the other, quietly say to his assistants, "I think that should do it."

A level of capability which is not within the realm of "automobile" as most people think of it. The car had to be back on Monday and this was our last weekend with it, the last chance to find out how fast it would go. There were two of us, the better to spot, uh, possible deterrents to the project. Freeway traffic was too crowded, but we found an alternate, a good side road with a nice straight almost a mile long and no traffic. There was a left-hand sweeper leading onto the straight, with a slight dip in the middle of it, and we had been going through a little gingerly, lifting slightly so that the speed through the sweeper was only 110 or so, and it wasn't allowing the big twelve to get really wound out before we ran out of room and had to shut off. As a result, the tach would show about 5800 and still be climbing. It was getting dark, sort of a now-or-never kind of thing, and we really didn't want to continue with headlights. Besides, we had made several passes already and were getting concerned about attracting attention from unsympathetic sources. So we went through once more, only this time the foot stayed down all the way. A bump caused a little bobble, but a quick glance showed the speedometer above 130 through the sweeper and the dip, and once on the straight the tach went to 5900, then 6000, then held at 6000. The Jaguar took care of the sweeper and the calculator took care of the arithmetic. The final number, with a slightly out of tune engine, is 141 mph, and you can believe it.

Crazy? Pointless? Maybe. But we think that quiet English engineer with his comfortable pipe would have understood, and maybe even been a little proud. ■

JAGUAR JOURNEY
To Munich by XJ-S Express

"IT'S A BLACK DAY for Stuttgart and Modena", cried British Leyland's advertising on the day they announced the Jaguar XJ-S. They might well have added Munich, West Bromwich and Newport Pagnell to the challenge. Would this ambitious new sporting (but not "sports") Jaguar live up to the standards to which it aspired? A drive in an automatic version through the Cotswolds prior to announcement convinced us that the challenge was viable, but a short journey within the restrictions of British B-roads was no confirmation that the XJ-S, described in detail in last October's MOTOR SPORT, would fulfil its purpose as a very high-performance Grand Touring car. That it is this and more we can now affirm, having given a manual version one of the toughest, fastest tests MOTOR SPORT has ever perpetrated, 2,435 miles, mostly on the Continent and mostly at speeds well in excess of the three-figure barrier.

That our actual Jaguar test car was one of the fastest production cars in the world had been confirmed by a 154 m.p.h. maximum speed check by a weekly contemporary a few weeks previously. This had been achieved two-up, with light luggage and with the air-conditioning switched off. Our own performance we feel emphasised the phenomenal performance of this latest Jaguar even more impressively: carrying three adults, nearly 20 gallons of Super, with its boot crammed full with luggage and with the air-conditioning switched *on*, this aesthetically ugly, aerodynamically brilliant, Browns Lane product wound itself to an incredibly easy 150.5 m.p.h. before traffic conditions halted the still rising speedometer needle. We know of no other car in the world which would fulfil these conditions with such smoothness and silence. It is the sort of performance which allowed us to leave Munich post-breakfast at 8.55 a.m., lose a couple of hours on the way for shopping and eating and another couple of hours crossing the

37

Channel and have the three of us deposited in our beds, though still quite fresh, as far apart as Hertfordshire, Essex and Berkshire, some 800 road-miles distant, by midnight the same day. That is GT motoring of the highest order. That its Lucas-Bosch fuel-injected 5.3-litre V12 engine should have averaged over 13 m.p.g. whilst obliging us with cruising speeds in the 130-140 m.p.h. region makes the XJ-S nothing short of incredible.

Our brief on this six-day expedition, for which British Leyland's Press Department had so willingly contributed this magnificent express, was to visit Ford Competitions Department in Cologne, Zakspeed at Neuheusel, Opel's Competition Department (sorry, Dealer Team) at Russelsheim, Porsche and Mercedes in the Stuttgart area, BMW in Munich and BMW tuners Schnitzer and Alpina elsewhere in Bavaria. My own journey began with a lift to collect the XJ-S from Browns Lane in a Ferrari 365 GT4—which disposed of the Modena side of the comparison and gave Jaguar's Chief Development Engineer David Fielden a chance to drive the opposition. The Ferrari had better lines and more rear seat room, but that beautiful music from the V12's Weber orchestra had become less enchanting by the time we'd traversed the M1. The XJ-S's silence felt golden. From Allesley we had a cross-country dash to the Donington Speed Show, the Jaguar more sure-footed on the greasy roads than the Ferrari, even with Gerry Marshall at the helm of the latter. Comparative road tests show the 34.5 cwt., 285 b.h.p. DIN, four-speed gearbox Jaguar to be faster on acceleration and maximum speed than the 33.5 cwt., 340 b.h.p. DIN, five-speed gearbox Ferrari. A case of prancing ponies against Warwickshire shirehorses, perhaps? The Ferrari is also £6,200 more than the XJ-S ....

However, all was not well with the XJ-S, which pulled half-way across the road under braking. A return to Jaguar gave new pads and re-routing of the radiator overflow pipe (already done on later production cars) to prevent it pumping anti-freeze mixture over the nearside caliper.

After 249 miles of running in England the yellow band of the vertical fuel gauge had fallen into the red area, requiring some 17 gallons replenishment for the 20-gallon tank to send the band out of sight at the top of the scale, an average of 14.6 m.p.g. A slow puncture was discovered and a wheel changed half-an-hour before we left for Dover, so we had a suspect spare for our entire Continental trip. However, the toolkit proved to be very comprehensive.

By the time J.W. and I had taken aboard our photographer in Essex, the deep and deceptively commodious boot was packed full, the pile of suitcases crowned by a plastic "wardrobe"; its position was a mistake, for, by the time we reached Cologne, a neat hole had been burned through it by the boot light in the lid. This dangerous light operated permanently with the sidelights; I'm glad to hear that courtesy operation is projected for future cars. Perhaps an underbonnet light will be fitted at the same time? And maybe a light too in the glovebox, the lockable lid of which needed a slam to close? Although the XJ-S's general equipment is exceedingly comprehensive, it is sadly lacking in some minor details, like the above. Why no k.p.h. markings on the speedometer of this transcontinental express? Why a door mirror which is not adjustable from inside the car (even the lowly Escort offers that as an option) and why no nearside mirror in a car which suffers from poor three-quarter rear vision? And why, as Jaguar have publicly thrown the gauntlet at the Mercedes 450SLC, are there no wiper/washers (or at least an option of) for those huge Cibie halogen headlamps, the European versions of which had a sharp dip cut-off and too short a main beam range, but didn't dazzle, while the UK version, which replaced them on our return to England, had superb range and spread, but annoyed other drivers? "Hmm, that should be leather," said Burkard Bovensiepen, owner of Alpina, Europe's biggest performance equipment company, echoing many other people's comments as he pointed at the cheap, convoluted plastic gear-lever gaiter. At the time we were driving up a country road in top gear with no throttle applied, though the pedal was about to be squashed to the floor without a hiccup from the engine—incredible engine flexibility—but still that detail caught his eye. Build a £9,527 vehicle, Jaguar, and a brilliant overall concept

Clear instruments and a telescopically adjustable steering column face the driver. Below, another 75 DM-worth of Super goes aboard, but no oil. Achtung Spitfeur! Wo ist Das? Eine Merlin V12?

won't necessarily overcome customer reticence brought on by poor detailing. Apart from the West of England cloth roof-lining, the plasticky interior looked down- rather than up-market from the rest of the Jaguar range. The Germans were pretty scathing about the exterior finish too, particularly the crudeness of the door window frames.

Perhaps readers will think I'm nit-picking. If so it is because I feel strongly that in its uncanny amalgam of "pace and quiet" the XJ-S is unequalled; it would be tragic to tarnish its appeal in fastidious markets such as Germany for want of fine detailing.

We were worried initially that the rear seat would be too cramped for long distances. However, even our 6 ft. photographer survived without too many complaints, though he had a valid moan about the (optional) rear seat safety belt buckles protruding through the middle of the seats to hamper his enforced side-saddle position. My own 5 ft. 7 in. were cosily comfortable sitting upright. The front seats felt much improved since the pre-production car J.W. and I drove last summer; we no longer slid forwards under braking, but the backrests would benefit from more lateral support. The right-hand handbrake, which falls to the floor out of the way when applied, no longer fought with the door arm-rest when pulled up.

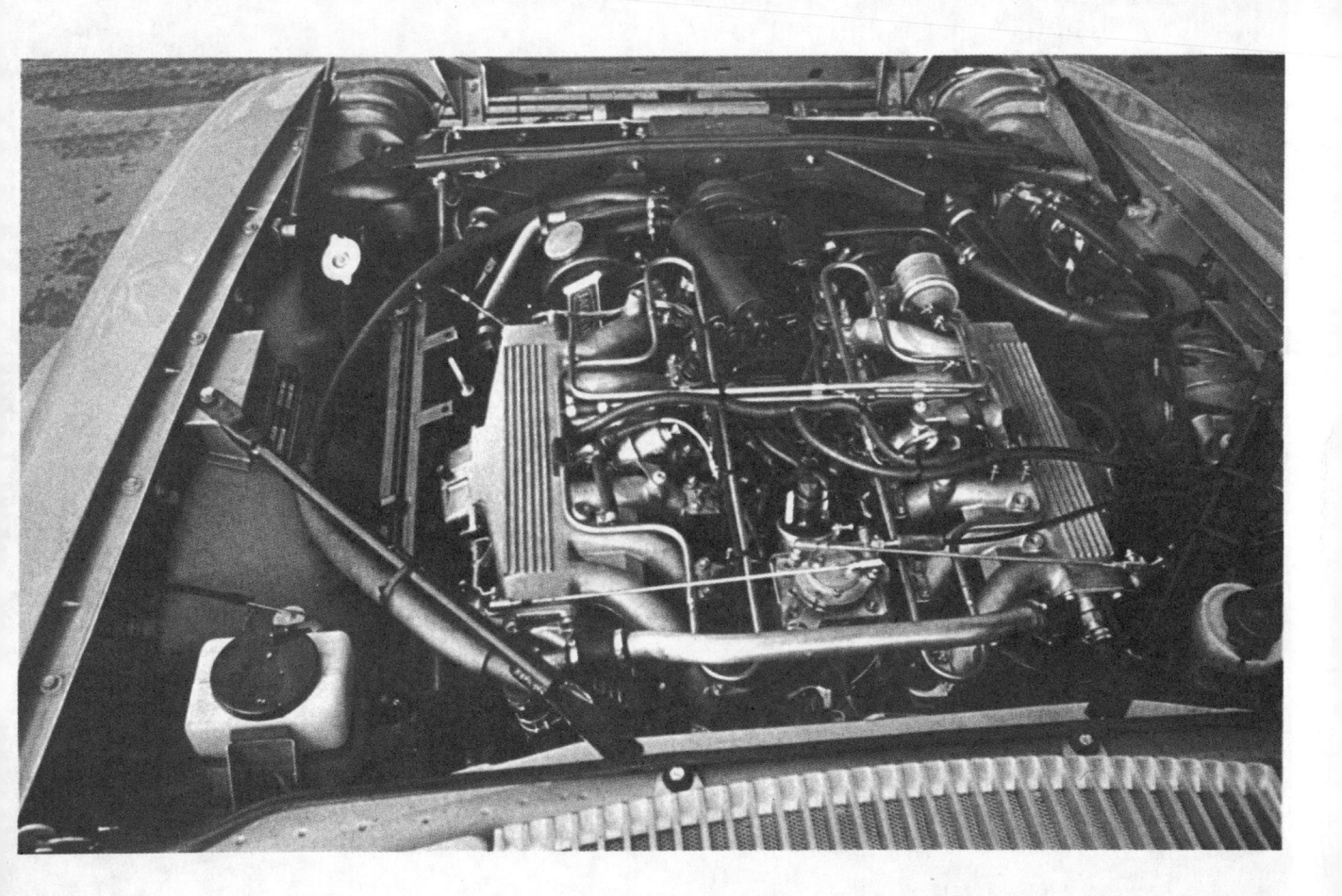

Once we'd eaten ourselves comfortably across to Calais aboard Free Enterprise VI, we were able to let the XJ-S have its head towards Cologne, free from endorsement worries.

Letting the XJ-S have its head from rest takes it to 50 m.p.h. in 1st gear, 84 m.p.h. in 2nd and 116 in 3rd, with 60 m.p.h. coming up in 6.7 sec. and 100 in 16.7 sec. The ratios are lower than need be for the 294 ft. lb. torque; 2nd will cope with most standing start conditions—well, so too will top. At Britain's 70 m.p.h. legal maximum, our passage was silent save for some roar from the 205/70 VR 15 Dunlop SP Super tyres. Some wind hiss emanated from a faulty nearside door seal, changing to a sudden roar at exactly 130 m.p.h. But engine noise was little more than a whirr, even up to the 6,500 r.p.m. red line. One hundred m.p.h. is roughly 4,000 r.p.m. in top, nice and lazy, but even on this gearing the engine felt to be so much on top of its work that the overdrive, which the handbook confirms will become optional in production, might have been a good economic asset. Nevertheless, our roughly 100 m.p.h. average from Calais to Cologne produced 13.57 m.p.g., all the benefits of which were lost in a lengthy search for our Cologne hotel. This pattern was to be repeated throughout our trip: next time we shall pack some decent street maps—they're cheaper than XJ-S performance.

Little effort was required to keep the XJ-S on course at high autobahn velocities; the Adwest power steering, a shade woolly and overlight for fast twisty-road work (though practice soon breeds familiarity), showed no vices or imprecision; there was absolutely no question of front-end lift as speed increased, no sign of wandering or decreasing stability. So long as the driver's brain was in gear the XJ-S remained unflurried, making 150 m.p.h. as undramatic as riding in a tube train. Such is the deteriorating standard of continental driving since the oil crisis, even in Germany, that any vehicle ahead, however innocent looking, must be treated with suspicion and occasionally full use had to be made of the ventilated disc brakes, which decelerated over

CONTINUED ON PAGE 55

"*The finest production engine in the world*", above. The silken, fuel-injected V12 produces an obviously very genuine 285 b.h.p. DIN at 5,800 r.p.m. Left, the XJ-S at rest outside the Intercontinental Hotel, Munich, reached after a 100 m.p.h. average from Calais.

Move over, Merc

...the big cat's quicker. But is it better? A convoy confrontation between two great cars—the Mercedes 450 SLC and Jaguar's XJ-S—ends in a photo finish.

Smoke pours from the inside wheel of the Mercedes as it exits from a side turning. It could do with a limited-slip differential

Rarely have two thoroughbreds been pitted in fiercer competition than Mercedes, 450SLC and Jaguar's XJ-S. The protagonists have clashed before, of course, their sports racing cars duelling at the Le Mans and elsewhere in the Fifties and their saloons tangling at the Nurburgring a decade later. Now, each strives to capture the limited but lucrative market for luxury, high performance coupes.

Both companies have retained a strong corporate identity in their designs. The Mercedes typifies that "carved from solid" appearance characteristic of the marque and all it represents, its quality being reflected by very high standards of engineering and finish. Equally conventional, if a little more cramped in layout, the Jaguar exudes an air of plush refinement that is the hallmark of the breed.

In some areas the thinking is virtually at one. Both cars have large capacity, fuel-injected, V-configuration engines, the Jaguar a V12, the Mercedes a V8. Their all-steel, monocoque shells are supported by fully independent suspension and the steering of each car has power assistance.

While the 350SLC is available with either manual or automatic transmission, its big brother, the 450, comes with a three-speed, self-change unit as standard. Although some 90 per cent of XJ-S customers are also likely to opt for automatic transmission, the only car available to us for this comparison was unfortunately a manual.

Manual or automatic, paying upwards of £9000 for a car you can expect lavish appointment. Neither the Jaguar (£9527) nor the Mercedes (£11,522) disappoint, though the approach differs.

So does the thinking on accommodation. The difference between a 2+2 and a 4-seater can be fine but significant. Its shorter wheelbase and lower roofline puts the Jaguar firmly in the 2+2 category while the Mercedes, with its superior leg and headroom, must be regarded as a full four-seater.

The Jaguar for refinement, the Mercedes for accommodation? Yes, but there's more to it than that. Pushing the three-pointed star through the air is an alloy-headed V8 of 4520cc. Its valves are operated by one overhead camshaft per bank of cylinders and its fuel metered by Bosch electronic injection (D-Jetronic on our car, K-Jetronic on the very latest imports). The output is a healthy 225 bhp at 5000 rpm. Jaguar employ an even larger, all-alloy V12 of 5343cc. It too has one camshaft per bank and electronic injection (by Lucas), and its net horse-power is no less than 285 bhp at 5500 rpm.

With such power on tap one could forecast vivid performance. Once

Fine dials on the Jaguar are partially spoilt by their cheap-looking, silver-painted rims

The Mercedes facia is enhanced by its superb instruments which are neatly framed by the rim of the large steering wheel

41

The coupes in full flight. Both handle extraordinarily well by normal standards. If you had to chose between them, the Jaguar with its limited slip differential, is the more progressive of the two

Functional and luxurious, the Jaguar's leather-trimmed cockpit is highly organised. Both the radio and air conditioning are standard

The heating and air-conditioning controls are a little scattered but most others are ergonomically positioned. The change-pattern for MB's automatic gearbox serves as a model to others

The Jaguar is very much a 2+2 with ample room up front but cramped accommodation in back. Right: Superior headroom is the key to the Mercedes' more practical layout

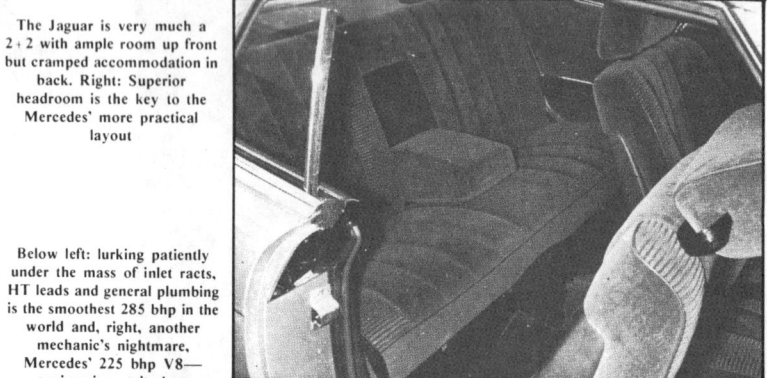

Below left: lurking patiently under the mass of inlet racts, HT leads and general plumbing is the smoothest 285 bhp in the world and, right, another mechanic's nightmare, Mercedes' 225 bhp V8— engineering at its best

again neither car disappoints, but it is here the more potent Jaguar takes a clear lead. Though, sadly, there was no chance to extend either steed to the limit, but for once we have few qualms about quoting the manufacturers figures, which we approached with ease on several occasions. For the Jaguar (manual) this means a staggering 155 mph and for the Mercedes a very presentable 134 mph.

Not until you put a watch on the Jaguar is its full potential revealed, so smooth and quiet is the engine, so effortless its output. With the ability to rush from rest to 60 mph in a mere 6.7 sec, this new generation Briton ranks as one of the fastest accelerating production cars in the world. Of course we are still talking of the manual version: the automatic is a little more tardy.

Not that the Mercedes could be regarded as a sluggard either, romping to 60 mph in only 8.5 sec and happily keeping pace with the Jaguar anywhere other than a long straight road.

The stodginess of the Jaguar's transmission does not encourage unnecessary changes, but the extreme flexibility of its lazy V12 engine successfully disguises the problem. The engine can be persuaded to pull from rest to 150 mph in top gear and certainly wastes little time getting into its stride, sprinting from 80-100 mph in a disdainful 8.0 sec. Once again the Mercedes is hot on its heels though, accomplishing that same transition in only 8.6 sec.

Both cars are first time starters, the Jaguar purring into action so silently that a glance to check if the tachometer is on the move becomes almost second nature. The Mercedes is quiet at idle too. And neither car suffers a hint of hesitation, flat spot or foible of any kind, their big engines spinning without temperament.

The V12, though muted even at 6500, adopts a more determined tone from around 5500 rpm onward—not that such revs are really required. The Mercedes is equally happy to rush up to its more modest 5800 rpm rev limit, but becomes a little hammery in the process.

Round one to Jaguar.

If you have to think twice before brimming either's car tank then you're living above your means. In both cases, the penalty of such performance is high thirst, the Jaguar completing no more than 12.8 miles on each of its 20 gallons. The Mercedes faired better, returning a 15.1 mpg average over 1500 miles. It, too, sports a 20 gallon tank and thus has the higher range of the two—300 miles.

Round two to Mercedes.

We see two good reasons for specifying the optional Borg Warner automatic transmission on an XJ-S; first, because self-change units befit the image of this type of car and second, because the standard of Jaguar's manual transmission does not quite match that of the rest of the car. The change is stiff and notchy when rushed and no better than mediocre however you treat it, while the clutch is heavy and a little unprogressive.

The Daimler-Benz three speed automatic unit fitted as standard to the 450SLC, on the other hand, is so good as to serve as a model for other manufacturers. Some automatics virtually defeat their own object by being over-eager to change up and reluctant to kick-down. Not so the Mercedes, whose manual over-ride is usually redundant thanks to a system that always seems to be in the right gear. Under part or full throttle conditions, the gearbox slurs almost imperceptibly from one ratio to another, and a commendably smooth kick-down is available up to 80 mph.

Given full throttle the Mercedes makes its two automatic changes at 4300 rpm (45 mph), and 5300 rpm (87 mph) respectively. However, for maximum acceleration it paid to hang on to each ratio a little longer and our best standing start times were achieved using 5500 rpm (giving 57 and 90 mph in first and second respectively).

The Mercedes takes the honours again.

Performance of the calibre exhibited by either vehicle would naturally be an acute embarrassment without handling to match. As it is the two thoroughbreds meet standards far in excess of most popular needs.

Similar thinking is seen in the design of their suspensions. For the Jaguar this means double wishbones at the front and the proven Coventry formula of a lower wishbone and drive shaft acting as the upper link on the rear. The Mercedes also sits on wishbones at the front, and for the rear the equally proven Stuttgart formula of semi-trailing arms are employed. Both cars feature coil springs all round and each has power assisted steering —rack and pinion on the Jaguar, recirculating ball for the Merc.

For years we have campaigned against the extreme lightness of Jaguar steering, so we were cheered to find the XJ-S sporting both a higher-geared rack and smaller steering wheel than its predecessors. The corresponding increase in resistance and extra precision they bring are very noticeable, and if not quite to the standard of the beautifully weighted Mercedes system, the Jaguar's is nevertheless good.

To pick a victor from the two in terms of handling calls for some very rapid driving. The Mercedes feels taut from the outset. Its firm suspension and relatively heavy steering combine to give that all important requirement: "feel" which breeds confidence, and it requires no special skills to rush the 450SLC along winding roads with a rewarding degree of precision. But it has its failings. For instance, its traction in the wet is distinctly poor, and even the most moderate of drivers can expect to be faced with the occasional tail-slide; likewise some spinning of the inside wheel, a phenomenon easily promoted by such power and the lack of a limited slip differential.

While the Mercedes' handling up to the limit of adhesion is exemplary, on the limit it is less forgiving. After mild understeer gives way to a slight tail out posture the semi-trailing rear suspension invariably allows the offside wheel to dig in, giving rise to a spinning inside wheel and an untidy cornering style.

With its lighter steering, softer ride, and uncanny lack of noise, the Jaguar is a more imposing proposition. However, the experience of a few hundred miles behind the Coventry Cat's lengthy bonnet shows that what it lacks in feel it makes up for with progressive, remarkably predictable, handling. Certainly it is the more forgiving of the two vehicles and even in the wet, hard-cornering is a delight, its tendency toward a final tail out pose being more reminiscent of a well-sorted rally car than a heavy coupe. The adhesion of the Jaguar's special Dunlop tyres is impressive in all conditions.

Round four to the Jaguar.

The golden rule of providing brakes to match performance seems too obvious to mention, yet so many manufacturers in the past have left stopping more to will power than brake-power—not least our friends across the Atlantic. Neither Mercedes nor Jaguar could be accused of such a misdemeanour. Both their cars sport twin servo-assisted hydraulic circuits operating on four hefty discs, and both stop with the minimum of fuss. If we were to be pedantic then the Mercedes' pedal is on the sensitive side for town work and the Jaguar's a little inert, but the effectiveness of both systems is beyond reproach.

A dead heat.

Presumably, if accommodation is your problem, you'll not be in the market for a coupe anyway. If you are, then the quantity and the sizes of the likely passengers could tip the balance in the Mercedes' favour, though it's a dead heat on luggage accommodation, both boots holding exactly 8.9 cu ft of our test cases.

Both cars have ample accommodation for those in front, and it is over rear seat accommodation that the divergent thinking is apparent. Armed with the knowledge that the Mercedes has a substantially longer wheelbase and a higher roof-line you'd expect it to be more capacious. It is.

With the belts unhooked from their retaining clips on the top of the seats, access to the rear is straightforward. As the safety-catches for the backrests are vacuum-operated and automatically recoil when the engine is switched off, there is no fumbling for manual latches.

Both head and legroom in the rear are adequate for a couple of average-sized adults and a third could probably find sufficient room for a short journey.

Reaching the rear of the Jaguar is harder work, and once installed you are faced with less legroom and minimal headroom—as well as a noticeably short cushion and an upright backrest.

The Mercedes wins the seat space race.

However, the Jaguar certainly wins on oddment stowage, with cubbies in the side panels, facia and console and pockets in the doors. Mercedes containers are confined to a glovebox, door pockets and the first-aid cabinet built into the rear parcel shelf.

Though identical in accommodation, the boots each have their advantages. The Jaguar's, with its lower lip, is easier to load with heavy objects while the 450SLC's squarer shaped one makes for easier stowage of bulky objects.

Swings and roundabouts?

A significant lack of roll, pitch, dive, squat and typical Jaguar resilience characterise the fine ride of the XJ-S. Apart from a slight tendency to lurch momentarily on fast, bumpy corners the big Jaguar copes with even the poorest of surfaces with aplomb.

Progress in the Mercedes is also most impressive, though the character of ride could not be further removed from the Jaguar's. At town speeds its tautness would not warrant criticism at all were it not for the simultaneous lack of sound absorption —bumps are heard as well as felt. At cruising pace, however, irregularities are smoothed out in the sort of silent, satisfying way one expects from a car of this calibre.

This round to the Jaguar.

Few among us would lavish in excess of £9000 on a shopping car and undoubtedly many XJ-Ss and 450SLCs will spend their working days pacing the highways of Europe. Not only can both steeds swallow up such road miles ad infinitum, but more importantly will do so without imparting strain to their drivers. The driving positions display different thinking but are both worthy of praise.

In the Jaguar you sink into a supple, well-fashioned, leather covered seat. Your hands fall on a perfect-sized, leather-bound steering wheel and your feet to well sited pedals. The gearlever is nicely positioned and the handbrake flops conveniently to the floor once applied. The minor controls are neatly and symetrically distributed between column-mounted stalks and the facia. Some will complain of a lack of cushion tilt adjustment, others of lack of grip from the slippery leather seats. But most will find the XJ-S an ideal long distance car.

In the Mercedes you are secured by a firmer cloth covered seat (leather is optional). Your hands fall on a larger, pimply steering wheel and your feet on a pair of massive but well-positioned pedals. The gear shift pattern is fool-proof and the lever well sited. The minor controls—featuring one multi-function stalk—are as neatly organised as the Jaguar's.

But we'd award round 10 to Jaguar; just.

The Jaguar's four minor instruments are contemporary and clever in design. They are arranged so that their needles move vertically, and with everything shipshape (and the fuel tank half full) they form a single yellow line across the panel so that only the briefest glance is needed to check that all is well. Like the neatly calibrated speedometer and revcounter they are surrounded by a thin silver painted line which, in our eyes, gives them an undeserved air of cheapness.

The Mercedes' crop of three large circular dials is neatly framed by the upper rim of the steering wheel. Each is clearly marked and exceptionally easy to read. Like the Jaguar's they are effectively lit at night. They may be staid in design but the result is excellent.

Round 11 to Mercedes; just.

Air conditioning is standard on the Jaguar, an option on the Mercedes. Both have comprehensive heating and ventilation systems. Of the two layouts the Jaguar's shows the most promise but in practice sufficient air to the face and adequate heat to the feet is difficult to achieve. The Mercedes, on the other hand, gives the best of both worlds with torrents of fresh air available through the face vents.

Another mark to the Merc.

Jaguar are emerging as masters of noise suppression. The XJ-S is a prime example of their expertise. Murmurs from its vast engine never really rise beyond a distant whine, road noise is virtually inaudible on many surfaces, minimal on others, and wind roar is seemingly no louder at 120 mph than at 70 mph. Bump-thump is very well suppressed. Being hypercritical you would fault the faint whine in first and second gears and the occasional hiss of the power steering. Otherwise you would just marvel at one of the most refined cars in the world.

Though relaxing, the Mercedes does not match these standards. The engine becomes a trifle insistent when exerted, there is constant wind noise from the large door mirror and the tops of the doors from 50 mph onward and considerable road roar on most surfaces. In this respect the Mercedes lags not only the Jaguar but most of its illustrious opposition.

Though both are lavish in appointment, the cars differ widely in specification. Both have fuel injection, electronic ignition, electric windows, a central-locking system and door mirrors (internally adjustable on the Mercedes) but from there on comparisons end. With your £9527 you acquire an XJ-S plus air conditioning, radio, alloy wheels, low profile tyres and adjustable steering. On the safety side there is an impact switch to isolate the fuel pump in the event of a collision.

If anything, safety features even higher in Mercedes' book, and the £11,750 (the price rose after our colour sheet went to press) 450SLC is equipped with under-bumper fog lamps, variable intensity rear lights, a headlamp wash/wipe system, a first aid kit, and a warning triangle as well as a sun-roof (useless above 50 mph due to lack of wind deflector), and automatic transmission. Both cars carry an excellent tool-kit.

We specify a dead heat.

West of England cloth in abundance, the unmistakable fragrance of real leather and the feel of quality pile carpet add up to unsurpassed luxury for the XJ-S. In comparison the injection mouldings and velour trimmings of the 450SLC are a little spartan to some eyes. But for many, the excellence of the fit of these mouldings, the undenied quality of the paintwork, and the total lack of rough unfinished edges, will ultimately impress even more. Both cars are luxurious, both cars are beautifully made. The Mercedes is the better finished of the two.

The final bouquet goes to Germany.

It's a dead heat on points. So how do you choose between two such magnificent motorcars? For some of course the clinching factor will be price—the Jaguar is £2223 cheaper after all—for others the decision may be more abitrary. If so, for us it would still be the Jaguar, that uncanny, unrivalled refinement tipping the balance in Coventry's direction, but only just.

JAGUAR		MERCEDES
• • •	Performance	• • • •
• •	Economy	• • •
• •	Transmission	• • • •
• • •	Handling	• • •
• • • •	Brakes	• • • •
• •	Accommodation (L)	• • •
• •	Accommodation (P)	• •
• • •	Ride comfort	• • •
• •	At the wheel	• • •
• • •	Visibility	• • • •
• •	Instruments	• • •
• •	Heating	• • •
• •	Ventilation	• • •
• • • •	Noise	• • •
• •	Equipment	• • •
• •	Finish	• • •

PERFORMANCE

Jaguar		Mercedes
155 †	Max speed mph	134 †
116 / 84 / 50 at 6500	Max in 3rd / 2nd / 1st rpm	— / 96 / 60 at 5800
6.7	0-60 mph sec	8.5
6.6	30-50 mph in top sec	3.3 (kickdown)
12.8	Overall mpg	15.1

† see text

SPECIFICATION

Jaguar		Mercedes
V12	Cylinders	V8
5343	Capacity	4520
90/70	Bore/stroke	92/85
Sohc per bank	Valves	Sohc per bank
9.0 : 1	Compression	8.8 : 1
285 bhp (DIN) at 5500	Max power	225 bhp (DIN) at 550
294 lb ft (DIN) at 3000 rpm	Max torque	278 lb ft (DIN) at 3000
4-speed manual	Gearbox	3-speed automatic
24.7	mph/1000 rpm	24.0
Unitary all steel	Chassis con	Unitary, all steel
Wishbones/coils/ anti-roll bar	Front suspension	Wishbones/coils/ anti-roll bar
Wishbone/driveshaft/ radius arms/coils/ anti- roll bar	Rear suspension	Semi-trailing arms/ coils/anti-roll bar
Rack and pinion-assisted	Steering	Recirculating ball-assisted
Disc all round	Brakes	Disc all round
205/70VR15 Dunlop	Tyres	205/70 VR14 Michelin
33.4 cwt	Weight	33.0 cwt

STANDARD EQUIPMENT

Jaguar		Mercedes	Jaguar		Mercedes
•	Adjustable steering		•	Laminated screen	•
•	Armrests	•	•	Outside mirror	•
•	Ashtrays	•	•	Petrol filler lock	•
•	Breakaway mirror	•		Radio	
•	Childproof locks		•	Reversing lights	•
•	Cigar lighter	•	•	Rev counter	•
•	Clock	•	•	Seatbelts: front	•
•	Coat hooks			rear	
•	Collapsible steering	•		Seat recline	•
•	Dual circuit brakes	•		Seat height adj.	•
•	Electric windows	•		Sliding roof	•
•	Fresh air vents	•	•	Tinted glass	•
•	Grab handles			Vanity mirror	•
•	Hazard warning	•	•	Wash/wipe combination	•
•	Head restraints	•		Wiper delay	•
•	Heated backlight				

road test
by John Bolster

The cornering power, which would be impressive in quite a light car, is extraordinary for such a heavy machine.

Jaguar XJS: nothing but the best

In 1961, or was it 1962?, I drove a 3.8-litre E-Type Jaguar to watch a Grand Prix race. I frequently exceeded 150mph on the way and was amused to find that none of the racers went as fast on the circuit. Yet, I broke no laws.

In 1976, I drove the new Jaguar XJS, which has 12 cylinders and 5.3 litres to go very little faster. So rapidly has the motoring situation deteriorated that one could be arrested like a criminal for extending the car on any gear except first. A short time ago, when we were told that petrol supplies were running low, it seemed unlikely that cars like this would ever again be built, and it's difficult to justify the cost of buying and running such a vehicle. Yet, almost incredibly, there is a rising demand for super-cars and this, by far the most elaborate and luxurious Jaguar ever sold, is assured of a warm welcome.

It had seemed that the pleasure of owning a good car was waning, and that those who were clever enough to keep some folding money away from the tax man were spending it on boats and 'planes. Somehow, neither the water nor the air was the answer and the road, for all its restrictions, still exerts its old fascination. If a man can afford a Jaguar or a Ferrari, that is what he will buy, even though he probably cannot tell you why. It's not a mathematical equation, but to compare 12 cylinder motoring with four-cylinder commuting is like asking a chap if he prefers the warm, mild beer at the local to Dom Perignon champagne.

So, I shall not attempt to make any excuse for using a ton and three-quarters of extremely complicated machinery to carry me about for a week. If you tell me that I could have ridden a moped for six months on the same amount of petrol, you are going to get a very rude answer. I may be a bloated capitalist of the extreme right, but I loved every moment of it, so there!

This Jaguar is not intended to be a sports car and that it happens to be extremely fast is purely incidental. The 12 cylinder, light-alloy engine is fitted with Bosch mechanical fuel injection, manufactured under licence by Lucas, but this again is not to secure ultimate performance but to obtain unrivalled flexibility and low-speed torque. Perhaps most important, injection gives a welcome reduction in fuel consumption, compared with the carburetters normally fitted to this engine. Though the electronic ignition is of racing type, its main advantage is to ensure exactly timed sparks in all cylinders without frequent servicing.

Another important feature of the engine is its clean exhaust, the vertical valves and Heron-type combustion chambers having better anti-pollution characteristics than the classical inclined valves of earlier Jaguar models. For a car aimed at the export market, this is vital. A synchromesh gearbox or automatic transmission may be chosen at the same price.

As is well known, there are two lengths of Jaguar saloon, but the XJS uses the floor pan of the shorter one, with the rear suspension moved forward 6.9ins to give a wheelbase of only 8ft 6ins. The body shell is lighter than that of the XJ saloon but a full 3 cwt heavier than the considerably narrower E-Type, though its better shape gives it less aerodynamic drag than that car. The V12 E-Type was not nearly so "clean" as the old, original E-Type, and the new body has the advantage of being designed from the outset to pass modern regulations, instead of having to be adapted, as the E-Type was through the years.

The XJS is a wide car, giving superb comfort for two people. The rear seats are too cramped for adults on long journeys, but they are ample for short trips and, of course, perfect for children. The bonnet is shorter than that of the E-Type but is still long enough to be somewhat awkward when entering a road from a gateway. The rear window is kept clean by the fins on either side of it, which advantage is somewhat cancelled out by the consequent blindness of the rear quarters. From an engineering point of view, the greatest virtue of the body is its excellent torsional rigidity, which is of the utmost importance for roadholding.

This body shell is suspended on wishbones in front, on the normal Jaguar system, though the rating of the coil springs and dampers is suitably adjusted. The Adwest rack and pinion steering is higher-geared than on the XJ saloons and the E-Type, though the power-assistance is similar. The front brakes have cast-iron ventilated discs of 11.8ins diameter, with 4-piston calipers on dual circuits.

The rear suspension is of the usual Jaguar type, each hub having two spring and damper units. A very rugged lower wishbone is paired with a fixed-length driveshaft, forming the upper link, with a radius arm for triangulation. The inboard brakes have 10.38ins cast-iron discs and two-piston calipers. Each disc has a damper ring recessed into its periphery to ensure silent operation.

Anti-roll torsion bars are used at both ends of the car, that at the rear being unused on a Jaguar. Insulated sub-frames reduce the transference of noise and vibration into the body. The brake application is assisted by a tandem vacuum servo and although entirely adequate braking could be obtained by using only one circuit, a warning light would inform the driver if either system was developing a lower pressure than normal. The handbrake lever is of the commercial-vehicle type, falling horizontally to the floor on the right of the driver's seat when the car has been parked. It must be raised and re-engaged before being released, so it's impossible to kick the button and let off the brakes by accident.

The driving position is comfortable and the steering column length is adjustable, though a

road test

short driver may find difficulty in fully depressing the clutch pedal. Although there are cars with heavier clutch operation, this one may become tiring at the traffic lights. I do feel that all powerful cars should have servo-assisted clutch pedals, as the Packard did before the war, if only to match the lightness of the brake application.

Proper round instruments are legible at a glance, a highly desirable feature of any really fast car. Most sensibly, the speedometer dial only reads up to 160mph, which gives a reasonable spread of calibrations over the useful working area.

Too many cars are fitted with 200mph speedometers, which means that the figures are crowded in so closely together that they can easily be misread at a hurried glimpse. This is a pretentious fashion that fools nobody.

Perhaps the first lesson the XJS teaches one is just how good fuel injection can be, for not only does it give a useful increase of top-end power but it utterly transforms the low-speed flexibility. This Jaguar starts at once from cold and can be driven in thick traffic straight away. It is the first modern car to equal the flexibility of my Silver Ghost, running evenly on top gear at a walking pace and accelerating smoothly from 4mph on that ratio. Where it differs from the 1911 Rolls is that it has approximately double the maximum speed, so there has been some progress in 65 years!

Really to enjoy this V12, it's best to have the manual transmission. Only then can one appreciate the steam-like flexibility of this wonderful engine. We have had engines in the past, even quite cheap American ones, that would accelerate strongly and smoothly from 4mph in top gear. Where this one breaks new ground is in combining such flexibility with outstanding performance at high revs. The Jaguar peaks at 5500rpm and remains completely smooth up to the recommended maximum of 6500 — indeed, it is capable of very much more.

A sensible choice of gear ratio has been made and though a slightly higher "cog" would give one the chance of breaking the 160mph barrier, the better acceleration of the existing arrangement is far more valuable. Where this car is so outstanding is in its extraordinary acceleration from 120 to 140mph, where most other very fast machines are starting to labour. It seems to level out at 6250rpm, which is about 154mph, so there is no danger of entering the red section of the rev-counter dial.

These speeds are somewhat academic nowadays and in case somebody accuses me of advocating such driving, perhaps I should utter a word of warning. Anybody who is reasonably competent should be able to drive safely at speeds up to 130mph, but though much higher velocities are not at all dangerous in themselves, they call for judgement which can only be acquired by practice. The man who seldom drives at much more than 100mph would think twice about going up to 150mph without some exploration of intermediate speeds. If things go wrong, it all happens a bit sharpish, so don't say that uncle Bolster didn't warn you!

The whole point of owning an XJS is that it is entirely unstressed at any normal speed, and stability which has been worked out for 150mph is incomparably better than that of ordinary cars. It is utterly safe, and in an emergency it can be braked to a standstill in an incredibly short distance, without trying to take charge. Like all recent Jaguars, it was designed in collaboration with the Dunlop engineers, who have developed a special tyre for it. As a result, it generates cornering power which would be impressive in quite a light car but is extraordinary in such a heavy machine.

Near-neutral handling and exceptional traction are its outstanding characteristics. It is ultimately possible to make the rear end break away on a test circuit, but only by driving much harder than would be in any way reasonable on the road. The behaviour continues to be safe and predictable on wet surfaces and the steering is very much better than the rather over-assisted setup on other Jaguars, though there is still not much feel of the road.

The engine is quiet enough to permit conversation in subdued tones at 140mph, yet it does not have the woolly sound of most well-silenced power units. There is a proper note, rising inspiringly as the big V12 reaches peak revs, but in normal use the XJS is a quiet car, if not totally silent; the gearbox has an audible whine on the indirect ratios, which would be lost in the general commotion of most cars. The gearchange might be a little lighter, but its quite acceptable.

Sumptiously appointed interior.

Really impressive, even by Jaguar standards, is the insulation of road noise. The wind becomes audible at 90mph or so, but curiously enough it seems less noticeable at higher speeds. The heating and ventilation work very well, with refrigerated air conditioning which will be appreciated in the sunny climates where the fortunate owners of such cars are likely to relax. That the windows rise and fall electrically and the doors lock from one point are nowadays minimum requirements among upper-crust cars.

From time immemorial, the enthusiast has bemoaned the tendency of manufacturers to turn away from "hairy" sports cars towards ultra-

The author holds a steady and unflustered 140mph....

luxurious GT coupés. The answer is that everybody longs to own a stark semi-racer, but nobody actually buys them. A few wealthy Englishmen and a lot of rich foreigners will rush to buy the XJS, and it will make a great deal of money for Leyland. As I have already said, a car of this price and performance is scarcely a logical buy, but as long as there is a demand for such *exotica*, Britain should endeavour to supply it. I've driven the German and Italian equivalents, and very nice they are too, but nobody who has tried them all could be in any possible doubt. The V12 from Coventry is the one, and if they can make them all as well as they built the car they lent me, Britain has a world-beater. ▷

road test

Above: the appearance of this most expensive Jaguar is really aggressive. Below: the specially developed Dunlop tyres grip superbly.

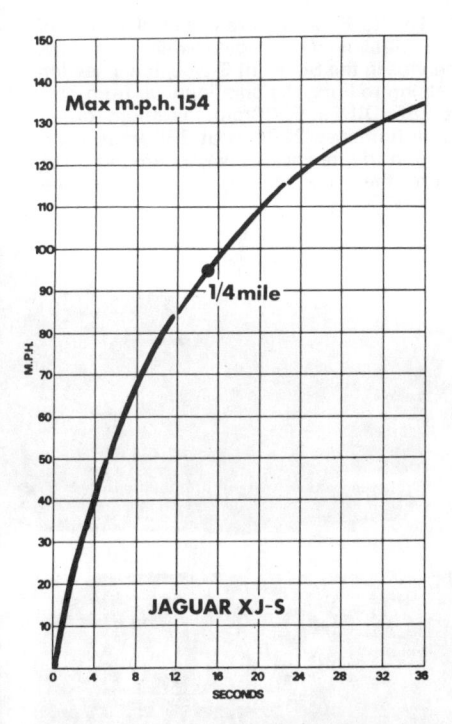

The impressive, and thirsty, fuel-injected V12 engine.

SPECIFICATION AND PERFORMANCE DATA

Car tested: Jaguar XJS 2+2 coupé, price £9527 including car tax and VAT.
Engine: 12 cylinders, 90 x 70mm (5343cc). Compression ratio 9 to 1. 285bhp DIN at 5500rpm. Single chain-driven overhead camshaft per bank. Lucas-Bosch fuel injection. Lucas electronic ignition.
Transmission: Single dry plate clutch. 4-speed synchromesh gearbox with central change, ratios: 1.0, 1.389, 1.905, and 3.238 to 1. Hypoid final drive, ratio 3.07 to 1.
Chassis: Combined steel body and chassis. Independent front suspension by wishbones, coil springs and telescopic dampers, and anti-roll torsion bar. Adwest rack and pinion power-assisted steering. Independent rear suspension by lower wishbones, fixed-length driveshafts, radius arms, and twin coil spring damper units, with anti-roll bar. Servo-assisted dual-circuit disc brakes with rear limiting valve. Bolt-on light-alloy wheels, fitted Dunlop Sport Super 205 70 VR 15 tyres.
Equipment: Speedometer. Rev-counter. Voltmeter. Oil pressure, water temperature, and fuel gauges. Clock. Heating, demisting, and ventilation system with refrigerated air conditioning. Heated rear window. Electrically-operated door windows. Electric door locks. Windscreen wipers and washers, two-speed plus flick-wiper. Flashing direction indicators. Reversing lights.
Dimensions: Wheelbase, 8ft 6ins. Track, 4ft 10ins/4ft 10.6ins. Overall length, 15ft 11.72ins. Width, 5ft 10.6ins. Weight, 1 ton 14cwt.
Performance: Maximum speed, 154mph. Speeds in gears: third, 116mph; second, 84mph; first, 50mph. Standing quarter-mile, 14.9s.
Acceleration: 0-30mph, 2.8s; 0-50mph, 5.0s; 0-60mph, 6.7s; 0-80mph, 10.4s; 0-100mph, 16.4s; 0-120mph, 25.7s.
Fuel consumption: 13 to 18mpg.

It is almost impossible to fault Jaguar's XJ-S, and to drive one on a long journey is still a memorable experience in one's motoring life no matter what one is used to driving. In the XJ-S we have a vehicle which will outrun virtually any sports car, and yet match the world's greatest luxury saloons for comfort, ride quality and silence. A quite amazing blend of contradictions and a unique technological achievement by Jaguar's small group of engineers.

The car is, of course, based on the XJ saloon (which Sir William has described as the Jaguar he is most proud of), the floor pan being that of the short wheelbase model but with the rear suspension assembly moved forward into a shortened rear seat pan to effect a reduction in wheelbase from 109in to 102in. The rear independent suspension is that developed from the E-type with its twin coil spring/damper units each side (reverting back to the E-type's rear anti-roll bar), and the front the twin wishbone coil spring XJ system with inbuilt anti-drive.

The XJ-S has a more powerful version of the all-alloy V12 engine which first saw service in the Series III E-type, fuel injection helping to bring the bhp figure up from 250 to 285 (DIN) at 5500rpm; maximum torque is an immense 293lb/ft at 3500rpm. Even though the newer cars weigh around 3cwt more than the last of the E-types, the power increase is enough to keep it ahead of the "E" throughout most of the speed range. To quote from *Motor's* road test acceleration figures, this translates into a 0 – 60mph time of 6·7, with 100mph coming up in 16·2; top speed is around 155mph.

The rapidity at which the XJ-S reaches such speeds as this may suggest an element of aggressive brutality in its character, but nothing could be farther from the truth. One can sit in the well shaped, highly comfortable leather seats and with no effort or worry achieve the highest cross country average speeds one is likely to put up driving any car in the world over the same route. It was quite possible to reach 140mph with reasonable ease in the E-type, but one was definitely conscious of the fact; in the XJ-S, 140mph becomes a commonplace, used-every-day speed, not

Unmatched in its combination of speed and comfort, the XJ-S must be the ultimate luxury touring car for two people. But is it a Jaguar truly wanted? askes **Paul Skilleter**

XJ.S

Above, power complex: covered in smogwear and induction tubing, the XJ-S's engine compartment resembles a plumber's shop ravaged by a wild gorilla; but it produces magnificently smooth power nonetheless. As for looks, the XJ-S is, at least, not wedge-shaped; twin round headlamps are fitted on US cars.

Opposite page, top: plenty of front seat adjustment but rear passengers are cramped on long journeys. Centre, instrument bezels are inappropriate and resemble the Mini Clubman's far too closely. Bottom, tradition lives - the old 'Jaguar' script is defiantly retained on the bootlid; and no 'Leyland' badges anywhere

to be particularly remarked upon — a figure on a dial rather than an obvious sense of achievement.

Needless to say, the car's brakes were well up to the job of dragging down 37cwt (laden) from three-figure speeds without the slightest hint of complaint, thanks to 11.2in ventilated discs at the front and 10.38in inboard discs at the rear; they were particularly outstanding in the wet too, and in fact *Motor* recorded a deceleration of 1g in rainy conditions! The steering is pleasantly power assisted and while not being imprecise, is marginally lacking in "feel"; power assistance is less than on the XJ saloons however, and the system is better for it.

Rarely is the handling of the XJ-S put fully to the test due to the sheer degree of roadholding obtained from the Dunlop tyres, but even near the limit the car is extremely safe if lacking the ultimate in finesse provided by such ultra-modern (and smaller) cars as the Lotus Elite. A gentle understeer is the usual characteristic, which under influence from the throttle can turn into an equally mild oversteer; one certainly has a feeling of excellent balance at all times and while the car is at its most impressive on very fast bends, it is still tremendous fun over twisty country roads. And even while you are driving fast, road and wind noise are both incredibly unobtrusive, and the suspension copes with the worst irregularities with indifference.

Yes, technically, the XJ-S gets almost maximum marks in every department you can think of, but perhaps the most fundamentally interesting point about it is the question of how it rates in the history of the sporting Jaguar, and whether it is completely the right product for the late 1970s.

It all goes back to the mid-sixties when plans were being made for a new Jaguar engine. Without realising it, the company had come to a fork in the road, as it could either have opted for a small, high efficiency unit or a large, relatively inefficient engine. The further variation, that of a large *and* highly efficient unit, was dropped when it became clear that the experimental four ohc, 4994cc, 500bhp engine designed originally for a racing application was too cumbersome and costly for use in a road Jaguar.

So the present 5.3-litre was evolved, any entirely new design owing nothing to the XK engine having a single ohc, and flat cylinder heads with the combustion chamber formed mostly by the piston; indeed horsepower comes from size rather than efficiency, and it was unfortunate to say the least that the launch of the V12 engine virtually coincided with the petrol crisis — everyone became aware that supplies of oil are neither infinite or cheap, and suddenly large thirsty engines appeared slightly antisocial. It wasn't that XJ12 owners couldn't afford to pay for their 11mpg beast's thirst, it was just that they became conscious of paying for it.

Almost certainly, in fact, it was originally assumed by Jaguar that within a reasonable period of time, the V12 would replace the faithful XK engine altogether (indeed the XJ saloon was designed for the V12 alone and it was with great reluctance that Sir William allowed it to be marketed at all with the six cylinder engine), and this might well have happened by now but for the assertion of strength by the Arabian states. But as things turned out, the XK engine has proved to be something of a saviour for the company, with its relative efficiency and economy, and there are now no signs of its being dropped.

Similarly, when the E-type "replacement" came to be discussed at Browns Lane, the decision was taken to tread the path of comfort and luxury rather than outright sporting appeal, and so we lost the traditional open two seater Jaguar sports car that had been a feature of the company's model range since 1935. While leaks from the factory had made it abundantly clear that the E-type was to be the last Jaguar two-seater, the size and character of the XJ-S when it appeared still took some people aback; and for the first time since 1949 when sporting Jaguars were marketed seriously in the States, a new Jaguar failed to get unanimous rave reviews in the specialised motoring press in its most important market — North America.

Not that anyone could really fault the way the XJ-S performed — it wasn't that. The reason didn't really lie with the car at all, but, to quote from what an American friend of mine wrote to me, "it is not that the XJ-S is an unworthy Jaguar (perish the thought!), but that for me it is the *wrong* Jaguar. And my sentiments are by no means unique".

It is a fact of life that at the time this is written, one can see classified advertisements in the West Coast newspapers offering brand new XJ-Ss at list price; no queues, no premiums, no excitement. You *always* had to wait three months for a Jaguar sports car before — but then the XJ-S isn't a sports car and wasn't intended to be. And while its economy is better than the V12 saloon's, the XJ-S's dimensions, price and styling do not pluck irresistibly at the customer's heartstrings as did the XK and E-type before it . . .

What then, with the huge benefit of hindsight, could Jaguar have offered either as well as or instead of the XJ-S? Quite simply, a true successor to the E-type, a genuine open two-seater (even with a built-in roll over bar if necessary), without so many of the XJ's luxury accoutrements but something like $5000 cheaper. It would have filled what has been called a "conspicuous hole" in the North American market, providing a high performance open sports car at a price affordable by the upper middle income man.

The price of the XJ-S definitely seems to be an inhibiting factor, and it represents a big jump up the scale for a sporting Jaguar. The price of Jaguar saloons has followed a quite normal pattern, and still equates with the price rise of such home produced cars as the Chevrolet in the same way that the E-type did; but the amount of dollars needed to buy an XJ-S suddenly elevated the car to a position out of proportion to all else, and for the first time a Jaguar cost the same as the equivalant Ferrari.

This was something new to the Jaguar-buying public, and it didn't like it all that much; especially when the car "looks like a Camaro" (though possibly US crash and safety regulations had something to do with the XJ-S's final shape — earlier prototype body designs with different bumpers, lights and bonnet line did look rather better). Added to which of course you couldn't get an open one, although it is possible that once again Ralph Nadar was responsible for this as during the car's development years it certainly looked as if the soft-top would be outlawed altogether; since then, it has been ruled unconstitutional that people might be prevented by law from killing themselves by having accidents in open cars.

So at what imaginary specification have we arrived for a Jaguar sports car that would sell like hot cakes in North America? A two-seater, or compact 2-plus-2, no longer than the old E-type but a little wider, most definitely available in open form, and powered either by the injected V12 engine or an entirely new high efficiency, four-valve-per cylinder three-litre unit; to sell at between $12,000 and $15,000.

We know that Jaguar have, and are, experimenting with smaller power units, including six-cylinder derivatives of the present V12, but there seems little chance that a genuine sports car is even under consideration at Browns Lane; after all, most of the money comes from selling the saloons, and the Jaguar sports car has always been a peripheral offering anyway.

Thus in summation, we would say that, if the XJ-S suits your needs, then buy it — there is no better luxury 2-plus-2 touring machine in the world and it will be a constant joy and satisfaction. But if you want a Jaguar sports car, with seats open to the skies and an urgent, cat-like agility and a spirit which is the embodiment of speed itself, then sit back and muse of the days of XKs and E-types and the howl of an unfettered straight-six; for the Big Cat has grown fat and lazy and only hunts indoors.●

AutoTEST
Jaguar XJ-S automatic

**Automatic transmission version of Jaguar's topmost sporting model.
Performance and economy suffer by comparison with manual version, yet still excellent in class.
Safe, predictable handling to high limits, and good brakes.
High standard of equipment and finish.
Smoothness, quietness and ability to cover ground quickly yet unobtrusively remain strongest points**

Though some people quarrel with the looks of the back end, few deny the XJ-S looks the part from the front. Halogen headlamps provide excellent illumination for quick night driving

HOW MANY PEOPLE who can afford to spend over £12,000 on a car are prepared to put up with manual transmission? Perhaps, if they are buying something totally exotic, low-slung and impractical; otherwise the inevitable heavy clutch and the bore of pushing a lever from slot to slot do not recommend themselves. Jaguar find themselves caught between two stools in this respect. The XJ-S can hold its own with almost any low-slung exoticar, despite being much more practical, so it must be offered with manual transmission. Yet it can only be a minority of customers who buy the XJ-S as a Lamborghini substitute, and most of the others will want automatic transmission.

The original *Autocar* Road Test of the XJ-S was carried out on a car with the four-speed manual gearbox. There were two reasons for this. In the first place, the car's ultimate performance had to be established, and there would have been a considerable penalty — as we now discover — with automatic. Second, the XJ-S was originally offered with the Borg-Warner Model 12 automatic as used in all XJ 4.2 and 5.3 models for some time. As we now know, it was Jaguar's intention to replace the Model 12 with the General Motors THM400, which they regarded as a superior transmission in many ways. For the first year or so of production, therefore, tests of automatic XJ-Ss were conspicuous by their absence.

There are no changes to the XJ-S other than those physically demanded by the automatic transmission. The final drive ratio, and hence the overall gearing in top, remains the same at 3.07-to-1; so direct performance comparisons are valid. The engine, fuel-injected, continues to produce its 285bhp at a peak of 5,500 rpm. Nor does the chassis differ in any respect, with all-independent suspension, power-assisted rack and pinion steering, disc brakes ventilated at the front and plain at the back, and 70-series Dunlop 205-15in. tyres carried on light alloy wheels.

Performance and economy

No matter how much power is available, a car with automatic transmission will always be slower than the same car with a manual gearbox — unless the overall gearing is fiddled in the automatic's favour, or the manual transmission cannot really cope with the task. As we have said, the Jaguar's gearing remains the same, and its manual transmission is extremely well matched to the engine; so the automatic pays the penalty.

This is not to say the XJ-S automatic is slow. It loses time to the manual car progressively as the speed increases from the standing start. As might be expected, nearly half a second is lost in the step-off from rest, 30mph coming up in 3.2sec rather than 2.8sec. An interesting point is that the automatic regains a fraction of a second to 60mph, because the manual car suffers an intervening gearchange while the GM transmission runs to 64mph at the red line in low ratio. After that, however, there is no contest, as witness the times to 100mph (18.4 against 16.9sec) and 120mph (30.4 against 25.8sec). All the same, it comes as a salutary reminder to students of acceleration that this apparently wide advantage is worth only half a second at the kilometre post.

It is more difficult to compare acceleration in each gear, because the lower ratios of the three-speed automatic do not correspond with those of the four-speed manual gearbox. In top gear, only the higher speed range can be studied because the GM automatic kicks down quite readily into intermediate up to 78mph. From 80 to 100mph in top takes the automatic 8.5sec, compared with 7.9sec in manual form. The increment from 100 to 120mph leaves a wider gap, 12.2 against 10.0sec.

The intermediate ratio takes the car all the way from 23mph (the kickdown point to low ratio) to its maximum of 108mph at the red line. Most of the time, it gives acceleration quicker than third gear, but slower than second gear in the manual version. Only at the top end does intermediate "run out of breath" compared with manual third gear, which of course has a higher maximum. The same picture is seen when comparing the automatic low ratio with manual second gear.

Where maximum speed is concerned, there is a far more obvious discrepancy in performance. While the manual car managed a mean 153mph, the automatic could only muster 142mph in near-ideal testing conditions. Even this speed takes the car beyond peak power, leading one to speculate what might be the effect of a higher (say 2.8-to-1) final drive ratio. It is also worth noting that with the air-conditioning compressor switched on, the mean maximum speed fell to 138mph, indicating the amount of power — perhaps 15bhp — swallowed by the system at high speed. A set of acceleration runs with the air conditioning on showed its rather smaller effect on standing-start times, a handicap of little more than a second to 100mph (19.6sec mean against 18.4sec with the air conditioning off).

At steady speeds, there is a deficit of something between five and ten per cent in fuel consumption, the penalty being heaviest at lower speeds where some torque converter slip is occurring, compared with the manual version. Not unexpectedly, the fuel consumption curve is much flatter than is usually the case in lower-powered cars, ranging only from 22mpg at a steady 30mph to 13.3mpg at 100mph. At first sight this suggests consumption should vary little however the car is driven, but the curve takes no account of the extra thirst brought on by rapid acceleration, which can have a profound effect. In the course of our longer than average test period the best brim-to-brim consumption we saw was 17.9mpg, for a long journey mostly on motorways. The worst, in typically thick commuter traffic, was 11.7mpg. These two figures neatly bracket our overall figure of 14mpg, exactly 10 per cent worse than we achieved in the manual car, and give some idea of the extremes likely to be encountered according to conditions and driving style.

Handling and brakes

The XJ-S has power steering as standard, and there is evidence that Jaguar (like Rolls-Royce) have now accepted that it is better to provide power assistance than a power take-over. As a result, it is no longer possible to wind the front wheels to full lock with one's little finger when the car is stationary; on the other hand, there is now sufficient feel in the steering to enable the car to be driven with confidence very close to its limit, even in tricky conditions. No longer, as in the early XJs, must the driver merely hope the front wheels have some grip left. The steering is also sensibly geared, with three turns of the wheel between extremes of a 36ft lock, by no means bad for this kind of car. It is quick enough to ease the problems of driving around town back-doubles, yet by no means so sensitive as to induce that go-kartish weaving encountered in some cars with over-direct steering.

Compared with the kind of tyres fitted by some of the Jaguar's competitors, the Dunlops are not wide. Almost certainly, considering what the car weighs, wider-section tyres would increase ultimate cor-

Considerable roll angle is evident in this cornering shot of the XJ-S on the MIRA handling track. The car is close to the limit here: note inside front wheel almost off the ground, but inside rear still with enough grip to prevent wheelspin

nering ability. Yet the use of tyres of "only" 205-section is a well-judged compromise which takes into account the need for reassuring stability (especially over roads of varying camber) as well as the excellent steering feel and reasonable turning circle. As things are, the XJ-S is very high in the league table of handling and roadholding, though by no means at the very top. What is more important, except to the sporting purist, is that the handling is sufficiently good-natured that even a moderately skilled driver can use most of it. There is no feeling in the XJ-S, as in some of its contemporaries, that one needs to have been to racing school to drive it anything but gently. Much of the credit for this must go to the supple suspension design, as important to the handling as to the ride, keeping all four wheels on the ground far more of the time than some rivals manage.

Certainly one would hardly credit that nearly 55 per cent of the kerb weight rests on the front wheels, for any understeer is hardly noticed. The power steering, of course, will do its bit towards disguising what little there is. To a large extent the imbalance of weight has been redressed by careful choice of anti-roll bars: the XJ-S has a rear bar as well as the saloon-type one at the front. What remains can so easily be balanced by judicious application of power, helped by the long and smooth accelerator movement. The limit is set by roadholding rather than loss of control, and it is no problem to reach and hold the point at which all four wheels are beginning to skitter sideways across the surface. Only a clumsy combination of steering and braking will lead to the front end sliding straight ahead.

The brakes themselves are extremely good in most respects, though with two minor reservations. One concerns their lightness. For the most part they are progressive, but the first tentative pressure on the pedal produces what might well be regarded as normal light check braking. A pedal effort of 60lb suffices to achieve the ultimate crash stop with a deceleration of well over 1g. With practice it is possible to feather the brakes very smoothly, but we wonder if such sensitivity is really called for. The effect is accentuated by the speed-sensitivity of the brakes, most evident in our fade test. During the first five stops, the driver had to ease off his pedal effort as the car slowed in order to hold the deceleration down to 0.5g. Not until the seventh stop had the brakes warmed to the point where the

effort stayed the same until the car reached a standstill.

Our other doubt about the brakes came when achieving the best possible stop. It became clear that the back wheels tended to lock slightly prematurely, leading to some mild control problems in a panic stop. We also noted a degree of wheel judder when stopping very quickly.

The hand-brake proved most effective — not always a "supercar" trait — and pulled 0.32g when used alone on the level. It had no trouble holding the car either way on the 1-in-3 test hill, and like most powerful automatic cars, the XJ-S treated the restart with absolute contempt.

Comfort and convenience

The most obvious question is: how far can the XJ-S be regarded as a four-seater? The back seats are apparently well shaped to provide support for two occupants, but the lack of headroom, and more particularly kneeroom, behind large occupants of the front seats means they are only suitable for medium-sized children. In this capacity the back seats serve admirably. A ride in the back of the test car was received with enthusiasm by a ten-year-old test subject, but he was getting towards the upper limit of those who would find the back comfortable (and even he was riding behind the front passenger rather than the driver). At best, then, the XJ-S qualifies as an emergency three-plus-one.

For those in front, the comfort is of a high order. There is plenty of rearward seat adjustment — no artificial limit here for the sake of a spurious gain in the back — and the seat shape provides adequate support for real comfort. It is reassuring to find that for once, the superficial appeal of the "showroom" seat has given way to the firm support of the seat that remains comfortable for driving (as we found) up to five hours at a stretch. The test car had leather upholstery, which continues to provoke argument among our test staff; suffice it to say that some of us would have preferred high-quality cloth.

Though the XJ-S wheelbase is shorter than that of the saloons, the ride feels hardly inferior, which means that by the standards of GT cars it is among the best. The long wheel travel enables the suspension to absorb a great deal of shock even on broken or unmade road surfaces, while the quality of the damping is such that there is little suggestion of upsetting "heave". Indeed, the XJ-S is arguably superior to the saloons in this respect. The only untoward effect we noted was some low-frequency vibration fed through from the road surface at certain speeds, which seemed to be noticed more by the driver than the passengers. Roll angles build up noticeably in hard cornering, yet again passengers seem less aware of them than spectators at the side of the road. Single large humps, like

A daunting prospect for almost everyone, the XJ-S engine compartment is tightly packed with equipment. Complicated fuel injection piping and induction system adds to crowding. So does air-conditioning plumbing, power steering pump and the usual collection of filters and reservoirs

AutoTEST
Jaguar XJ-S automatic

canal bridges and level crossings, are crossed without drama: it is only over the most precipitous that the nose-heaviness of the car tends to make it land rather heavily front wheels first.

From the driver's point of view, there are plus and minus points. On the plus side are the steering, already discussed, the smooth accelerator action and the lightness (once one is accustomed to it) of the brakes. The minus points concern the transmission selector, the minor controls and the pedal layout. We have often criticised particular gating arrangements in automatic transmission selectors, but the gate in the XJ-S was so vague and awkward in its operation that it was far from easy to determine what the arrangement was supposed to be. Certainly the lever positions did not line up with the PRND21 indications on the quadrant (on the wrong side of the gate for the home-market driver). The handbook indicates that there is free movement of the lever only between 1 and 2, and between D and N; but our inability to "feel" the gate made life extremely awkward, and we lost count of the number of times we overshot D and found ourselves revving in Neutral when returning from the 2 position.

The GM transmission itself does all that is asked of it in the way of rapid yet smooth changes of ratio, except perhaps when a manual change-down is selected on the overrun (which is always the most difficult case to meet). Drivers familiar with the former Borg-Warner transmission will note three changes of consequence. A safety-change on low ratio brings an automatic shift to intermediate at 6,500rpm even when one is selected. The 2 position permits automatic changes between low and intermediate, rather than simply holding intermediate as before; and the speeds at which low ratio may be engaged are markedly lower. If the driver selects 1 while on the move, the transmission will not shift to low until the speed is down to 15mph, if it is left to itself. Even the kickdown switch will not bring about the change above 23mph. To all intents and purposes, therefore, it is not possible to use low ratio once the car is properly on the move. Granted there may be plenty of performance there without using low ratio; yet in our more exuberant moments we several times felt frustrated by our inability to do so.

The instruments and controls are generally well laid out, once one is used to the novelty of vertical-quadrant presentation for the minor instruments rather than the traditional Jaguar dials. In the test car, both the speedometer and rev-counter over-read in some small degree, the latter showing the 6,500rpm red line at a true 6,250rpm. Two annoying oddities are the trip reset knob — tucked away under the facia and completely impossible to operate on the test car — and the use of special-to-Jaguar column stalks (rather than the perfectly good Leyland "corporate" ones) for lights, indicators and wipe/wash. All our drivers commented adversely on the need to depress a separate little button carefully hidden behind the end of its stalk in order to operate the screen washer; and all were intrigued by the presence of a matching button on the other stalk which apparently had no function at all. The screen wipers themselves are probably the worst single feature of the XJ-S, for they are small, slow and hesitant — almost eccentric — in their manner of operation. In no sense do they match the performance of the car, whereas the splendid halogen headlamps provide remarkable range on main beam, and a well-controlled spread when dipped.

Another point which bears comment is the pedal layout. The brake pedal is larger than that in the manual car, presumably with the object of allowing left foot braking if the driver wishes. But on several occasions we found that when braking with the right foot, we could not brake as hard as we wished because the *left* foot, resting on the floor beneath the pedals, became trapped and prevented further pedal travel. A proper rest for the left foot might be the best solution.

The air conditioning system, fitted as standard in the XJ-S, generally works well and responds quickly to any change in the setting of the temperature control knob which permits the selection of any

Main instruments are speedometer (left) with total and trip odometers, and rev counter, with vertical-scale water, oil pressure, fuel tank and voltmeter instruments between them. Press-in switches are for hazard warning, heated rear window, map and interior lamps. Lights switch is behind wheel spoke on left. Air conditioning control on left or radio/cassette player, with distribution on right. Central locking override by transmission selector; electric window lifts are to the rear of this. Versions of Leyland's corporate column levers are used

Specification

ENGINE
Cylinders	Front; rear drive
	12, in 60 deg vee
Main bearings	7
Cooling	Water
Fan	Viscous and electric
Bore, mm (in.)	90 (3.54)
Stroke, mm (in.)	70 (2.76)
Capacity, cc (in³)	5,343 (326.1)
Valve gear	ohc
Camshaft drive	Chain
Compression ratio	9-to-1
Octane rating	97 RM
Fuel injection	Lucas/Bosch electronic
Max power	285 bhp (DIN) at 5,500 rpm
Max torque	294 lb ft at 3,500 rpm

TRANSMISSION
Type	General Motors THM 400 three-speed automatic with torque converters

Gear	Ratio	mph/1000rpm
Top	1.000	24.74
Inter	1.485	16.66
Low	2.485	9.96

Final drive gear	Hypoid bevel, limited-slip
Ratio	3.07-to-1

SUSPENSION
Front — location	Double wishbones
springs	Coil
dampers	Telescopic
anti-roll bar	Yes
Rear — location	Drive shafts and lower wishbones
springs	Twin coil
dampers	Telescopic
anti-roll bar	Yes

STEERING
Type	Rack and pinion
Power assistance	Yes
Wheel diameter	15½ in.

BRAKES
Front	11.2 in. dia. disc
Rear	10.4 in. dia. disc
Servo	Vacuum type

WHEELS
Type	Aluminium alloy
Rim width	6.0 in.
Tyres — make	Dunlop SP
— type	Radial-ply tubed
— size	205/70-15in.

EQUIPMENT
Battery	12 volt 68Ah
Alternator	60 amp
Headlamps	Halogen, 110/220 watt (total)
Reversing lamp	Standard
Hazard warning	Standard
Electric fuses	16
Screen wipers	2-speed
Screen washer	Electric
Interior heater	Air conditioning
Interior trim	Leather seats, pvc headlining
Floor covering	Carpet
Jack	Screw cantilever
Jacking points	Two each side
Windscreen	Laminated
Underbody protection	Bitumastic overall

MAINTENANCE
Fuel tank	20 Imp. galls (91 litres)
Cooling system	37 pints (inc. heater)
Engine sump	20 pints SAE 20W/50
Transmission	16 pints ATF Dexron 2
Final drive	2¾ pints SAE 90EP
Grease	14 points
Valve clearance	Inlet 0.012-0.014in. (cold) Exhaust 0.012-0.014in. (Cold)
Ignition pick-up	0.020-0.025in. gap
Ignition timing	10 deg BTDC (stroboscopic at 750 rpm)
Spark plug—type	Champion N10Y
— gap	0.025in.
Tyre pressures	F26; R24 psi (normal driving)
Max payload	720 lb (325 kg)

Maximum Speeds

Gear	mph	kph	rpm
Top (mean)	142	228	5,740
(best)	143	230	5,780
Inter	108	174	6,500
Low	64	103	6,500

Acceleration

True mph	Time secs	Speedo mph
30	3.2	31
40	4.5	41
50	5.9	51
60	7.5	61
70	9.5	72
80	11.9	83
90	14.7	94
100	18.4	104
110	23.1	114
120	30.4	124

Standing ¼-mile: **15.7 sec, 93 mph**
kilometre: **27.8 sec, 117 mph**

mph	Top	Inter	Low
10-30	—	—	2.3
20-40	—	—	2.7
30-50	—	3.9	2.7
40-60	—	4.4	3.0
50-70	—	4.6	—
60-80	—	4.7	—
70-90	—	5.4	—
80-100	8.5	6.1	—
90-110	9.5	—	—
100-120	12.2	—	—

Consumption

Overall mpg: 14.0
(20.2 litres/100km)
Calculated (DIN) mpg: 16.4
(17.3 litres/100km)

Constant speed:

mph	mpg
30	22.0
40	21.6
50	20.9
60	19.6
70	18.0
80	16.4
90	14.8
100	13.3

Autocar formula
Hard driving, difficult conditions 12.7 mpg
Average driving, average conditions 15.4 mpg
Gentle driving, easy conditions 18.2 mpg

Grade of fuel: Premium, four-star (97 RM)
Mileage recorder: 1.4 per cent over reading

Oil
Consumption (SAE 20W/50) Negligible

Brakes

Fade (from 70 mph in neutral)
Pedal load for 0.5g stops in lb

	start/end		start/end
1	28/20	6	32/28
2	28/18	7	32/32
3	28/22	8	32/30
4	28/22	9	32/32
5	30/22	10	33/33

Response from 30 mph in neutral

Load (lb)	g	Distance (ft)
20	0.30	100
30	0.48	63
40	0.70	43
50	0.88	34
60	1.05	29
Handbrake	0.32	94

Max. gradient 1 in 3

Test Conditions

Wind: 8-12 mph
Temperature: 5 deg C (41 deg F)
Barometer: 29.60 in. Hg
Humidity: 65 per cent
Surface: dry asphalt and concrete
Test distance 2,247 miles

Figures taken at 4,600 miles by our own staff at the Motor Industry Research Association proving ground at Nuneaton, and on the Continent

All Autocar test results are subject to world copyright and may not be reproduced in whole or part without the Editor's written permission

Regular Service

Item	3,000	6,000	12,000
Engine oil	Check	Change	Change
Oil Filter	—	Change	Change
Gearbox oil	—	—	Check
Spark plugs	—	Clean	Change
Air cleaner	—	—	Change
Total cost	£17.05	£42.02	£57.04

(Assuming labour at £5.50/hour)

Parts Cost

(including VAT)

Brake pads (2 wheels) — front	£19.45
Brake pads (2 wheels) — rear	£19.55
Exhaust system	£280.53
Tyre — each (typical advertised)	£60.00
Windscreen	£33.75
Headlamp unit	£23.33
Front wing	£60.05
Rear bumper	£31.97

Warranty Period
12 months/unlimited mileage

Weight

Kerb, 34.5 cwt/3,866 lb/1,754 kg
(Distribution F/R, 54.7/45.3)
As tested:
38.6 cwt/4,321 lb/1,961

Boot capacity: 15.0 cu. ft.

Turning circles:
Between kerbs L, 36ft. 6in.; R, 35ft. 6in.
Between walls L, 38ft. 6in.; R, 37ft. 6in.
Turns, lock to lock 3.0

Test Scorecard

(Average of scoring by Autocar Road Test team)

Ratings: 6 Excellent
5 Good
4 Above average
3 Below average
2 Poor
1 Bad

PERFORMANCE	5.17
STEERING AND HANDLING	4.50
BRAKES	4.60
COMFORT IN FRONT	4.33
COMFORT IN BACK	2.86
DRIVERS AIDS	3.75
(instruments, lights, wipers, visibility etc.)	
CONTROLS	3.86
NOISE	5.17
STOWAGE	4.00
ROUTINE SERVICE	2.80
(under-bonnet access: dipstick etc.)	
EASE OF DRIVING	4.90
OVERALL RATING	**4.42**

Comparisons

	Price £	max mph	0-60 sec	overall mpg	capacity c.c.	power bhp	wheelbase (in.)	length (in.)	width (in.)	kerb weight (lb)	fuel (gal)	tyre size
Jaguar XJ-S	13,200	142	7.5	14.0	5,343	285	102	192	70½	3,866	20	205/70-15
Aston Martin V8	16,599	146	6.2	12.4	5,340	—	103	183	72	3,893	21	GR70-15
Maserati Khamsin	17,960	130	7.5	15.1	4,930	320	99	172	69	3,454	20	215-15
Mercedes 450SLC	13,950	136	9.0	14.1	4,520	225	111	186½	70½	3,605	20	205/70-14
Rolls-Royce Corniche	33,134	120	9.6	11.9	6,750	—	120	203½	72	5,174	23½	235/70-15

Jaguar XJ-S automatic

interior temperature between 65 and 85degF. The system is generally quiet in operation, though there can be a disturbing burst of fan noise when driving away from a cold start: it also suffers the drawback of all but the most sophisticated systems, of causing a rapid though temporary misting-up in certain conditions. It also seems slightly primitive that one should have to individually close each face-level air inlet to divert the greatest flow of warm air into the footwells.

Quietness remains perhaps the greatest single attribute of the XJ-S, in common with its V12 stablemates. Apart from the smoothness and quietness of the engine itself, few engineers have been able to match the XJ's standard of road noise insulation, and the S is almost as good as the saloons in this respect. Attention to sealing (helped no doubt by the doors which are lower and stiffer than those of the saloons) keep wind noise to an equally low level. As a result, normal conversation is possible up to at least 120mph, while at 70mph one is hardly conscious of any noise at all.

Living with the XJ-S

One of the nicest things about the V12 Jaguar engine is its lack of temperament. The XJ-S started immediately whatever the conditions, idled smoothly, pulled readily from cold and warmed up quickly. The car's thirst may be prodigious, but the fuel tank holds 20 gallons to give a safe range of 250 miles or so, and filling to the brim (via a cap beneath a lockable flap in the left rear "fin") is less tedious than in many modern cars. The fins themselves, sweeping down from roofline to tail, are the only serious obstruction to the driver's vision, partly offset by the provision of an internally adjustable door mirror.

The fuel tank lives forward of the boot, which in consequence is deep enough to house the big spare wheel vertically against its front bulkhead, alongside the battery for which there is no room under the crowded bonnet. The remaining space for luggage is more generous than is often found in cars of this class, and most suitcases can be stowed upright, making things easier to organise. Interior stowage for smaller items is generous, with a central locker between the front seats, pockets in both doors and a glovebox under the passenger side of the facia. There are even small inset pockets in the trim beside the back seats, while the rear parcels shelf has a deep lip to hold things in place. Electric windows and centralised locking (in other words, of the passenger's as well as the driver's door) are standard. In the test car, the driver's door window was so slow in operation that we spent most of the test fearing it would fail altogether. There are no less than five interior lights, including the central one in the roof.

Above: Rear quater view shows "sails" which assist airflow but obstruct driver's vision. Fuel filler cap lies beneath flap visible in nearside sail. Right: Front seats fold forward to five access to back, and long doors make entry easier – though one must watch out for front safety belts to avoid tripping. Below: Back seat seems comfortably shaped and occupants well provided for with armrests and oddment recesses; but lack of head and knee room mean those occupants must be small if they are to relax

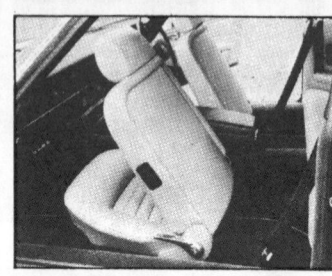

Right: Depth of boot enables spare wheel (with protective cover) to be stowed upright alongside battery, here shown with cover removed. Excellent tool kit suffices for most owner-driver tasks. Remaining space takes lots of luggage. Below: Driver's door mirror is internally adjustable. Below right: Fusebox is underneath facia on driver's side. Fuses are identified on cover

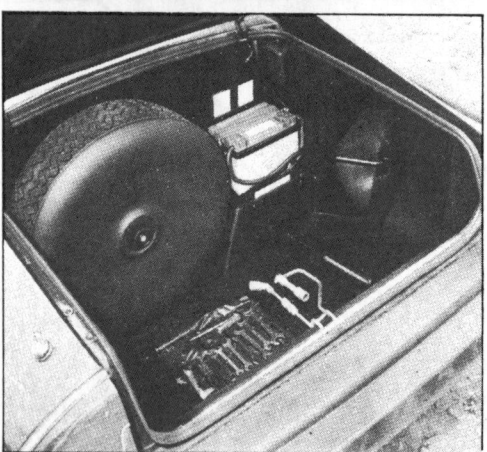

The scene under the bonnet is enough to daunt all but the most confidently qualified. There is not only the superb engine with its fuel injection and anti-pollution equipment; the air-conditioning system, power steering pump and other accessories add to the crowding. Despite this, the few points that need regular checking are by no means difficult to reach. The dipsticks, for engine and transmission, are easier to use than most, while the reservoirs are translucent. Extracting the sparking plugs is more of a problem, though the excellent tool kit includes an articulated socket spanner designed for the job. Fortunately the electronic ignition system needs no regular re-setting; a good deal of the fuel system plumbing needs to be shifted before the distributor can be reached. A minor irritation is that the bonnet release is on the passenger's side of the car.

The fusebox, containing all the fuses except those protecting the headlamps, is inside the car, low down under the driver's side of the facia and needing some contortion to investigate; a diagram inside its cover clearly identifies each fuse.

In conclusion

There are few enough cars at any price which offer as much as the XJ-S. If one is pedantic enough to insist on 12-cylinders and automatic transmission, there are only two, one of which costs half as much again, and the other, almost twice as much. In other ways, the Jaguar finds more logical competition in cars like the Mercedes 450SLC, which is priced very close to it and offers more room though less sheer performance. Perhaps the greatest asset of the XJ-S is that it can be whatever the owner wants it to be: one of the most capable and quick genuine GT cars, or a civilised, undemanding and incredibly refined carriage.

MANUFACTURER:
British Leyland UK Ltd.,
Grosvenor House,
Redditch, Worcestershire

PRICES
Basic	£11,282.00
Special Car Tax	£940.17
VAT	£977.77
Total (in GB)	**£13,199.94**
Seat Belts	(Standard)
Licence	£50.00
Delivery charge (London)	£40.00
Number plates	£10.77
Total on the Road (exc. insurance)	**£13,300.71**
Insurance	Group 7

EXTRAS (inc. VAT)
*Stereo radio/cassette unit	£298.07
Non-standard paint	£136.91
Whitewall tyres	£48.00

*Fitted to test car

TOTAL AS TESTED ON THE ROAD £13,598.78

On the road

CONTINUED FROM PAGE 13

Jaguar, and thus virtually any manufacturer, has produced before. This has really been achieved in the XJ-S's unique blend of speed, comfort and silence – I can't think of any other car in the world that is endowed with all these attributes in one package, Rolls-Royce and Mercedes included. Even better, the XJ-S handles in a manner which doesn't let you down over the twisty bits and has a far greater resistance to roll than the XJ saloons – you can't exactly say that it handles like a pure sports car, but it appears to get as close to that objective as is possible with a car of that size and character. To be quite frank, I didn't get the car near the limit during my half-hour's drive what with a Jaguar man sitting at my left elbow, though in a way this supports what I've said – the limits of the vehicle are such that to initiate any sort of breakaway above 60mph corners means that you have to be motoring extremely quickly. Definitely a case of visibility and nerve limiting cornering speeds rather than the car in most circumstances. All right, some of us may prefer even more direct steering but quite honestly it's marginal, and on balance I'd probably say that steering "feel" couldn't be much improved.

Seating and driving position are both excellent, the seats having a cushion which effectively grips the driver having a slightly harder outer diameter with a softer inner section. Room in the back is passable though not spacious by any means if the front seats have been pushed back – but then if you want to carry four people regularly and over long distances, you buy the (cheaper) XJ 5·3; horses for courses. Because of the generally low noise levels and the effortless ease with which the car performs, speed becomes less exhilarating than you thought it was; the XJ-S accelerates in the surging but undramatic fashion of the XJ 5·3, only rather more quickly and even more quietly – a feat indeed!

I think it goes without saying that the XJ-S will sell – in fact I'm fairly certain that the first year or so's production must already be sold, if those people who've said they've had an order for the E-type's successor on the books for the past X number of years really meant what they said. In any case, and in spite of a new 2,000 foot assembly track alongside those of the XJ saloons, the XJ-S is hardly being made in quantity – about 15 a week to begin with, building up to about 60 in around a year, so I understand. Reaction to the XJ-S from the States will be interesting – that country loves the Triumph TR7 so I wonder if the XJ-S will also receive the traditional rapturous welcome that a new sporting Jaguar usually enjoys in that country.

Even though the XJ-S will become the most expensive Jaguar yet sold, the small number which will be made ensures that it will by no means contribute a particularly large slice to Jaguar's profit figures. And while Jaguar have been and are one of British Leyland's few profit making divisions, making the right quantities has been one of several headaches – it is interesting but a little disheartening to learn that production of the XJ saloons did not increase until five years after the car's introduction; the uplift came when Geoffrey Robinson's new broom swept through Browns Lane. The story of Robinson's brief career at Jaguar is a fascinating one, but too complex to relate here; sufficient to say that during the few months he was at the factory, he became a Jaguar man and the Jaguar workforce became Robinson men, and one can only regret that he was forced to resign when Jaguar were integrated even more into Big L. But for the time being at least, Jaguars are still Jaguars and without question the XJ-S continues the grand tradition. ●

CONTINUED FROM PAGE 39

2 tons of laden Jaguar with astonishing rapidity. The wide tyres then sometimes jiggled a little in lorry ruts, but the car slowed on course. It is difficult to convey the serene sense of security with which the whispering XJ-S ate up the autobahn miles, disposing smugly of flat-out Mercedes S-series cars with the same surge of phenomenal acceleration which is present from zero to 150 m.p.h.

Off the autobahns the XJ-S had magnificently high standards of roadholding. In damp conditions it had a tendency towards understeer on slow roundabouts, when the Powr-Lok limited slip differential, a boon otherwise for traction, was of no particular help. But set up correctly, this big, heavy Jaguar cornered very quickly indeed, wet or dry, drifting towards oversteer on the limit, with a vague sensation as though the outside wheel was tucking in. If the XJ-S feels more stiffly sprung than the XJ saloons it is not totally immune to wallow, which is why it retains a comfortable ride, but it manages to respond adroitly to quick changes of direction. Anti-dive geometry robs the brakes of some feel as speed falls to the lower regions; this, and a deceptive impression of slower speed than reality, provoked a few instances of brake locking on damp roads.

Automatic mixture control by the "brain" of the electronic injection normally ensured instant hot or cold starting, one of the main improvements over the old carburetter engine. Unfortunately, while cold starting remained perfect, hot starting deteriorated during our trip: a one-way valve fitted to the fuel pump outlet to ensure that pressure in the fuel line is maintained wasn't seating properly, underbonnet vaporisation resulting from the drop in pressure. Ironically, in view of our destinations, the valve is made not by Jaguar or Lucas, but by Bosch in Stuttgart, who are modifying the design. A slight misfire, sometimes at low speeds, more normally when reapplying throttle after decelerating to 80-90 m.p.h., was thought to be a loose connection on an injection temperature sensor. Neither of these minor faults delayed us nor slowed us.

Where the XJ-S wins hands down over any other performance car is in its amalgamation of supreme high-speed performance with docility, flexibility and a total lack of temperament at low speeds. We encountered heavy traffic in all the German cities we visited and later used the XJ-S for London commuting. In these conditions it was almost totally silent save for the clicking in and out of the standard Delanair automatic air conditioning; no sign of plug wetting nor overheating. The big pedal pad eased the effort of operating the fairly heavy clutch and in any case the V12's flexibility and smoothness of throttle control made this a two-gear car (1st and 3rd or 2nd and top) in town. Top gear could be used practically all the time if one was cruel enough. This laziness of town driving convinced us firmly that the manual car is a totally superior proposition to the Borg-Warner automatic alternative: the manual's major attributes are vastly superior acceleration, particularly for safe overtaking and far better car control for fast road driving away from motorways. Gear-change movements were easy, but a fierce clutch made smooth changes difficult from 1st to 2nd.

Some other details we didn't like were the lame-dog slowness of the wipers lifting off the stop for a single screen wipe, the difficulty of operating the shrouded washer button in the end of the steering column stalk (Lucas are engineering a modified version, say Jaguar), the need to carry four keys for various locks and the uselessness of the central locking system in this two-door car. To make this worthwhile it should be possible to have simultaneous locking and unlocking of the opposite door from the key-lock of one door, Mercedes-wise. On the other hand, the electric windows wound satisfactorily quickly, the optional Philips RN 642 stereo radio/cassette player/recorder with automatic aerial was simply superb (a Radiomobile 1085S radio is standard equipment) and on one occasion the horrendous hydraulic bumpers proved to be effective and worthwhile after all.

A rear hub oil seal went when nearly back in England, the nearside door lock linkage fell apart (production tolerances will be improved as a result), the air conditioning control illumination occasionally died (the bulbholder specification will be changed) and we had that noisy door seal (a new seal with a thicker lip section is going into production). We feel that these minor failings in this early production car do not detract from its overall reliability on a journey upon which it was thrashed consistently and unmercifully at speeds which most XJ-S owners will never aspire to, both by us and by Mercedes' Chief Test Driver, who put it through its paces on the Stuttgart test track. In spite of three-figure average speeds on the Continent, this XJ-S achieved an overall 14.1 m.p.g., inclusive of use in Britain, and the engine, which an eminent German engineer described as "the best production engine in the world", used less than one pint of oil throughout.

In its diverse performance attributes the XJ-S is unequalled. With more attention to detail finish and appointments and some styling revisions it could so easily be the best car in the world.—C.R.

If the first thought that comes to your mind when someone says "Group 44" isn't "the thundering white horde from British Leyland," obviously you haven't been following SCCA national racing the past 11 years. Because since 1965 when Group 44 was formed by Bob Tullius and Brian Fuerstenau, the team has campaigned TR-3s, TR-4s, Spitfires, MGBs, Midgets, TR7s, Jaguars and virtually every other British Leyland product with four wheels to countless SCCA national victories including 14 national titles. Incredible.

During 1976, Group 44, composed of Bob Tullius driving an E-Type Jaguar and an XJS coupe, Brian Fuerstenau in an MGB, John McComb in a TR7 and John Kelly in an MG Midget, compiled an impressive record of 29 victories in 45 starts. Four of those wins were accrued by Tullius in the Jaguars, three with the E-Type and one with the late-arriving XJS. Don't let the coupe's sole victory in 1976 lull you into thinking the XJS is less competitive than any of Group 44's previous race cars. In the car's first race, the Mosport Trans-Am, Tullius captured the pole and turned the fastest lap in Category I and finished 4th in class. At its next race, an SCCA national at Lime Rock, Tullius and the XJS won the pole, the race and set a B production lap record in the process. At the Indianapolis national race the Jaguar didn't start because of a dry-sump problem but when Tullius's three wins in the E-Type were added to his one victory with the coupe he still had enough points to qualify for the annual Road Atlanta runoffs. And there, against the best B production cars in the country, Tullius qualified the XJS almost a second quicker than the fastest of the Corvette contingent. But a 1st-lap spin and a 7th-lap fire put the immaculately prepared Jag coupe on the sidelines and denied Tullius his 6th national title.

Although Jaguar won its stripes and its reputation with racing successes all over the world during the Fifties, the past several years there's been hardly a meow heard from those cats from Coventry. So we've been following Tullius's exploits with the E-Type and XJS with more than usual interest. Joe Rusz, R&T's Motor Sports Editor, happened to be talking to Paul Brand of Group 44 and casually mentioned our interest in the Jaguar coupe. The idea of a production-vs-race car comparison test hit Joe and Paul simultaneously and a few weeks later we had our bags, test equipment and cameras loaded on a plane headed for Road Atlanta to test Tullius's racing XJS.

As the car rolled out of the enormous Group 44 Quaker State trailer, Tullius cautioned us not to be too critical of some of the detailing. "This car is only a prototype," Bob said. "Production versions will be a lot more sanitary." Tullius needn't have worried. The workmanship is every bit worthy of the name Group 44.

TWO JAGUAR COUPES

Group 44's XJS is more than the cat's meow

ROAD & TRACK TRACK TEST

PHOTOS BY JOE RUSZ

The transformation of the XJS into a race car starts with a total teardown. Every nut and bolt is removed and the body is stripped of all paint and undercoating. Construction starts with the installation of the rollcage. "The stock body is quite stiff," Tullius said, "but the demands of racing are severe so the cage is located to stiffen the chassis and work as an integral part of the structure of the car."

Although the SCCA allows the use of wheel flares of an alternate material, Group 44 has not taken advantage of this rule, opting instead for the strength of the stock steel as opposed to the lightness of fiberglass.

According to Tullius, Group 44's racing successes have never resulted from superior horsepower. Rather it's handling that separates the Group 44 cars from the also-rans. For the XJS this means rebending the steering arms to eliminate bump steer and shortening the steering rack so the steering-arm lengths from pivot to pivot are correct. Power assist is retained and no changes are made to the steering gear. However, the pump is slowed down considerably; the stock pump turns faster than crankshaft speed, but on the race car it turns about one third as fast. All rubber in the suspension is replaced with solid bushings and the front spring rates are quadrupled and Koni double-adjustable shocks are fitted. For the anti-roll bar, Group 44 buys 1.25-in. diameter solid stock of the correct length from Holman-Moody. The bar comes with splined ends and Latton "Lanky" Fouchee, Group 44's chief mechanic and chassis designer, fabricates the end links.

At the rear the production lower arms are retained but are beefed up. The stock trailing arms are replaced by new semi-trailing arms to eliminate a geometry conflict Jaguar gets away with on the production XJS by having the entire rear subframe rubber mounted. This conflict causes the suspension to bind when solid bushings are substituted for the stock rubber units.

Stock XJS cornered at 0.726g; race car generated 1.000g.

Unlike the front coil springs, the rear's (two per side) are adjustable. Each spring has a rate of approximately 275 lb/in. (or 550 lb/in. per side) compared to about 1300 lb/in. for each front spring. Rear shock absorbers are double-adjustable Konis (also two per side) and the adjustable anti-roll bar is 0.69-in.-thick Holman-Moody solid stock.

"When it came to picking brakes we looked for the biggest thing that would fit inside the wheels," Fuerstenau said, "because we knew that a nose-heavy car like the XJS always needs a lot of front brake." The system consists of Lockheed rotors and 8-piston calipers up front and Girling rotors and stock calipers from the front of an E-Type at the rear. Brake ducts aid cooling at the front but not at the rear so the rear calipers are mounted at the rear of the discs for better air flow onto the rotors.

"Over the years our brake systems seem to get simpler and simpler," Fuerstenau said. "We use a single master cylinder from a Ford pickup truck and we don't even have a brake balance bar to adjust front/rear brake bias." Minilite 15 x 10 magnesium alloy wheels are fitted front and rear and the Goodyear Blue Streak racing tires are 25.0 x 10.0-15 front and 27.0 x 11.0-15 rear.

Although Lanky Fouchee and Brian Fuerstenau collaborate on chassis design, Brian is solely responsible for engine development. As a tribute to Brian's expertise Tullius says he has never had an engine problem in three years of running the V-12 except for a few cracked valves and connecting rods that were caught during normal overhaul. Much of the engine is stock; in fact, only the Arias pistons, the Crane camshafts and valves springs and the rod bolts aren't Jaguar components. The engine is blueprinted and normal detailing work such as magnafluxing is carried out. According to Fuerstenau, most of the cost of the $10,000 V-12 is in hand labor. "If you had a wrecked V-12 you could build a competitive race engine for around $3000 in parts," Brian figures.

Because of the number and size of the bearings in the V-12, a lot of time and thought went into the oiling system. The dry sump is a Fuerstenau design using a Weaver pump and having a capacity of 30 quarts. Normal oil pressure is 85 psi, and if the

PERFORMANCE COMPARISON
Production & Racing Jaguar XJSs

	Production	Racing
Acceleration:		
Time to distance, sec:		
0–1320 ft (¼ mi)	16.5	13.6
Speed at end of ¼ mi, mph	88.0	115.0
Time to speed, sec:		
0–30 mph	3.8	2.9
0–60 mph	8.6	5.0
0–80 mph	13.9	7.4
0–100 mph	22.2	10.3
Fuel economy:		
mpg	13.5 (normal)	4.0 (race)
Handling:		
Speed on skid pad, mph	33.0 (100-ft radius)	31.8 (67.5-ft radius)
Lateral acceleration, g	0.726	1.000
Speed thru 700-ft slalom, mph	52.7	na
Brakes:		
Minimum stopping distances, ft		
From 60 mph	166	154
From 80 mph	291	260
Pedal effort for 0.5g stop, lb	32	60
Fade, % increase in pedal effort, 6 stops from 60 mph @ 0.5g	19	nil
Interior Noise:		
Idle in neutral, dBA	54	99
Maximum 1st gear	79	120
Constant 50 mph	64	na
70 mph	70	104
90 mph	76	111

pressure drops below 60 psi an oil pressure safety switch shuts off the engine. "The V-12 needs lots of oil coolers," Brian said. "On the XJS I've used two aircraft surplus coolers—one on the scavenge side of the system and the other on the pressure side—and each has a thermostat to maintain oil temperature on cool days."

If there's one area where Jaguars are notorious for having problems, it's the cooling system. Not so, Tullius says, and to prove his faith the XJS uses the stock radiator.

The induction system consists of an intake manifold from British Leyland UK that was designed for the E-Type. "With this manifold there's really not enough hood clearance in the XJS so we've had to compromise on the length of the velocity stacks fitted to the six twin-choke Weber 441DF carburetors," Tullius explained. "We gave a great deal of thought to fuel injection," Fuerstenau continued, "but there's no system available. The English system is too tall and won't clear the hood." A 30-gal. fuel cell resides in the trunk and has Holley pumps at each side. "We use one pump until the tank is about three-quarters empty," Brian explained, "then we switch on the additional pump to scavenge the fuel left in the tank."

Proof that it doesn't take a lot of exotic equipment to win races, Tullius admitted that of all the engines Group 44 has built and raced, only two, the TR7 and the Jag V-12, have ever been on a dynamometer. "From past experience, I knew the engine didn't have much torque because of its short-stroke design," Fuerstenau said, "so I first concentrated on reliability and then on gaining horsepower at higher rpm. Total dyno time was minimal compared to most engine development programs."

The first thing you notice when the engine is fired up is that it doesn't have the melodious V-12 note for which Jaguars are known. Rather the exhaust sounds rough and harsh. "I don't like the sound either," Tullius said, "but the choice was running the pipes out the back and interfering with rear brake cooling to have the engine sound like a typical Jaguar, or running separate pipes out behind each door and having the engine sound like a 6-cyl Datsun with twin pipes. The choice was simple."

To fire the V-12, Fuerstenau uses a Speedatron optical CD system with a rev limiter. No one makes a distributor timing rotor to fit the V-12 engine so Brian machines this himself.

Behind the engine is a lightweight competition flywheel from Tilton Engineering, a 3-plate Borg and Beck racing clutch and a 4-speed gearbox. In the U.S. the XJS is only sold with a 3-speed automatic so Group 44 swaps this for the manual gearbox with close-ratio gears that comes in the European XJS. The differential is stock except it's locked by welding. Final drive ratios range from 3.77 to 4.55:1. Cooling for the transmission and the differential is provided by MGB oil coolers and Holley electric pumps.

The morning test session at Road Atlanta was devoted to determining the race car's straight-line performance in acceleration and braking, noise and handling on the track's 67.5-ft radius skidpad. For comparison we had figures from a test of the XJS in January 1976. As is often the problem with race cars set up for rolling starts, the XJS was reluctant to move from rest, demonstrating its displeasure by bucking and stumbling until the revs built up. Once underway, the Jaguar hurried down the Road Atlanta back straight in impressive fashion with our new computer recording times of 5.0 sec to 60 mph, 10.3 sec to 100 mph and the quarter-mile in 13.6 sec at 115.0 mph. This compares to 8.6 sec, 22.2 sec and 16.5 sec at 88.0 mph for the stock XJS tested with our old equipment. However, any disparity caused by the different testing techniques was minimized because the Engineering Editor rode shotgun in the racing XJS during all the performance tests.

Although a race car doesn't often brake to a complete stop, Tullius's Jaguar also won this test with distances of 154 ft and 260 ft from 60 and 80 mph respectively, versus 166 and 291 for the production XJS. As expected, the race car brakes suffered no fade, but the street car's were nearly as good showing only a 19 percent increase in pedal effort after the sixth stop.

Is the race car noisy? You bet. At idle the noise-level reading was 20 dBA higher than the maximum reading obtained with the stock Jaguar and during wide-open-throttle acceleration in 1st gear the needle swung to 120 dBA. That's loud.

With that portion of the test completed, Tullius changed into his driving suit and took the street XJS out for a few laps on the demanding 2.5-mi Road Atlanta course. The first timed lap was 2 minutes 0.14 seconds; three laps later Tullius was down to 1:58.61 and came in to explain how the car was acting.

Tullius was impressed: "Even though it's a street car, this car does what it's supposed to do. It doesn't understeer a whole lot and it doesn't oversteer a whole lot; in fact, you can get it to do a broad slide which is quite unusual for a stock street machine. It handles much like the race car if you can interpolate the difference between the need for the race car to be stiff and the street car to be soft. The only real difference between it and the race car is the throttle application. With the street car you can stand on the throttle and it won't spin the wheels and send you roaring into the woods somewhere. When the car goes into a

Front brakes consist of Lockheed vented rotors and 8-piston calipers. Flexible hose routes air onto the rotors to aid cooling.

Rear suspension of racing Jaguar retains dual coil springs and dual shock absorbers of production car. Shocks are double-adjustable Konis and springs are adjustable for changing ride height. Heim-jointed lower anti-roll bar link can be moved into various holes of horizontal arm to vary rear roll stiffness.

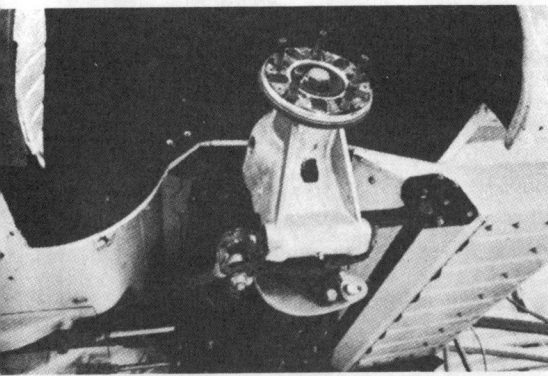

Heim-jointed semi-trailing links replace the stock Jaguar trailing arms. Precise geometry is maintained by attaching the inner pivot points to rollcage braces.

Major portion of trunk is consumed by a 30-gal. fuel cell.

SPECIFICATIONS COMPARISON
Production & Racing Jaguar XJSs

	Production	Racing
Price	$19,000	$45,000
General:		
Weight, lb	3830 (curb)	3130 (race)
Weight distribution (with driver), front/rear, %	56/44	53/47
Track, in.	58.0/58.6	60.0/60.6
Width	70.6	74.8
Height	49.7	47.5
Ground clearance	5.5	3.0
Usable trunk space, cu ft	13.6	3.9
Fuel capacity, U.S. gal.	23.0	30.0
Engine:		
Compression ratio	7.8:1	11.5:1
Bhp @ rpm	244 @ 5250	475 @ 7600
Torque @ rpm	269 @ 4500	360 @ 5500
Carburetion/fuel injection	Lucas electronic	six Weber (2V)
Fuel requirement	unleaded, 91-oct	premium, 100-oct
Drivetrain:		
Transmission	automatic; torque converter with 3-sp planetary gearbox	4-sp manual
Gear ratios:		
4th		1.00
3rd	1.00	1.28
2nd	1.45	1.65
1st	2.39	2.14
Final drive ratio	3.31:1	3.77:1
Chassis:		
Brake system	11.8-in. vented discs front, 10.9-in. discs rear, vacuum assisted	12.0-in. vented discs front, 11.2-in. vented discs rear
Swept area, sq. in.	448	457
Wheels	cast alloy, 15 x 6JK	Minilite, 15 x 10
Tires	Dunlop SP Sport, 205/70VR-15	Goodyear Blue Streak; 25.0 x 10.0-15 front, 27.0 x 11.0-15 rear
Steering, turns, lock-to-lock	3.3	1.5
Front suspension	unequal-length A-arms, coil springs, tube shocks, anti-roll bar	unequal-length A-arms, coil springs, Koni adjustable tube shocks, anti-roll bar
Rear suspension	lower A-arms, fixed length halfshafts, trailing arms, dual coil springs, dual tube shocks, anti-roll bar	lower A-arms, fixed-length halfshafts, semi-trailing arms, dual adjustable coil springs, dual Koni adjustable tube shocks, adjustable anti-roll bar
Instrumentation:		
Instruments	160-mph speedo, 7000-rpm tach, 99,999 odo, 999.9 trip odo, oil press., coolant temp, voltmeter, fuel level, clock	9000-rpm tach, oil press., oil temp, coolant temp, fuel press.
Warning lights	oil press., brake sys, hand brake, voltmeter, low fuel, ignition, low coolant, rear-window heat, seatbelt, hazard, lights on, brake light failure, side light failure, major fault, secondary fault, high beam, directionals	none
Accommodation:		
Seating capacity, persons	4	1
Seat width, front/rear, in.	2 x 20.5/2 x 19.0	14.0
Head room, front/rear	36.0 x 33.0	39.5
Calculated data:		
Lb/bhp (test weight)	17.2	7.1
Mph/1000 rpm (top gear)	22.5	22.2
Engine revs/mi (60 mph)	2670	2700
Piston travel, ft/mi	1230	1240
Brake swept area, sq in./ton	214	270

corner it tends to understeer just a bit until you get the throttle on and then it gets up on its haunches and does a little broad slide or the tail hangs out a little. It's exactly like the race car only softer and gentler."

When queried on the performance of the stock Dunlop tires, Tullius said, "The tires work fine. They aren't overheating or going away. But they are rolling over a lot, especially at the front, and additional air will help that."

With the pressures increased from 32 to 40 psi all around Tullius climbed back into the Jaguar for two more laps, the fastest being a 1:56 flat. Just that one change reduced his lap times by more than 2.5 sec.

"When I went out the first time the tires felt soft," Tullius said. "Putting more air in made the handling a lot more positive and more like the race car. The car was a lot less wallowy, especially in the esses where you have to make quick changes of direction."

How does Tullius like driving a car with power steering? "I'll tell you a little story about that," Bob said. "At first I didn't want the power steering, and I complained bitterly to Brian and Lanky. But they finally convinced me to try it and I got in the race car and drove it for half a lap and forgot the power assist was there. I don't think any race driver would ever go back to conventional steering after he got used to power assist.

"Probably the most surprising thing about the XJS is its aerodynamics," Tullius said. "At an indicated 130 mph on the back straight there was absolutely no indication that the speed was affecting the stability. According to the sales brochures, Jaguar did a lot of wind-tunnel testing of the body design, but I figured half of that was BS and the other half was somebody dreaming. Well let me tell you, now I'm a believer. To see if Jaguar's claims were anywhere near right we decided to keep this prototype XJS aerodynamically stock. You'll notice that we even use the production front spoiler. When I ran at Daytona I could effortlessly drive up and down the banking at 180 mph."

The race car was still fitted with Daytona gearing but that didn't stop Tullius from cracking off laps as fast as those he turned at the SCCA runoffs the previous November, the quickest being a 1:32 flat. So the race car, even on a cold track that didn't allow the tires to achieve optimum temperatures, was 24 sec faster than the street car.

Corvette racers who worried that the XJS would be the scourge of B production in 1977 are probably right now heaving a collective sigh of relief, because this year Tullius has his sights on Category I of the professional Trans-Am series instead of the SCCA nationals. But that only means that when the checkered flag drops on the last race of the 1977 season, there'll probably be a bunch of disgruntled professional racers who will have tasted defeat at the hands of Bob Tullius and the thundering white Jaguar from Group 44.

Cockpit is typically race-car stark. Steering column retains telescopic adjustment feature of stock XJS.

Top portion of Jaguar V-12 is dominated by six twin-choke Webers. Front carburetors lack velocity stacks because of limited hood clearance.

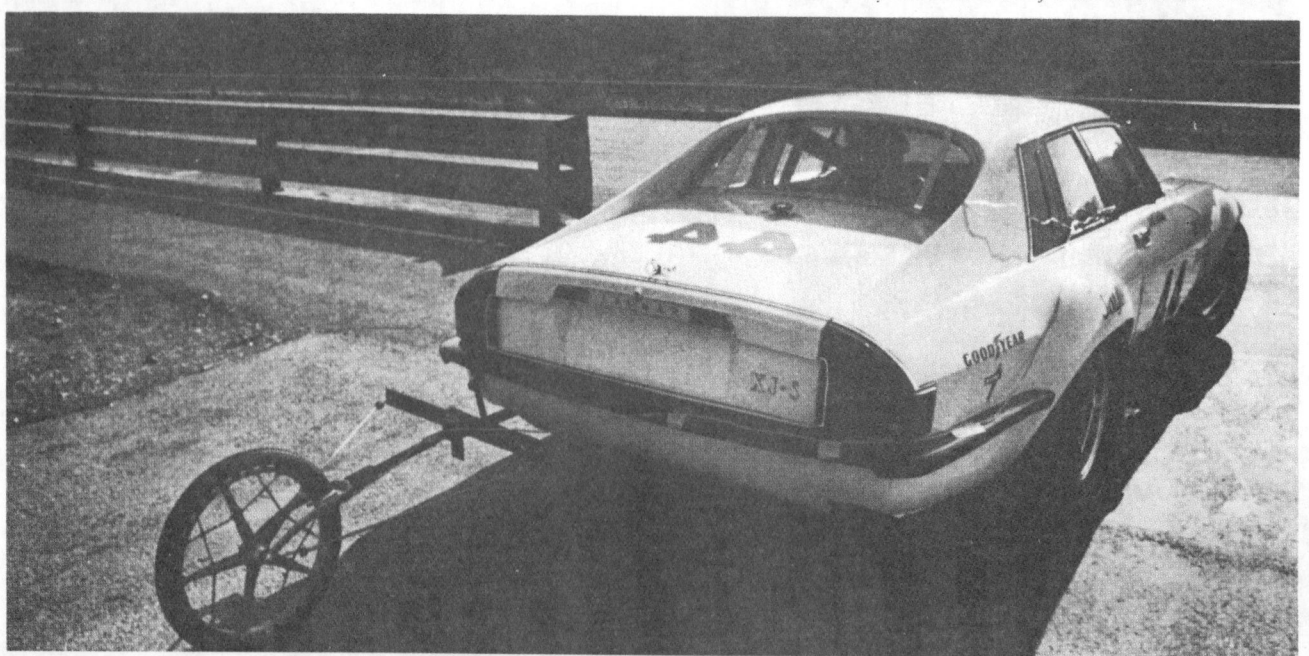

ROAD TEST
TULLIUS JAGUAR XJS

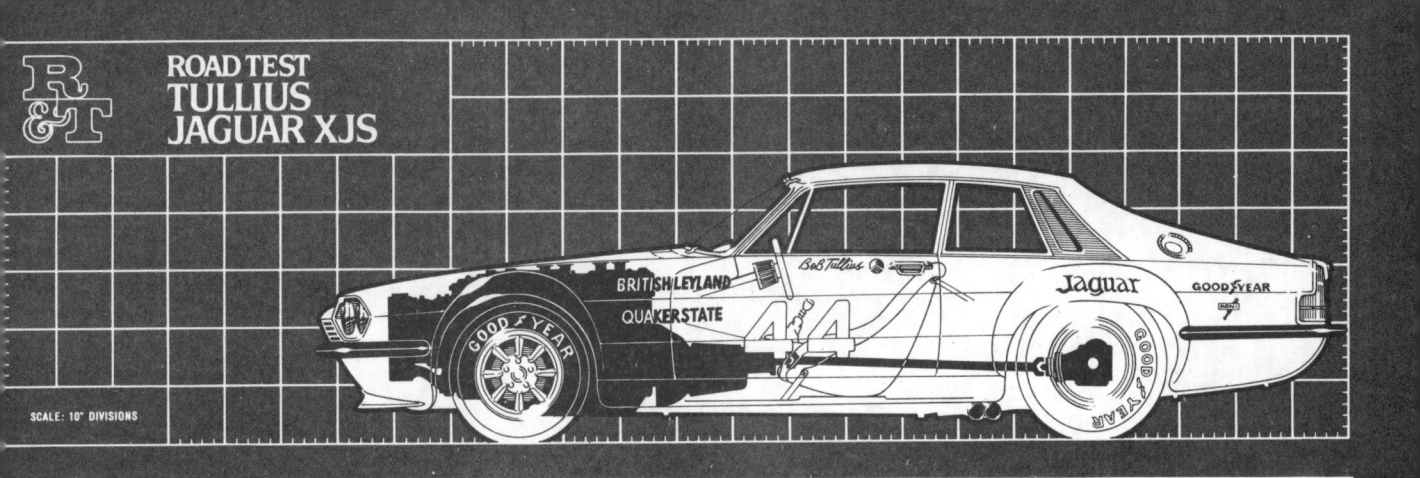

SCALE: 10" DIVISIONS

PRICE
List price $45,000

MANUFACTURER
Group 44 Inc
308 Victory Dr
Herndon, Va. 22070

GENERAL
Curb weight, lb	3130
Test weight	3380
Weight distribution (with driver), front/rear, %	53/47
Wheelbase, in.	102.0
Track, front/rear	60.0/60.6
Length	191.7
Width	74.8
Height	47.5
Ground clearance	3.0
Overhang, front/rear	39.9/49.8
Usable trunk space, cu ft	3.9
Fuel capacity, U.S. gal.	30.0

DRIVETRAIN
Transmission	4-sp manual
Gear ratios: 4th (1.00)	3.77:1
3rd (1.28)	4.83:1
2nd (1.65)	6.22:1
1st (2.14)	8.07:1
Final drive ratio	3.77:1

CHASSIS & BODY
Layout	front engine/rear drive
Body/frame	unit steel
Brake system	12.0-in. vented discs front, 11.2-in. vented discs rear
Swept area sq in.	457
Wheels	Minilite, 15 x 10
Tires	Goodyear Blue Streak; 25.0 x 10.0-15 front, 27.0 x 11.0-15 rear
Steering type	rack & pinion, power assisted
Overall ratio	17.5:1
Turns, lock-to-lock	1.5

Front suspension: unequal-length A-arms, coil springs, Koni adjustable tube shocks, anti-roll bar

Rear suspension: lower A-arms, fixed-length half shafts, semi-trailing arms, dual adjustable coil springs, dual Koni adjustable tube shocks, adjustable anti-roll bar

INSTRUMENTATION
Instruments: 9000-rpm tach, oil pressure, oil temp, coolant temp, fuel pressure
Warning lights: none

ENGINE
Type	sohc V-12
Bore x stroke, mm	90.0 x 70.0
Equivalent in.	3.54 x 2.76
Displacement, cc/cu in.	5343/326
Compression ratio	11.5:1
Bhp @ rpm, net	475 @ 7600
Equivalent mph	152
Torque @ rpm, lb-ft	360 @ 5500
Equivalent mph	113
Carburetion	six Weber (2V)
Fuel requirement	premium, 100-oct

ACCOMMODATION
Seating capacity, persons	1
Seat width	14.0
Head room	39.5
Seat back adjustment, deg	0

CALCULATED DATA
Lb/bhp (test weight)	7.1
Mph/1000 rpm (4th gear)	22.2
Engine revs/mi (60 mph)	2700
Piston travel, ft/mi	1240
Brake swept area, sq in./ton	270

ROAD TEST RESULTS

ACCELERATION
Time to distance, sec:	
0-100 ft	3.4
0-500 ft	7.8
0-1320 ft (¼ mi)	13.6
Speed at end of ¼ mi, mph	115.0
Time to speed, sec:	
0-30 mph	2.9
0-40 mph	3.7
0-50 mph	4.3
0-60 mph	5.0
0-70 mph	5.8
0-80 mph	7.4
0-100 mph	10.3
0-110 mph	12.4
0-120 mph	15.2

SPEEDS IN GEARS
4th gear (8200 rpm)	180
3rd (8000)	133
2nd (8000)	103
1st (8000)	79

FUEL ECONOMY
Race driving, mpg	4.0
Cruising range, mi (1-gal. res)	116

HANDLING
Speed on 67.5-ft radius, mph	31.8
Lateral acceleration, g	1.000
Speed thru 700-ft slalom, mph	na

BRAKES
Minimum stopping distances, ft:
From 60 mph	154
From 80 mph	260
Control in panic stop	very good
Pedal effort for 0.5g stop, lb	60

Fade: percent increase in pedal effort to maintain 0.5g deceleration in 6 stops from 60 mph..nil
Overall brake rating..........very good

INTERIOR NOISE
All noise readings in dBA:
Idle in neutral	99
Maximum, 1st gear	120
Constant 30 mph	na
50 mph	na
70 mph	104
90 mph	111

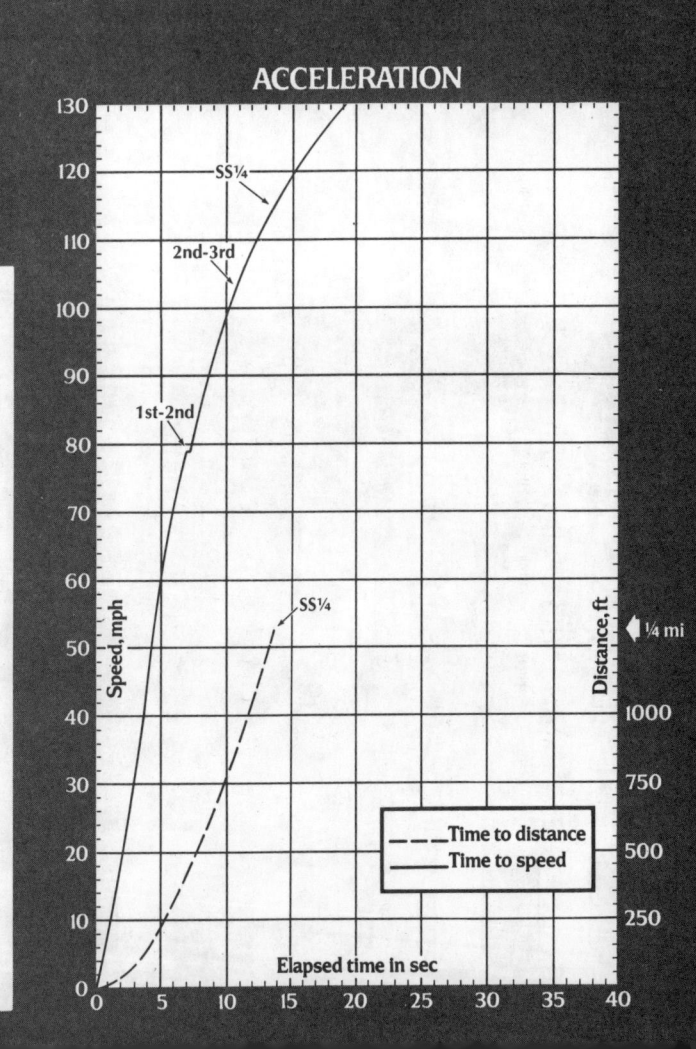

ACCELERATION

MOTOR Test

The Top Cat!

It's rumoured that Linda Blair rehearsed her neck-twisting routine for The Exorcist from watching a Jaguar XJ-S drive past. Truth is, it's not the fastest road car, nor the prettiest, nor the cheapest to run. But it sure gets the roses for being the best value in the supercat set . . .

HUMAN NATURE can be a strange animal. Legend has it that it turns mice into men and makes women melt at the knees — all because of a big hunk of metal and leather.

The XJ-S Jaguar transcends all the normal criteria on which cars are summarily judged with a mere flick of the wrist, and it refuses to accept the character of the driver, no matter how strong, because it is indeed far greater than any mortal man.

The whole XJ-S image begins with the shape of the car. It's conservative yet aggressive. It has been likened to the finer Grand Tourers of the Sixties, a comparison that is in no way unjust or derogatory.

The importance of the XJ-S is that while those cars suffered from a bout of the post-Sixties fashion blues, the XJ-S has managed to place itself on the Seventies list of 'most desirable cars' and is likely to remain there for some time.

An examination of the reasons why the car is so sought after and so highly regarded by the motoring fraternity, and those affluent enough to be able to purchase and maintain an XJ-S, comes up with some interesting conclusions.

The XJ-S is not the best performer in its class. Nor is it the most economical to run; the cheapest to maintain; the most practical or the most reliable. But it's a Jaguar and, what's more, it's a Jaguar coupe of the first class.

On the open road, the XJ-S maintains all the legends of Jaguar performance. The 5.3-litre V12 is incredibly smooth and willing to perform, and though purists would shudder at the thought of a grand touring car with automatic transmission, the Jaguar proves that, if you can afford the necessary $32,750, life was indeed meant to be easy.

Jaguar's decision to replace the Borg-Warner box with a General Motors unit seems to have been a wise one. The changes are smooth and precise, even when the big cat is being pushed hard, and while there is a manual overide on the gearbox, using it is a rather pointless exercise because there seems to be limitless power for overtaking manouevres and superb brakes for slowing the beast down. The gate also proved to be far too stiff and vague for accurate manual changes.

The automatic transmission also makes the big car easier to push around city streets, where its size could otherwise cause a few problems.

Another great help in this area is the power-assisted steering which is light enough to make driving the car a simple matter, yet sensitive enough to make driving at speeds a real pleasure.

The choice of tyre size on the XJ-S has been a careful compromise because, while the tyres are not as wide as some fitted to competitors, they maintain a good turning circle and light steering effort, and do not detract from the Jag's incredible road holding and handling characteristics.

There are a few reasons why it is best not to consider the Jag as a city car. First it has been designed as a high-speed tourer. Its suspension and performance have both been geared to these requirements and that's really what it does best. That's not to say the Jag can't handle the hassles of peak hour insanity — it can. In fact, it does it a lot better than many of its close rivals. It has better visability than a Lamborghini or Ferrari and is a more pleasant car to push through heavy traffic because of its well mannered behavior and its superbly low level of

The front seats feature ample leg room for giants.

The V12 and its associated plumbing fill the engine bay.

interior sound.

Then, of course, there's the consideration of fuel consumption, and it's here that the Jag really suffers from the city enviroment.

On the open road, the XJ-S can manage 15 litres per 100 kilometres but the toll of constant stop/start driving whittles that down to an expensive 25 litres per 100 kilometres around the traps.

Even so, there is the toll of the climate control air conditioning to be taken into consideration, and that can mean a drain of 8-12 kW or, in fuel consumption terms, a loss of between 5 and 10 percent.

With 212 kW under the bonnet, the XJ-S can manage some very impressive performance figures, considering its 1760 kg of weight.

One passenger jumped at the opportunity of being taken for a quick spin and foolishly suggested that we "open her up and see what this baby can do." Needless to say, after the run he had a glazed look in his eyes and could only mumble incoherent phases littered with words like "necksnapping." Since then, he has become periodically morose, claiming that life no longer has the same meaning because he doesn't own an XJ-S.

True, the straight-line performance of the cat is impressive, but of more importance is the way it handles high-speed country driving.

The normal stretches of blacktop in Australia are a far cry from the autoroutes of Europe, yet the XJ-S handles them with surprising ease and comfort, and it is here that the car proves to be a real masterpiece.

While the XJ-S has been given a shorter wheelbase than its sedan-bodied brothers, it doesn't seem to suffer in the ride and handling areas at all.

The long wheel travel soaks up the most vicious bumps, and broken edges or pot-holed surfaces never arouse a feeling of discomfort inside.

In the seating department, the XJ-S is wide open to criticism on a couple of points, yet praise on others. The car has been designed as a four seater, or at least as a 2-plus-2, but in effect the rear seats are lacking in sufficient legroom to provide enough accomodation for four fully grown adults.

This leaves us with a 2-plus-2 that can easily take two adults in the front but only children in the back, both because of the lack of legroom and the lack of headroom for taller-than-average passengers.

The seats themselves are really comfortable, both back and front, and have been carefully moulded to give good support over long distances.

There were comments about whether they should have been cloth trimmed instead of leather, but the leather has been used for a good many reasons. It will last as long as the rest of the car if looked after properly, whereas cloth trim, no matter how good, will wear out and need replacing.

Seat travel is excellent, and can accomodate the most formidable giants.

Unfortunately, the rear seats have been set so low, in an effort to give some measure of headroom for backseat passengers, that they tend to hide the rear passengers behind the C-pillar and cut off their outward visibility.

Driving controls and instruments on the XJ-S are discreet, conservative and efficient. The only black mark here is the placement of the light switch on the dash behind the steering wheel next to the steering column. It's a tricky place to hide it and would take a bit of getting used to.

One great step forward in the area of improving driver comfort has been to put a leather glove on the steering wheel. But while the leather covering is an improvement, the wheel still needs to feel a bit beefier.

The high beam/indicator/wiper/washer stalks on the steering column are easy enough to use but the operation of the windscreen wipers is totally out of keeping with the car.

While they worked well, they were too small, and had a ridiculous 'park' phase, where they went into limbo after use before returning to a station that was far too high up the windscreen.

On such a delightful and well engineered car as the XJ-S, this was a great disappointment, as we had come to expect greater things.

Lighting on the XJ-S was amongst the best we have come across, and set standards that could well be adopted by other manufacturers.

Now comes the question of interior sound levels, and it's here that Jag makes so many other cars look positively sick.

Even at speeds of more that 150 km/h, it was possible to carry on a normal conservation with passengers — including those in the back seat.

There was also no need to turn up the cassette player to drown out road or engine noise. There wasn't even much wind noise at high touring speeds, and that's an important consideration in long-distance touring.

It means that even after six or eight hours of driving, both pilot and passengers can arrive feeling relaxed. They're not hoarse in the throat from shouting or deaf from the high noise levels.

While on the subject of music, Jaguar has made the incredible omission of a power radio aerial — and that truly is a mysterious oversight, as there is plenty of space in the rear for such an aerial to be mounted.

That brings another mark in favour of the XJ-S. The luggage space in the boot is enormous. The spare wheel has been mounted vertically across the back of the boot next to the battery, allowing extremely good use of the boot depth yet fairly good access to both the spare and the battery.

The boot also contains a fairly comprehensive tool kit, but heaven help anyone who has to do more than change a tyre on the XJ-S. Under the bonnet, the expansive V12 and all its associated plumbing for air conditioning, power steering and fuel injection is an absolute nightmare.

Oh, it's well laid out and pretty neat, but totally confusing for anyone but a mechanic.

With an engine as sophisticated as a fuel-injected V12, there is going to be the need for regular, and no doubt expensive, visits to the dealer for a tune-up and service — and Jaguars are notoriously costly in this area.

But, then, people have always said that if you can afford to buy one, you can afford to maintain one, and there is little doubt this would apply to the XJ-S.

In all, the $32,750 Jaguar XJ-S is a truly magnificent piece of machinery. It is desirable in every way as a safe, high-powered touring car with superb handling and arguably the best interior comfort and sound levels of any grand tourer on the market today.

PRICE

Dollars x 1000

JAGUAR XJ-S	~32
DE TOMASO PANTERA GTS	~35
BMW 633 CSi	~39
FERRARI 308 GTB	~42
MERCEDES-BENZ 450SL	~37

Scale: 5, 10, 15, 20, 25, 30, 35, 40, 45

FUEL CONSUMPTION

Litres per 100 kilometres

JAGUAR XJ-S	~16
DE TOMASO PANTERA GTS	~20
BMW 633 CSi	~15
FERRARI 308 GTB	~17
MERCEDES-BENZ 450SL	~14

Scale: 22, 20, 18, 16, 14, 12, 10, 8, 6

ACCELERATION

0-100 km/h, seconds

JAGUAR XJ-S	8.3
DE TOMASO PANTERA GTS	~6.2
BMW 633 CSi	~8
FERRARI 308 GTB	~7
MERCEDES-BENZ 450SL	~9

Scale: 14, 13, 12, 11, 10, 9, 8, 7, 6

WEIGHT TO POWER

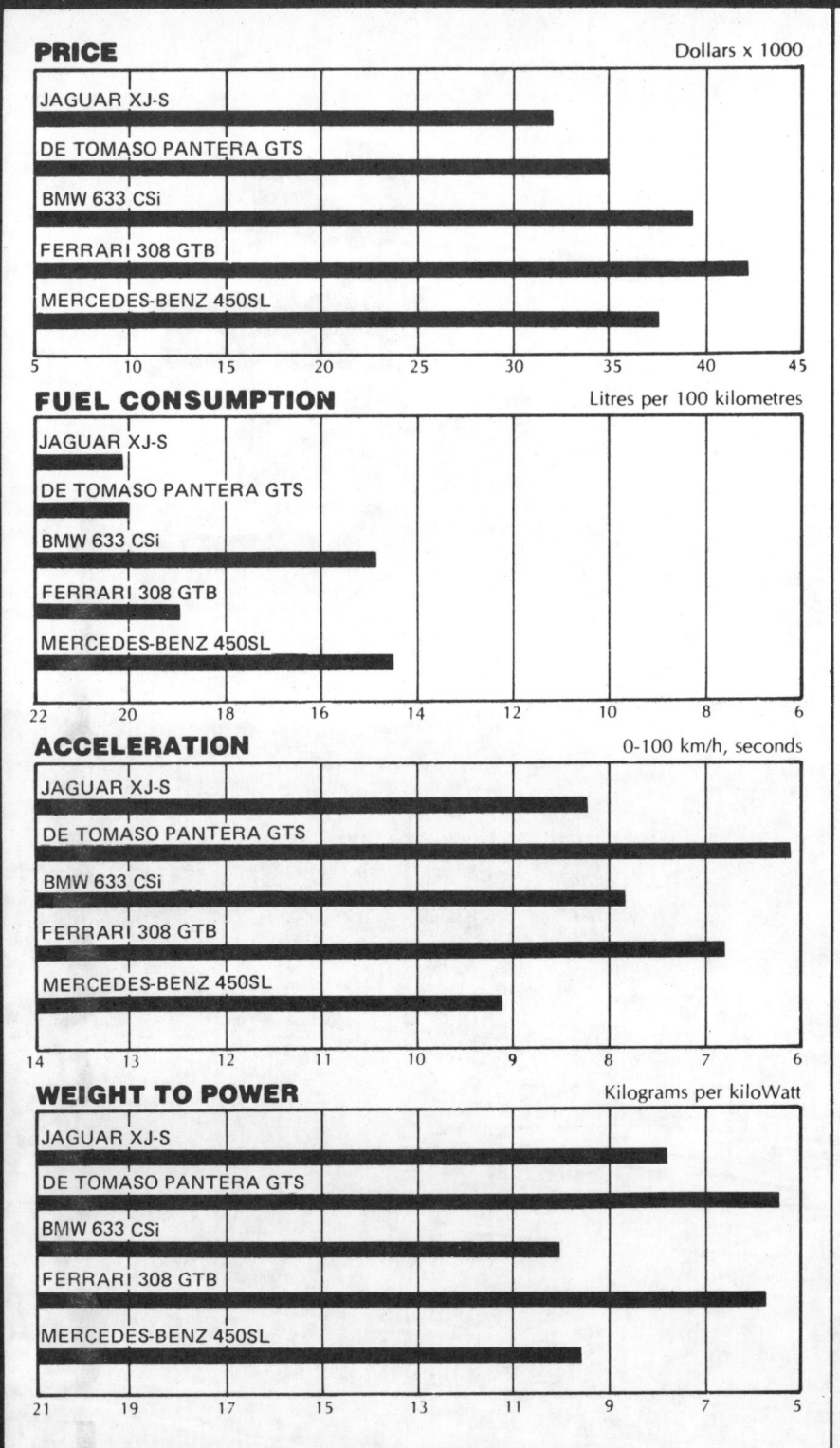

Kilograms per kiloWatt

JAGUAR XJ-S	7.95
DE TOMASO PANTERA GTS	~5.5
BMW 633 CSi	~8
FERRARI 308 GTB	~6
MERCEDES-BENZ 450SL	~9

Scale: 21, 19, 17, 15, 13, 11, 9, 7, 5

THE FIVE-STAR TEST

Comfort	★★★★★
Handling	★★★★
Brakes	★★★★
Performance	★★★★★
Quietness	★★★★★
Luggage capacity	★★★★

CHECKLIST

Adjustable steering	Yes
Carpets	Yes
Cigarette lighter	Yes
Clock	Yes
Day/night mirror	Yes
Hazard flashers	Yes
Heater rear window	Yes
Laminated screen	Yes
Petrol filler lock	Yes
Radio	Yes
Tachometer	Yes
Intermittent wipers	No
Rear window wiper	No

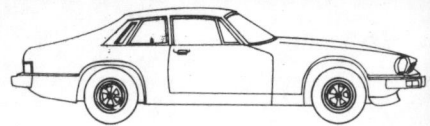

JAGUAR XJ-S ROAD TEST DATA

ENGINE
Location...Front
Cylinders...12
Bore x Stroke.................................90 x 70 mm
Capacity...5343cc
Carburation...................................Fuel injection
Compression Ratio......................................9.0:7
Fuel Pump..Electrical
Valve Gear...DOHC
Maximum Power....................212 kW at 5500 rpm
Maximum Torque ....399 Nm at 3500 rpm

TRANSMISSION
Type......................Borg-Warner three-speed automatic
Driving wheels...Rear
Gearbox ratios
First..2.39
Second..1.45
Third...1.00
Final Drive Ratio..3.07

SUSPENSION
Front.....................Independent by coils and wishbones, anti-roll bar
Rear.....................Independent by coils, lower wishbones, radius rods, anti-roll bar
Shock Absorbers..................................Telescopic
Wheels...6JK x 15
Tyres..205/70UR x 15

BRAKES
Front..284 mm disc
Rear...263 mm disc

STEERING
Type....................Power-assisted rack and pinion
Turns, Lock to Lock..3.0
Turning Circle......................................12.0 metres

DIMENSIONS AND WEIGHT
Wheelbase..2590 mm
Front Track..1470 mm
Rear Track...1490 mm
Overall Length.......................................4870 mm
Overall Width.......................................1790 mm
Overall Height.......................................1260 mm
Ground Clearance....................................140 mm
Kerb Weight..1687 kg

CAPACITIES AND EQUIPMENT
Fuel Tank...91 litres
Cooling System....................................20.5 litres
Engine Sump...9.1 litres
Battery...12V 68Ah
Alternator..70A

CALCULATED DATA
Weight to Power................................7.95 kg/kW
Specific Power Output....................39.67 kW/litre

FUEL CONSUMPTION
Average for Test........................16.2 litres/100 km
Best Recorded..........................15.00 litres/100 km

ACCELERATION
0-60 km/h..4.4 seconds
0-80 km/h..5.8 seconds
0-100 km/h..8.3 seconds
0-110 km/h......................................13.0 seconds

OVERTAKING TIMES
40-60 km/h.......................................4.8 seconds
60-80 km/h.......................................4.2 seconds
80-100 km/h.....................................4.5 seconds
100-120 km/h...................................4.5 seconds

STANDING 400 METRES
Average...15.0 seconds
Best Recorded..................................14.8 seconds

SPEED IN GEARS
First..80 km/h
Second..146 km/h
Third..239 km/h

Cat Tales in the Sunset

Jaguar's XJ-S just may be the last of a breed

Road Test

by Fred M.H. Gregory

After a pleasant couple of weeks and a thousand or so miles of driving a Jaguar XJ-S, I'm tempted to have a tiny sign painted on it that says: Endangered species: get it while it lasts.

It's hard for me to imagine that Jaguar, or any other company for that matter, will produce a car like the XJ-S in the future. Everything is against the possibility: the price and availability of gasoline, the trend toward lighter and more efficient cars, and new ways of making horsepower without relying on gobs of cubic inches and lots of cylinders. The XJ-S is contrary to all these things: it is heavy, inefficient and mechanically complex.

And those are about the worst things you can say of the car; nearly every other facet of it is praiseworthy. The XJ-S is exactly what it's supposed to be—plush, fast, nimble, good looking and expensive.

Though it has changed only slightly since its introduction in 1975 (Jag has gone from using a Borg-Warner to a GM-built HydraMatic transmission), the XJ-S has held up well, remaining one of the best luxury GT cars on the market.

Compared to its competitors—the BMW 633i, Mercedes-Benz 450SLC and, though I'm stretching the comparison a bit here, the Porsche 928—the XJ-S has the most attractive price. At $26,000, it is $2000 cheaper than the BMW, $6000 less than the Porsche and a whopping $10,000 under the Mercedes. Also, Jaguar's price includes everything, since there are no options.

Unless you're one of the people who slipped out of Iran, just before the hammer fell, with a satchel full of numbered Swiss bank books, 26 grand is still a lot of bucks. But for that you get some things required for the good life. There are a couple of cows-worth of sinfully exquisite Connolly leather (specially tanned for Jaguar by the British firm) stretched over some well-formed, exceedingly comfortable seats. The appointments and workmanship throughout the interior are first rate and embellish a clean and simple design. Power assists abound, easing the operation of the steering, brakes, windows and door locks. The car's finish and the fit of its panels are in keeping with the fine craftmanship found elsewhere, and its V-12 engine is a one-of-a-kind among production cars of today.

The XJ-S's looks, which got mixed reviews when it first came out, can be evaluated more objectively today. They are striking; the XJ-S stands out among other cars like Robert Morley in a marathon. That is, it looks somewhat chubby but not displeasingly so. From the waistline down, the XJ-S's lines flow voluptuously in great rounded sweeps, a welcome relief from the angular styling so much in vogue today. The roofline follows this motif until it reaches the rear window, where the symmetry breaks down. The way Jaguar has handled this upper rear-quarter area has never set well with me. It seems to contrast unfavorably with the rest of the car's looks. There's a plastic panel just behind the rear side windows that appears to have been an afterthought. It departs from the concept that the rest of the styling follows. This roof treatment makes a recessed rear window necessary, creating annoying blind spots in both rear quarters. Still, this is a relatively minor flaw in an otherwise impeccable design. It just takes a little getting used to.

The same can be said of driving the car. When you first slide behind the wheel of an XJ-S, you feel a little crowded: the door sill seems a bit close to your head; the footwell appears somewhat confining; the headliner hovers too near. But the feeling soon dissipates and is replaced by one of reassuring snugness similar to that of putting on a favorite old suit.

The ambiance of the Jag is different from that of its German counterparts. In those cars there's an impersonal, mechanical aura, the kind usually created by know-it-all engineers. But in the Jag there's evidence of traditional British coachwork.

The XJ-S also follows Jaguar's performance tradition. With the exception of the 928, it is the class of its field in this area. Its 244-horsepower V-12 responds instantly and impressively throughout its full range, with no hesitation or flat spots. Though the car registered superior quarter mile times, its most satisfying charge comes in the mid-range speeds. A quick jab of the throttle and you're around slower-moving obstacles in a flash. The XJ-S cruises most comfortably, I'm sad to say, in the taboo 80-mile-per-hour stratosphere. It shakes a little over 90 and runs out of muscle just past 120. At useable—and legal—speeds, though, it leaves little to be desired. The temptation is to always call upon the car's substantial power, but this takes its toll in gasoline, pushing the mpg down to around 10. Sensibly driven, the XJ-S can manage mileage around 16.

The flaw in the Jag's performance is its braking. Even with 4-wheel disc brakes, its nearly 4000-pound bulk is reluctant to come to a smooth, rapid stop under the most extreme conditions. In normal driving, braking was adequate

Cat Tales

and presented no problems. But during track tests, where we try to bring the car to the quickest possible halt, we encountered a lack of stability and the need for very deft brake pedal manipulation.

The XJ-S's handling, on the other hand, is everything you would expect from such a car. The leather-bound steering wheel, which is adjustable a couple of inches fore and aft, is perfectly proportioned to the amount of effort the power-assisted rack-and-pinion unit demands—which is also just right; there's enough resistance to let you know exactly what the front wheels are doing but not so much that any substantial amount of effort is needed to control the car. It maneuvers with ease and agility and remains flat and neutral in hard cornering. During normal highway driving, the XJ-S goes where it's pointed and requires little attention to keep in a straight, true line. Its suspension is not overly firm and produces a quiet and stable ride. Whatever compromises had to be made between handling and comfort were done with little damage to either. The XJ-S can be driven fast with safety and comfort.

By definition, the car is a 2+2. But, like most of the cars bearing those numbers, it's somewhat lacking in the +2 department. Though a couple of adults can be hauled around back there for short spins, long-time occupancy would surely require the eventual ministrations of a chiropractor.

As a 2-passenger car, however, the Jag is ideal. Each front seat is fully adjustable and padded just-so (the belts are unobtrusive and easy to wear). Both give good side support and cradle the body well. A long trip seems like a jaunt up the block. The trunk is large enough to handle all the luggage that any pair of people might require, even for a two-week vacation.

Of course, there are some things

about the XJ-S that are not so laudable. Jaguar makes a big deal about everything on the car being standard. That's fine as far as it goes, but it doesn't mean that the car has everything a person who's going to peel off a substantial stack of bills might want (unlike the new Series III sedans). For example, there is no sunroof available, or power seats or power side mirrors. You can't get cruise control, and the level of gadgetry on the XJ-S, as compared to even a 280ZX Datsun, is primitive. And of the luxuries provided, some are less than you might expect. The power windows can be opened only one at a time; if you press both buttons, nothing happens. As for the stereo, though its sound is sufficient, you could easily do better by having any one of a number of moderately priced units on the market custom installed.

In a sense, this is nitpicking, but there is one area where the XJ-S deserves to be roundly criticized: Its ventilation system is nearly nonexistent. If you want fresh air, the only choice you have is to open the windows. Otherwise, you're at the mercy of air conditioning. Floor-level vents would be nice, since there's a noticeable amount of heat generated in the footwells, and though there are a couple of triangular panes of glass in the doors that would make useable vent windows; they're immovable.

But, in spite of its flaws, the Jaguar XJ-S remains one of the finest mass-produced cars in the world. It has personality, presence and a trait that is becoming rare in automobiles; character. Perhaps that is because of its British origin. After all, just about everything that comes from the U.K. is in some sense a reflection of the national character. There's that respect for tradition and fine work which no amount of decay seems to totally erase. Like the British, the XJ-S is a stiff-upper-lip car, and if like the British it has a few eccentricities, so what? /MT

ROAD TEST DATA
Jaguar XJ-S

SPECIFICATIONS

GENERAL
Vehicle type	Front-engine, rear-drive, 2-door, 2+2 coupe
Base price	$26,000
Options on test car	None offered
Price as tested	$26,000

ENGINE
Type	V-12 SOHC (2)
Bore & stroke	3.45 x 2.76 in./90 x 70 mm
Displacement	326 cu. in./5343 cc
Compression ratio	7.8:1
Fuel system	Lucas-Bosch fuel injection
Recommended octane number	91 unleaded
Emission control	Catalyst
Valve gear	OHV
Horsepower (SAE net)	244.4 at 5250 rpm
Torque (SAE net)	269 lb.-ft. at 4500 rpm
Power to weight ratio	16.21 lb./hp

DRIVETRAIN
Transmission	3-speed automatic
Final drive ratio	3.31:1

DIMENSIONS
Wheelbase	102 in.
Track, F/R	58.62/58.65 in.
Length	192.25 in.
Width	70.6 in.
Height	47.8 in.
Ground clearance	4.5 in.
Curb weight	3963 lb.
Weight distribution, F/R	N.A.

CAPACITIES
Fuel capacity	24 gals.
Crankcase	11 qts.
Cooling system	5.5 gals.
Trunk capacity	10 cu. ft.

SUSPENSION
Front	Independent, semi-trailing wishbones w/coil springs, shock absorbers, anti-roll bar
Rear	Independent, driveshafts form upper links, radius arms, shock absorbers w/coil springs, anti-roll bar

STEERING
Type	Power-assisted rack and pinion
Turns lock-to-lock	3.25
Turning circle, curb-to-curb	36.4 ft.

BRAKES
Front	11.8-in. ventilated discs
Rear	10.38-in. discs

WHEELS AND TIRES
Wheel size	15 in.
Wheel type	Light alloy
Tire make and size	205/70 SP Super Sport Dunlop radial
Tire type	Radial
Recommended pressure, F/R	26/24 psi

TEST RESULTS

ACCELERATION
0-30 mph	3.2 secs.
0-40 mph	4.6 secs.
0-50 mph	6.2 secs.
0-60 mph	8.4 secs.
0-70 mph	10.4 secs.
0-80 mph	13.1 secs.
Top speed	125 mph (approx.)
Standing quarter mile	16.1 secs./89.6 mph

BRAKING
30-0 mph	46 ft.
60-0 mph	195 ft.

FUEL CONSUMPTION
EPA city/highway	10/13 mpg
MT 73-mile test loop	16.1 mpg

SPEEDOMETER
Indicated	30	40	50	60
Actual mph	30	40	50	58

CONTINUED FROM PAGE 18
from the accelerator. And there is no dead pedal or foot rest of any sort for the driver's left leg; to make this even worse, the carpet in that area is covered with a slippery vinyl anti-wear pad, so you can't even dig in a heel.

Finally, there is the matter of the driving position. Perhaps the telescoping steering column confuses the issue, because the first impulse is to leave the seat alone and pull the wheel back until it's an easy reach. Then, with nothing to brace your left leg against, you find yourself sliding down out of the seat. Putting the steering wheel clear forward and running the seat ahead until your left leg touches the firewall is a much better solution. Only the proximity of your head to the windshield pillar seems wrong then, but you get used to that after a few hundred miles.

Certainly there would be no difficulty in getting used to the rear seat if you are short enough to live with the low ceiling. Kneeroom is reasonable for a coupe, and there is elbow room to spare. Only the ceiling is low. But unlike other manufacturers who tend to assume that the rear in a 2+2 is primarily decorative, Jaguar has gone to the trouble of providing a very comfortable seat. The padding is soft yet deep, the cushions offer good support and there are even little storage bins on each side.

Even so, all of this discussion of comfort, upholstery materials and fender lines seems wide of the mark. The XJ-S is remarkable for one quality; its extravagant and completely unusable performance. You're always reminded of it—the way it springs forward when you press the accelerator, its cat-like grace on the boulevard, its stability and balance when the speedometer climbs. It's a car made for epic journeys. Since those opportunities are rare, it attempts to appease its driver with power accessories and automatic temperature control. And it does a fair job. Later in the year, a four-speed manual transmission will be available—an unprecedented novelty in a car of this stature—and it will give serious drivers a stronger vote in the proceedings. But it won't change the car much.

The XJ-S is a dark and mysterious product of England's tortured auto industry, fantastically overqualified for today's driving conditions—a rich man's car for an opulent age that might have been but now clearly never will be. That is reason enough to want one. •

FULL TEST **JAGUAR XJS COUPE**

Cat amongst the pigeons!

David Segal finds that, although its a thirsty feline, the softened successor to the famous E Type is a smooth and satisfying highway cruiser.

AS WE ROCKETED down another endless hill, I wondered what he did for a living. A well-stocked back seat told me that he was probably a travelling salesman. Or perhaps a well to do farmer, who knows?

His Fairlane lurched heavily as it negotiated the tight left hander; building up speed again as the road unwound ahead. Suddenly the Fairlane peeled off to the right and into the turn lane. On went his brakes hard as he washed off speed, at the last moment swinging onto what, according to the sign, was the road to Cooma.

As I drew the Jag alongside, he looked across and gave me a wave. I returned the acknowledgement and turned my attention back to the road ahead. It was just reward, I thought, for the 150 km of speedy, safe and quite enjoyable nose to tail driving we had just experienced.

I had come across the Fairlane around an hour earlier. It was during one of those 'special' open-road tests that today are only spoken about in whispers.

That morning I'd flown into sunny Sydney to collect a shiny new Jaguar XJS from Leyland. The plan was simple - return, via the Hume, to hometown Melbourne.

I knew as I left Sydney, encased in the supple leather seat, with the V12 purring quietly away up front, that it was going to be a memorable trip. It was.

Actually it was relatively uneventful in the early part, save for a brief dice with a BMW 320. Than I came across the Fairlane.

He was travelling just five km/h slower than I, and normally I would have gone straight past. But I could see he was using a radar detector, something I had inadvertadly forgotten to bring. In a case like that, it's always worth slowing slightly to take advantage of the detector's police far-sightedness!

There began a very enjoyable hour's motoring; motoring which was certainly at high speed, but was just as certainly safe.

My co-traveller quickly caught on to my intentions and responded by quickening his pace slightly. Our cars and ages may have been worlds apart (he was almost old enough to be my father), but our on-road repartee amazed me.

As he headed for who knows what towards Cooma, I was truly sorry to see him go.

Shortly after our parting I became aware of a rattle somewhere distant from the comfort of the XJS's controlled climate cockpit.

It was a 'a rattle' that was to become THE RATTLE.

At first I could barely hear it above the Clarion tape deck, but as these things always do, it became progressively louder.

Naturally I did what any self-respecting motoring journalist does — I turned up the tape deck! However my theory of "ignore it and it'll go away" didn't do any good. Eventually I stopped to check it out.

The noise sounded like a loose exhaust hitting against the body somewhere, but a look revealed nought. I drove on.

Now it was very loud and starting to drive me slightly crazy. In between cursing everyone from the boss of Leyland downwards, I decided to stop at the dealer in Holbrook — if there was one.

An enquiry at the first garage brought an interesting response. "Oh, you've got one of those things. The local Doc's got one, we've tried to work on it but it's nothing but trouble," said the attendant, whom I discovered later drives a Datsun 260Z anyway!

Are there any Leyland dealers in town? "No, there's one in Albury, but he won't look at it, you'll just have to keep going to Melbourne."

Disillusioned, I had a second look for the problem, determined to find it this time.

I did.

The heat shield, which surrounds the two separate exhaust systems, had come adrift on one side. A screw had dropped out and the continuous vibration made the aluminium sheet bend and rattle against one of the exhaust pipes. Simple.

Much relieved, I swung the big cat into another Holbrook garage, this one, as I discovered, occupied by two women.

They looked like mother and daughter as they came outside to assist and to ogle the Jag. "It's terrible having a rattle in such a beautiful car — it is really lovely isn't it" said mum. Lovely young daughter remained silent during my explanation of the problem, but was clearly impressed.

They couldn't help, but pointed up the road a bit to someone who could. Again the reaction was the same. The pump jockey, in his early teens, couldn't believe it. I had to speak to him twice before he heard me wheruopn he ran out the back to fetch a mechanic.

The mechanic, another teenager, a few years older though, listened as I explained the problem and said simply "No worries."

Up went the hoist, the Jag looking like a beached whale on top of it, and to work went the mechanic. He cut a small piece of tin to act as a washer, drilled it, and after several attempts, found the right sized selftapper. The whole thing was fixed in about ten minutes.

Although it sounds terribly corney, he actually refused to take any of the money I offered — although he did ask to look at the engine!

Back on the road with the problem fixed, never to reoccur, the rest of the trip was totally uneventful. A bit of rain slowed down my progress for a while, but I was still back in town in time to join my friends at our weekly pub get-together.

What the final part of the journey did do was to underline just how good the Jaguar XJS really is.

Although those who are familiar with Lamborghinis and Ferraris over similar distances tell me that the XJS isn't even in the same ball park, I've got to say that it provided me with the most comfortable and least tiring long distance jaunt I can remember.

Trying not to appear to be stating the obvious, the V12 engine is the heart of it all. It is so silky smooth, so quiet, yet so powerful. Although virtually every motor noter from here to Pommieland has said it, it's true; the motor almost purrs along.

The 5.3 litre fuel injected all-alloy 12 puts out 212.5 kw at 5500 rpm, enough to punch it along to a top speed of 245 km/h. A redline of 6500 rpm means there is power to spare at all but the very top of the performance spectrum. And that's the beauty of the V12, at any time, at any speed, power is instantaneous, but delivered in the quietest most refined way. No sudden exhaust howl or lurch forward here. It wouldn't be cricket don't you know!

Naturally, the penalty for such performance is a thirst for fuel, and plenty of it. The best figure achieved on the "Coathanger" to "Upside down river" run was 20.1 1/100 km (14 mpg), whilst around town it dropped to a miserable 27 1/100 km (10.5 mpg). Those figures fall a bit short of Jaguar's claimed figure of 17.6 1/100 km (16 mpg), but at least the 90 litre fuel tank gives the XJS a decent touring range.

Backing up the superb engine is an equally capable three speed auto (a four speed manual is available o/s). Naturally it is operated by a centre console mounted T-bar, the shift itself being difficult to use manually thanks to a jagged offset change pattern. Left in drive the box is silky smooth, quiet, and with change points which are all but indiscernable.

During the long drive I found the steering to be the most difficult of all the controls to become familiar with. The rack and pinion system is power assisted, and consequently is extremely light and sensative. It was this sensativity that was the problem.

Once the driver was "dialled in," the Jaguar became a pleasure to point. The steering provided sufficient feel but more importantly was very precise. Point it and it goes there, no questions asked. The power also means it's light for easy around city manouvering.

The steering wheel itself is surprisingly large, and the rim too thin. The column is adjustable for reach though not height.

The silver test car had a problem in the steering area. It developed an annoying vibration at any speed above 120 km/h. Leyland had effected some temporary repairs to it on the morning of pick up; perhaps they had something to do with it. One suspected that it was more of a tyre balance problem.

Showing their worth time after time were the brakes. You just can't go past a good four wheel disc system. They just stop and stop and stop, no fade, no lock up, no fuss. The pedal is progressive in its action but doesn't really offer much feel — it just does its job.

Suspension specifications are as equally impressive as everything else. All independent, coil springs, Girling shocks and twin anti-roll bars. Race car stuff.

It all works as well as it sounds. The car's limits are high indeed, you really have to be very brave (did I hear someone say "or stupid"?) to find them. Once found, the characteristic is oversteer rather than the opposite, this naturally being even more prevalent in the wet.

At no time does the car feel like it's a handful or that it's going to get away from you. It's predictable and progressive in its actions, bumps all certainly disturb it, but not throw it.

What I did find surprising was the body roll. There's more than you'd expect from such a low, squat vehicle. It doesn't appear to effect handling unduly though.

The ride more than matches the handling. Apart from the occasional series of bumps which will catch the suspension out, and which you hear rather than feel, it is oh-so-smooth and completely quiet. No road noise, no contact with the outside world. You just float serenely on.

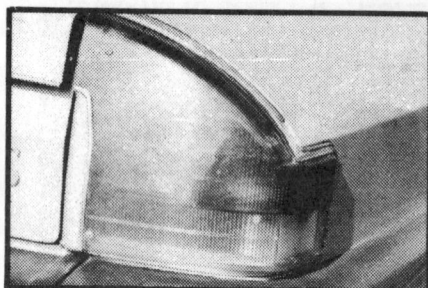

This on road suppleness is assisted by the leather bound luxury interior. Climb down into the well shaped bucket and the first thing you notice is the softness of the seat. You really sink in to it, a condition which is normally bad news over long distances. But not in the Jag. After my long run, I'd never been fresher and less tired.

When seated behind the wheel you face a neat instrument cluster. On the extreme left is the easily read speedo, complete with tripmeter. To the right is the 6500 rpm red-lined tacho. Between the two are four vertical barrel-shaped strip gauges for water temp, oil pressure, fuel level and voltage. All are a little difficult to read at first, and while the voltage gauge remains so, the rest become easy as the miles roll by.

The top edge of the binnacle has a strip of warning lights for indicators, park brake, high beam, brake fail and so on.

TOP: To change the plugs, first locate them! You'd have to say that the under bonnet area is crowded — a mechanic's nightmare in fact.

ABOVE: Effective as the brakes are, there's still quite a lot of lock up on the rear under emergency stopping conditions. This happened on virtually all the test stops from 110 km/h.

On the dash, to the left on the steering column, is the four position light switch. The switch is awkwardly out of view most of the time and is operated, after a while, more by feel than by sight. High beam and flasher mode for the quartz halogen lights are on the right hand turn indicator stalk.

A second, left hand stalk, operates the windscreen wiper/washers.

Flanking the optional sound system (yes, it's actually optional on this $40,000 motor

MOTOR MANUAL
ACTION ANALYSIS
Jaguar XJS

MODEL.....................Jaguar XJS
MANUFACTURER Jaguar ; Rover - Triumph
BODY TYPE/SEATS   Two door Coupe/2 plus 2

PRICE –
  Basic...................................$38,290
  As tested...........................$38,540

  OPTIONS FITTED (and cost) Clarion stereo radio/cassette player ($250)

ENGINE:
  Location................................... Front
  No. of cylinders........................... 12
  Capacity (cc)............................ 5343
  Bore and stroke (mm)............ 90 x 70
  Block....................................Alloy
  Head.....................................Alloy
  Valve gear.................... single overhead
  Induction.......................Fuel injection
  Compression ratio.................... 9.0:1
  Max. power (kw/bhp) 212.5/285 at 5500 rpm
  Max. torque
    (Nm/ft albs) ...... 348.5/294 at 3500 rpm

TRANSMISSION:
  Type........................ Three speed auto
  Shift location............................. Floor
  Ratios –
    First.................................... 2.39:1
    Second............................... 1.45:1
    Third................................... 1.00:1
    Final drive........................... 3.07:1

BODY/CHASSIS:
  Construction........................... Unitary
  Panel material............................. Steel

SUSPENSION:
  Front: Independent, semi-trailing arms, coil springs, anti-roll bar.
  Rear: Independent with drive shafts acting as upper links, twin coil spring and shock units, anti-roll bar.

STEERING:
  Type........ Power assisted rack and pinion

BRAKES:
  Type   Power assisted four wheel discs
                          pressure limiting valve
  Front................. 284mm ventilated discs
  Rear..................... 263mm solid discs

WHEELS AND TYRES:
  Wheel type............. GKN Kent alloy rims
  Diameter/rim width (in)................. 15/6
  Tyre make/type............. Dunlop SP Sport
                                       Steel radials
  Dimensions.................... 205/70 VR15

DIMENSIONS: Weight (kg)...... 1687
  Length (mm)............................ 4870
  Width..................................... 790
  Height................................... 1260
  Wheelbase............................. 2590

Front track............................... 1470
Rear track................................ 1490
Turning circle (m)..................... 11.05
Fuel tank capacity (litres)................90

PERFORMANCE:
  Speedometer error (km/h)
    Indicated............ 60   80   100
    Actual.............. 59.7   78.7   98.2

  Maximum speeds in gears (km/h)
    First....................................... 100
    Second.................................. 165
    Third..................................... 245

  Acceleration from rest to –
    60 km/h..............................4.8
    80 km/h..............................6.9
    100 km/h..............................8.9
    120 km/h............................ 11.5

  Acceleration from 60 km/h to –
    100 km/h...............................5.4

  Standing start to 400 metres –
    Elapsed time...................... 16.8
    Terminal speed (km/h)............... 145

  Braking –
    110 km/h to rest (average)........ 58.3m

  Fuel consumption (litres per 100 km/mpg)
    Driven hard........................ 27/10.5
    Driven normally.................... 20.1/14

*Performance figures recorded using SILICONIX ET 100 digital stop watches from Smiths Industries P/L, Technical Sales Division, 132 Bank Street, Melb. 3205

Cases used are Sampsonsite "Saturn" supplied by Namco Industries (Vic), Princes Hwy., Harrisfield.

Interior space for two is in the super luxury class, but we wouldn't be too keen on a long trip seated in the rear. Well formed as the rear seats are they're still strictly for kids.

car!) are the air conditioning controls. "Climate control" would be a better term.

One of the two dials is for temperature selection — just choose your favorite climate (in fahrenheit I should add) and then set the second knob to 'auto'. The system does the rest. And it constantly maintains that temperature.

Two other settings are provided on the second control for more varied and/or quicker atmosphere alterations.

Above those controls are a series of push buttons for hazard warning lights, heated rear window, map reading light (on the passenger's side) and the interior lights.

On each side on the T-bar auto shift are ashtrays, whilst just behind are the cigarette lighter, electric central locking switch and the two electric side window controls. These console mounted controls lead up into a shallow bits and pieces bin, the hinged lid of which acts as the central armrest.

Further storage is provided by a reasonably sized lockable glovebox and side bins in the rear compartment.

I've left mention of the handbrake to now for a specific reason. It is mounted down beside the driver's seat on the door side and applying it is simple. Not so releasing it. Despite verbal instructons from Leyland and written instructions in the form of a transfer on the front quarter window, I had trouble time and time again.

The only way to release it is by using both hands, pulling the lever up with all your might, then pushing the safety button in. If you are lucky, it MAY release. If not, try, try again.

Surprisingly, taking into account the XJS's price tag, there are a few pieces of equipment it lacks. The biggest omission is of course a sound system. Leyland says that Jaguar buyers like to have their own individual systems installed, so there is no point putting one in. That's the excuse they give anyway.

Equally surprisingly there's no intermittent wiper setting (although there is a single sweep mode), no exterior mirror on the passenger's side, no electric remote control (it's manual) on the exterior mirror, no wipers on the headlights and so on. I'm not trying to detract from the XJS, but rather point out that there are a few cheaper vehicles with a lot of these features. The LTD and Commodore SL/E spring to mind immediately.

Jaguar makes no pretentions about the XJS. It's a 2 + 2 - for that read two adults and two children.

With the front seats forward a little there is room to carry adults in the back, but only for short distances. Access to the back is limited to begin with, and if the driver is to be comfortable, rear seat passengers sit in the 'knees around your ears' position. The back seat, in itself is reasonably comfortable, but is children only territory.

With empty back seats the driver and passenger have heaps of legroom (in terms of length) but are a bit restricted in width. Two up, the cockpit is a very luxurious sort of environment, a personal luxury cocoon.

This 'cocoon' like environment is enhanced by the Jag's small glass area. You sit very low in what is already a low slung machine, and look out through a letter box slot windscreen. Equally small are the rear and side windows.

A big problem restricting vision to the side is the Cathedral Buttress-like 'C' pillar design which sweeps from roof to tail. These create a large blind spot which requires careful attention if you are not to wear a fellow traveller.

The quartz halogen headlights turn night into day well, easily facilitating fast travel. Not so good are the wipers which, strangly for a British car, are set for left hand drive. Their effectiveness is just acceptable, that's all.

'Beauty is in the eye of the beholder' and other famous catch-phrases, can be applied to the XJS. From the moment of its release there has been controversy over its styling. Those who are knowledgeable on such matters say Jaguar got it all wrong, particularly at the back, which even to the casual observer, looks awkward.

However the fact remains that the XJS has a good drag co-efficient, a figure of 0.38 in fact. And that's the best of all the Jaguars.

Despite the good air penetration though, there was an annoying wind whistly around the 'A' pillars of the test car.

One disappointing aspect was the test car's finish. Leyland certainly aren't regarded as world leaders in this area, but the XJS was plain disappointing. Panel fit was ordinary, although paint was reasonable. However it fell down badly inside. The glovebox didn't line up with the rest of the dash, bits of trim didn't meet where they should, even the plastic gauge surround looked terribly cheap.

Some of the minor controls and finishes also didn't measure up to what you'd expect from such an expensive (and exclusive) car. Sure, perhaps it is nit-picking, but all this detail stuff does add up.

The actual pricetag is $38,290. That's big bread but doesn't include on road costs. The excellent Clarion system fitted to the test car costs around $250. So we are really talking very close to 40 grand. Um, er . . . maybe next week?

The price puts the XJS right in the thick of things, for a few grand less you could buy a Lotus Espirit or Elite, or perhaps a Porsche 924, 911 SC or 911 Targa. And there are countless luxury saloons also for the taking at or below that figure.

But it is a fickle market area. Personal fancy dictates what will be bought, needs are secondary and price is a long last. □

SHORT TAKE

Jaguar XJ-S

The fastest production car in America.

Vehicle type:	front-engine, rear-wheel-drive, 4-passenger, 2-door sedan
Price as tested:	$30,000 (base price: $30,000)
Engine type:	V-12, aluminum block and heads, Lucas/Bosch fuel injection
Displacement	326 cu in, 5340cc
Power (SAE net)	262 bhp @ 5000 rpm
Transmission	3-speed automatic
Wheelbase	102.0 in
Length	192.3 in
Curb weight	3930 lbs
Zero to 60 mph	7.9 sec
Zero to 100 mph	22.0 sec
Standing ¼-mile	15.9 sec @ 89 mph
Braking, 70–0 mph	204 ft
Top speed	143 mph
EPA fuel economy, city driving	13 mpg
C/D observed fuel economy	11 mpg

• Between one point on the map and another, you will find out about the heart of the matter. Given a long run, virtually anyone who sets a heavy foot into the backside of the XJ-S will leave its cabin with a puckered-up wallet and a feeling of glowing camaraderie for the V-12 engine. Somewhere in the middle ground between the warmth and the wallet breathes a wonderful truth about the heart of the matter.

It is a truth that means different things to different people. Some say the XJ-S is irreverently and unforgivably out of step with the times in its thirst for fuel. Others, landed gentry with pucker-proof wallets for the most part, say, Pah, pay it no mind because it gives back plenty in exchange for what it takes. The XJ-S gives back more top speed than any other production car you can buy in this country: 143 mph. And if that doesn't skewer the heart of the matter dead to rights, there is no hope for the high life.

The world's supply of V-12s is falling through the cracks. The market for intricate engines with infinite horsepower has dried up, cracking open like barren ground after the evaporation of the wet stuff that once gave it life. A few stout-willed holdouts see the virtues of a thing as exciting as the XJ-S, a thing that lives at the hot point of single-minded trajectories—and damn the mainstream if it flows the other way. We will celebrate the XJ-S, thank you very much.

Not that we are taken with the idea of feeding this wonderwagon day in and day out. Rated at a spongelike 13 mpg by the EPA, the XJ-S gave us only eleven miles for every $1.30 or so (plus or minus the greed of individual station owners). Our 11-mpg average is both a testimonial to the freedom of the open road, where we usually leveled off above 100 mph, and an indictment of those who say the open road doesn't exist anymore. Just look for it. It's there.

Anyhow, it's just as well that not every car on the road is a Jaguar. That would suck up a bit too much fuel. But let's leave room among prejudices for a few free-spirited exceptions, because the world gets gray without them. And gray we can't use when America seems in need of a king-sized jug of Geritol and a lifetime supply of multiple vitamins.

Vitamins are what the veddy British have found more of. Rated horsepower for the 1980 XJ-S is up from 244 at 5250 rpm to 262 horsepower at 5000 rpm. The V-12's compression ratio has been pressured up to 9.0:1 from 7.8:1, and the wonders of modern science are running rampant all around the engine in the form of Lucas/Bosch electronically controlled (D-Jetronic) fuel injection, which uses closed-loop feedback and a three-way catalyst to fry emissions.

For a two-ton car, the XJ-S really gets rolling. Off the line it's nothing to peg an applause meter, but then the woebegotten three-speed automatic doesn't really dig in until 35 mph. You'll want to shift it yourself, because the transmission goes wimpy and overanxious, shifting up at an understressed 4500 rpm, far short of the scrambling 6500-rpm redline and also below the 5000-rpm power peak. Manhandling the shift lever is also the best way to set up a passing situation. Even with the twelve-cylinder torque, a downshift is a nice thing to have underfoot when you want by; but the automatic just won't give it to you at road speed, so it's hands on the levers, boys, and we'll be home in time for tea.

Once the revs are up, finding a ready point of reference in the rear-view mirror is a sure sign of a quick eye, because whatever was there a moment ago will have shrunk with the suddenness of a kaleidoscopic image. *Foop!* it goes, and things get way back there—except for wind noise around the windshield pillar, which comes right along, and a slight shiftiness in the steering. The noise isn't obtrusive at anything under 80 mph, but this is a car built to lope at a hundred. The steering is fine on long straights or lengthy sweepers, but using it hard feels like putting your arm around a willowy girl only to feel her bend at the waist. Quick changes of path bring a hefty shifting of weight and an impalpable hunting for stability. Some of the cure for both is simply to throw more power to the rear wheels, which make a fine rudder.

The sculpturing of Jaguar's coupe is beginning to show its age, though it still snares its share of stares, and the inside is a place of leather-lined comfort and convenience. The controls look a little stodgy, especially the thin-rimmed steering wheel and wispy shift lever. The front seats are quite good for long, fast hauls, less appealing in rubber-eating encounters with tight corners. The all-independent suspension and four-wheel disc brakes eat up corners, bumpy or not, with a most accommodating appetite, but the brakes contribute considerable instability when you push the pedal hard enough to wake them up. In the rear compartment, as back seats go these aren't much. Jaguar's sedans have the real ones.

The XJ-S is for getting down to business, for confronting the heart of the matter in a most gentlemanly way. And in doing so, it is of no small importance that the XJ-S will make the rate of your trips and the rate of your heart ever so much faster. —*Larry Griffin*

Rarity: Jaguar XJS convertible

COME TO THE CABRIOLET

Many think that Jaguar's XJS should always have been a roadster. A professional conversion has arrived at last

To understand the thinking behind the Lynx Engineering Jaguar XJS it is necessary to delve into Jaguar history and trace the big V12 coupé's ancestry.

Even ten years after its rapturous 1961 introduction the E-Type was still enormously popular, and the addition of the technically magnificent V12 powerplant only served to give a further boost to status of the curvacious coupé and roadster. Yet there were signs of age: the E-Type had mellowed considerably since its earlier out-and out performance youth; despite the superb V12 engine and its 250 bhp, the Series 3 car was little faster than the initial 3.8, six-cylinder model, and the provision of luxury items such as automatic transmission and power steering had softened the car's image significantly.

Everyone knew that a replacement was on the way, but when the XJS was finally presented in September 1975 it was the first new Jaguar model whose styling — coincidentally the first not to show the direct influence of the gifted Malcolm Sayer's work — failed to capture the public imagination. Performance and refinement it had a plenty — even at 150 mph the big coupé was almost silent — but though the XJS undoubtedly played the part, it was the first Jaguar whose looks did not match its performance. The styling was bulky and particularly awkward at the rear, where ungainly 'flying buttresses' trailed unfashionably from the edges of the rear window.

Worse still, in many eyes, was the absence of an open-topped roadster version, suspicions that Jaguar were abandoning the performance sports car market having been confirmed by the discontinuation of the E-Type in December 1974. It seemed a tragedy that the effortless but nevertheless extremely rapid performance of the V12 engine could only be enjoyed in the cocooned isolation of the coupé's air-conditioned, leather-clad 2x2 cabin, and that its sound would only be heard by passers by.

When it became clear after some years had elapsed that Jaguar were not intending to produce the car the sports world was waiting for, several private specialists turned their hands to satisfying the demand. First to come up with an effective conversion were an American firm, Royal Carriage Motors, of Washington State, with London-based Lynx Engineering's highly professional XJS Spyder following a matter of months later.

TWO YEARS DEVELOPMENT

The prototype XJS Spyder is the result of nearly two years' development work, for as everyone knows the removal of a unitary construction car's roof deprives the vehicle of much of its rigidity. Additional strengthening panels are welded in the sill and door pillar areas, as well as behind the rear seat, though despite the presence of a bulky, power-operated hood, rear-seat shoulder width is only fractionally reduced and both headroom and legroom (what there ever was of it, at least) are unaffected. Much of the credit for the neatness of the XJS conversion must go to Lynx's experience with their convertible Jaguar XJ6 and XJ12 coupés, production of which was halted not because of any problem with the conversion but because of the contracting market inevitable once BL had discontinued coupé production.

Even with the hood up few would argue that the Lynx Spyder is considerably better-looking than the standard XJS; release the two locking pins on the screen header rail, open the glovebox and operate the concealed switch controlling the hood-retracting rams and the XJ is immediately transformed into one of the most elegant and stylish roadsters ever seen.

Gone are the awesome flying buttresses: Lynx have cleverly sliced off the roof at the top of these structures, removed their inner skins and folded the outer layers over and down into the boot recess, thus ingeniously eliminating any seams on the top of the rear wings. In place of the narrow tunnel leading to the coupé's miniscule rear window is uncluttered rear deck, much in the style of the Alfa Romeo Spider, and all-round vision is significantly improved. From the side the car has a much more balanced profile — even hood-up — with none of the coupé's tail-heaviness, and the rear aspect at last gives some hint of the stylishness Jaguar so obviously planned for their prestige express.

SUPERB WORKMANSHIP

Lynx have taken the opportunity to operate the rear quarter windows electrically (they rotate, rather than rise or fall), providing an instant pillarless coupé or convertible at the touch of the respective switches. The standard of workmanship and design is superb — just about the only aspect of the XJS Spyder that can be criticised from an aesthetic viewpoint is the rather thick windscreen header rail — so obviously a left-over section of the coupé's roof.

In our all-too-brief test session we were clearly unable to test the Spyder at anywhere near the XJS maximum of some 150mph: at just over half his speed, normally considered the tolerable long-distance maximum for open-topped driving, the Spyder gave remarkably little turbulence or buffeting. Some scuttle shudder occures — this is sensed rather than felt — but some loss in rigidity has clearly come about somewhere along the line. With the velour-lined hood up the Spyder becomes, to all intents and purposes, a coupé, but again, we cannot vouch for the soft-top's behaviour at the very high speeds that the XJS's smoothness and silence encourages.

Silence with the roof up, that is. For with the top retracted the splendour of the vast V12 — which many prominent engineers from reputable rival firms consider the best production car engine ever designed — is able to delight the ears as well as please the parts that no other engines can possibly reach.

To Lynx's credit, the rest is pure XJS. Lavish, long and low, 12mpg being the penalty for 100mph in less than twenty seconds. It's so deceptively silent and fast that it needs a warning buzzer to sound off speed limits as they are successively and contemptuously demolished.

The power steering is still too light, the wheel too big and thin, and the controls too cheap-looking — but that's how Jaguars have always been, and one soon becomes accustomed.

Lynx have managed to come down on their initial £8950 estimate for the conversion job, the cost now running nearer £6500 and taking approximately ten weeks to complete.

With the fuel crisis hitting hard at XJS sales, new cars are being heavily discounted. The recipe for today's most magnificent roadster is thus simple: a £14,000 coupé, plus 50 percent for the conversion, plus VAT, equals £21,425 — as always with Jaguar, a price at which it knows no rivals.

As with all conversions, the vexed question of safety and approval rears its ugly head. At present Lynx are converting only used XJSs — exempt from Type Approval regulations — but cars for the German market will need TV certification, whereas those destined for the antipodes are to carry a special certificate giving the body's deflection under certain stresses.

According to Lynx's Guy Black — responsible for the design work — his firm are confident that the Spyder does not infringe any construction and use regulations, and even BL were prompted to say that they wished they were doing it themselves.

Although Jaguar have not given the XJS Spyder full factory approval, they have followed the project with great interest despite officially not going beyond supplying 'encouraging noises'. Perhaps that's because Jaguar's new chairman is, to quote one source, "moving things along more quickly." Could this mean that he has plans of his own ∎

Few would dispute that the XJS Spyder looks cleaner and more attractive than the standard coupé from almost any angle; the rear deck, in particular, shows the lithe elegance Jaguar must always have intended the XJS to have. The beautifully-trimmed hood (above) is raised and lowered electrically and there is no intrusion on precious passenger space; the superb V12 engine (opposite) continues to provide unrivalled performance

Road Test

JAGUAR XJS

Despite its unfashionable size and thirst, the Jaguar XJS can still be counted among British Leyland's proudest achievements

IF "EFFICIENCY" is the watchword for the 'eighties, what hope is there for the Jaguar XJS? Desirable beyond the standards associated with its comparatively modest price, BL's five-year-old flagship is nevertheless conspicuously large, heavy and thirsty for a two-plus-two in the austere '80s and in danger of being upstaged by mostly more fuel and space-efficient rivals from the likes of Porsche, BMW, and Lotus.

The technology for extracting more miles-per-gallon from Jaguar's magnificent fuel-injected 5.3-litre V12 without compromising its prodigious power is already at BL's disposal in the form of the May Fireball head (see *Motor* w/e April 26), though when (or even *if*) the company will play this particular card it's not saying.

Meanwhile, the XJS we test here has taken its first step towards greater economy as a result of running specification changes that will feature on all V12-engined Jaguars for 1981. These stem from the use of a microprocessor to control the fuel injection system, which not only provides better economy through more flexible and accurate fuel metering but also cuts exhaust emissions. Jaguar's engineers have, in turn, exploited this cleaner running by raising the compression ratio and, hence, boosting power and torque by 5.4 and 8.2 per cent respectively — a worthwhile gain which helps to erode the performance differential between the automatic and the old manual XJS (dropped early '79).

In other respects, the XJS continues virtually unchanged: it is a supercar in the classic Gran Turismo idiom, its front-mounted V12 engine driving to the rear wheels through the GM 400 three-speed automatic transmission. In fact, most of its running gear is lifted from the current XJ 12 saloon though it uses the floorpan of the now defunct shorter-wheelbase Coupe, with its wheelbase shortened still further.

The front suspension is by double wishbones which are angled to give anti-dive characteristics using coil springs and with concentric dampers. At the rear, the familiar Jaguar independent system employs lower tubular transverse wishbones, fixed length drive-shafts acting as upper links and two radius arms running forward to mounting points on the body. Anti-roll bars are fitted front and rear. Braking is by ventilated discs all-round (inboard at the rear) and the rack and pinion steering employs Adwest power assistance.

At £19,187, the XJS remains the "bargain" of the supercar market. Driving this latest version has only served to confirm that while it is possible to buy comparable performance for similar money, it is impossible to buy it with greater refinement and comfort at any price. And it is this unrivalled *combination* of qualities that makes the Jaguar so special even by the elevated standards of the supercar class (though a more sporting ride/handling compromise would make it even better). There's no shortage of worthy rivals: Porsche's 928 automatic (£21,827), Maserati's Kyalami (£33,999), BMW's 635 CSi (£18,950) and Aston Martin's V8 (£34,498) are, like the XJS, cast in the conventional front engine/rear drive mould and offer varying degrees of +2 accommodation (the BMW being the most successful in this respect). Ferrari's 308 GT4 (£17,534) has its V8 engine mounted amidships though it, too, has "occasional" rear seating.

With its new digital Lucas fuel-injection and raised compression ratio (from 9:1 to 10:1), the familiar, all-alloy 5343 cc V12 now develops 300 bhp (DIN) at 5400 rpm and 318 lb ft of torque at 3900 rpm. These outputs compare with 285 bhp and 294 lb ft of torque for the previous fuel-injected version of this engine. Despite its formidable bulk, power and torque of this order leave no doubt about the Jaguar's exceptional performance. Though we were unable to record a true maximum speed, the ease with which 140 mph could be summoned — even within the confines of MIRA — suggests that Jaguar's 150 mph claim is not overly optimistic, though we think 145 mph is a more realistic figure.

No less impressive is the Jaguar's acceleration, with 0-60, 0-100 and 0-120 mph coming up in 7.6, 18.0 and 29.7 sec respectively. These figures were achieved by leaving the gearbox in Drive, there being virtually nothing to gain by manually overriding it. Few cars with automatic transmission are quicker, even the significantly more powerful Aston Martin V8 taking 7.5, 16.4 and 25.7 sec to cover the same increments. Quick as the automatic XJS is, however, we shouldn't forget that both the old (and slightly less powerful) manual version of this car and some of the Jaguar's manual rivals could show it a clean pair of heels. The manual XJS we tested in 1976, for instance, accelerated from 0 to 60 mph in just 6.7 seconds.

Unless all your driving is in town, standing starts are of limited interest and it's on the move that the engine's tremendous torque and the automatic transmission combine to stunning effect. Using kickdown we recorded 30-50, 50-70 and 100-120 mph times of 2.8, 3.5 and 11.7 respectively: enough to accomplish overtaking manoeuvres that drivers of many brisk sporting 2-litre cars couldn't even contemplate.

What no figures can convey, however, is the *quality* of the transport the XJS provides. One of our testers commented that its V12 engine must surely be the quietest and smoothest production engine ever made, and it's

Above: leather-trimmed front seats look thinly padded but are actually well-shaped and comfortable. Above right: space in the back, however, is very cramped. Below: facia is neatly presented but perhaps a little bland. Steering wheel adjusts for reach.

hard to think of one that even compares. Equally remarkable is the sheer breadth of its powerband and total lack of temperament. All qualities which combine to create an experience you only ever encounter in a V12 Jaguar: super-quiet, silky and effortless *potency*. It's addictive, be warned.

Addiction will certainly prove expensive as our overall consumption of 13.5 mpg shows. It is encouraging that even with the efficiency losses introduced by automatic transmission, this figure still improves on the 12.8 mpg achieved by the manual XJS we tested in 1976. But apart from the Aston Martin (10.7 mpg), all our selected rivals are more economical, the Porsche 928 Automatic by a considerable margin (18.5 mpg) in its latest form.

Our touring consumption of 16.1 mpg computed (in the aftermath of our fuel-flow meter failing) from steady-speed figures supplied by BL, better illustrates what the average owner might reasonably expect to achieve with some restraint on a long run. This gives a range of around 322 miles on each 20 gallon tankful of 4-star.

Since the gearchange of the manual XJS was both stiff and notchy, we'll shed no tears over its demise. The GM 400 3-speed automatic gearbox is, in any case, smooth and responsive and enhances the astonishing refinement of the engine. If anything part-throttle kickdown is a little too responsive — especially since the engine's massive torque often renders a downchange unnecessary — though under most circumstances the box operates unobtrusively. Using the lever and revving to 6500 rpm gives theoretical maximum speeds in the intermediates of 65 and 109 mph (in practice the 1st to 2nd change takes place automatically at 6300 rpm), but there is little to be gained from doing this and one might as well allow the box to take care of the changes, which it does at around 56 and 93 mph.

The old criticism of the Jaguar's Adwest power-assisted steering being too light and feel-less still applies to a small extent though its crisp responses and sensible gearing do help compensate, creating a sense of wieldiness that belies the car's size and weight. As before, the XJS feels superbly balanced, its natural tendency towards mild understeer changing to progressive oversteer on the limit of adhesion or under power on a tight bend. Only on bumpy secondary roads does the handling deteriorate to any degree, the car's soft damping (set up for silence and comfort rather than ultimate handling) allowing some float and lurching when travelling quickly, though even then the levels of grip remain high.

The brakes are progressive and tremendously powerful, capable of hauling the heavy XJS to rest from high speed without drama or fade, wet or dry.

While the XJS is billed as a 2+2, it's not really a comfortable four-seater and, in its Gran Turismo role, is far

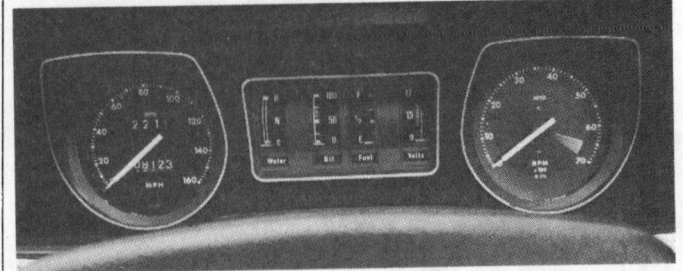

Instrumentation is clear but spoiled by dials' silver edging

better suited to two people and their luggage. Room in the front is ample, with a wide range of adjustment for the well-shaped, leather trimmed seats. But the rear is very cramped — both in leg and headroom — and with the front seats positioned to suit tall passengers, the back is really only suitable for children. To be fair, though, few of the Jaguar's supercar rivals are significantly roomier. Boot capacity is adequate at 8.4 cu ft, as is interior stowage which features an averagely-sized glove box, a large bin beneath the central armrest and small knick-knack pockets mounted on the doors and each side of the rear seat.

Apart from mild small-bump harshness at low speeds and a tendency for the suspension to check over sharp surface irregularities, the ride is up to the usual very high XJ standards. It's shown in its best light over lumpy surfaces which it soaks up with aplomb and no trace of sogginess, though there is still too much float at speed over undulations.

The driving position prompted much praise from our testers who appreciated the well-sited major controls, and adjustable (for reach) steering, though the column stalks are hard to reach if you have small hands. As before, the instrumentation comprises a large speedometer and matching rev counter which flank four smaller gauges for water temperature, oil pressure, fuel level and battery charge. Above the dials is a battery of on fewer than 18 warning lights. Although clearly presented, and calibrated, and free from serious reflections, all our testers agreed that by surrounding the instruments with a thin, painted silver line, BL have made an otherwise attractive display look cheap.

The standard air-conditioning (which is controlled by two rotary switches on the centre console) is good inasmuch as a frosty blast of air is available to order for hot days, but poor in its inability to provide proper bi-level heating/ventilation; the flow from the vents at each end of the facia is weak which leads to stuffiness when the heater is on, despite an adjustable temperature differential between the upper and lower circuits.

True to expectations, the XJS proved to be fabulously refined in every department, the engine note never rising above a distant hum and road and wind noise barely noticeable, even at very high speeds. Our dB reading of 66 at 70 mph is the lowest we have yet recorded (the previous record was held by the Mercedes 380 SEL) and the XJS must be one of the very few cars in which you can listen to the radio at 30 and 120 mph on the same volume setting.

Apart from the cheap-looking instruments, the XJS is trimmed and finished to a high standard — as indeed you would expect it to be for nearly £20,000. Traditional wood cappings are conspicuously absent but the interior is both subtle and functional in its style and finish, and we detected neither creak nor rattle during the test.

And despite lacking a cruise control, the XJS is also very well equipped with air conditioning, leather trim, power windows, central locking, a radio, alloy wheels, adjustable steering, an impact switch to isolate the fuel pump in the event of a crash, power steering, tinted glass and a multitude of interior lights (one for everyone) among other things, all standard.

The flying buttress-style rear pillars are far enough back not to create the blind spots you'd expect, yet surprisingly the windscreen pillars do get in the way. Even so, all-round visibility is quite good and judging the car's width when parking created few problems even for the shortest of our testers, despite the fairly low seating position. The Cibie Biode headlights are powerful on main beam, well-defined on dip (and not too sharply cut-off) while the two-speed wipers clear a large area of the windscreen without smearing.

Magnificent to look at, a nightmare to service: Jaguar's fabulous V12

MOTOR ROAD TEST NO 42/80 ● JAGUAR XJ-S

PERFORMANCE

CONDITIONS
Weather	Wind 0-15 mph
Temperature	66°F
Barometer	29.9 in Hg
Surface	Dry tarmacadam

MAXIMUM SPEEDS
	mph	kph
Banked Circuit	See text	
Terminal Speeds:		
at ¼ mile	94	151
at kilometer	117	188
Speed in gears (at 6500 rpm):		
1st	65	105
2nd	109	175

ACCELERATION FROM REST
mph	sec	kph	sec
0-30	3.3	0-40	2.5
0-40	4.6	0-60	4.4
0-50	6.0	0-80	6.0
0-60	7.6	0-100	8.0
0-70	9.6	0-120	10.6
0-80	12.0	0-140	13.7
0-90	14.6	0-160	17.9
0-100	18.0	0-180	24.3
0-110	23.2		
0-120	29.7		
Stand'g ¼	15.8	Stand'g km	28.2

ACCELERATION IN KICKDOWN
mph	sec	kph	sec
20-40	2.6	40-60	1.7
30-50	2.8	60-80	1.8
40-60	3.0	80-100	2.0
50-70	3.5	100-120	2.6
60-80	4.4	120-140	3.1
70-90	5.0	140-160	4.2
80-100	6.0	160-180	6.4
90-110	8.6		
100-120	11.7		

FUEL CONSUMPTION
Touring*	16.1 mpg
	17.5 litres/100 km
Overall	13.5 mpg
	20.9 litres/100 km
Govt. tests	12.7 mpg (urban)
	21.9 mpg (56 mph)
	18.6 mpg (75 mph)
Fuel grade	97 octane
	4 star rating

Tank capacity	20.0 galls	
	91 litres	
Max range	322 miles	
	518 km	
Test distance	1203 miles	
	1936 km	

*Consumption midway between 30 mph and maximum less 5 per cent for acceleration (see text).

NOISE
	dBA	Motor rating*
30 mph	58	7
50 mph	62	9
70 mph	66	12
Max revs in 2nd	75	23
(1st for 3-speed auto)		

*A rating where 1 = 30 dBA and 100 = 96 dBA, and where double the number means double the loudness.

SPEEDOMETER (mph)
Speedo	30	40	50	60	70	80	90	100
True mph	31	42	51.5	61.5	71	80	90	100

Distance recorder: 0.9 per cent fast

WEIGHT
	cwt	kg
Unladen weight*	34.3	1743
Weight ax tested	38.0	1930

*with fuel for approx 50 miles

Performance tests carried out by Motor's staff at the Motor Industry Research Association proving ground, Lindley.

Test Data: World Copyright reserved; no unauthorised reproduction in whole or part.

GENERAL SPECIFICATION

ENGINE
Cylinders	12 in vee
Capacity	5343 cc (325.8 cu in)
Bore/stroke	90/70 mm (3.54/2.76 in)
Cooling	Water
Block	light alloy
Head	light alloy
Valves	Sohc per bank
Cam drive	Chains
Compression	10.0:1
Fuel system	Lucas-Bosch digital electronic injection
Bearings	7 main
Max power	300 bhp (DIN) at 5400 rpm
Max torque	318 lb ft (DIN) at 3900 rpm

TRANSMISSION
Type	3-speed automatic

Internal ratios and mph/1000 rpm
Top	1.000:1/24.3
2nd	1.48:1/16.7
1st	2.48:1/10.0
Rev	2.08:1
Final drive	3.07:1

BODY/CHASSIS
Construction	Unitary, steel
Protection	Phosphate dip and electrophoretic primer; wax spray in box section cavities and inside doors; fully undersealed; localised zinc-oxide primer at bolt-on joins

SUSPENSION
Front	Independent by double wishbones, coil springs; anti-roll bar
Rear	Independent, with lower wishbones, fixed length driveshafts acting as upper links; radius arms; twin coil-spring/damper units per side

STEERING
Type	Rack and pinion
Assistance	Yes

BRAKES
Front	11.2 in ventilated discs
Rear	10.4 in inboard discs
Park	On rear
Servo	Yes
Circuit	Split, front/rear
Rear valve	Yes
Adjustment	Yes

WHEELS/TYRES
Type	Alloy, 6JK × 15
Tyres	205/70 VR 15 Dunlop
Pressures	26/24 psi F/R (normal) 32/30 psi F/R (full load/high speed)

ELECTRICAL
Battery	12V, 68 Ah
Earth	Negative
Generator	60A alternator
Fuses	12
Headlights	
type	Cibie halogen
dip	110 W total
main	220 W total

Make: Jaguar
Model: XJ-S
Maker: Jaguar Rover Triumph Ltd, Browns Lane, Allesley, Coventry CV5 9DR
Price: £15,401.00 plus £1,283.42 Car Tax plus £2,502.66 VAT equals £19,187.08

The Rivals

Other rivals include Bristol's 412 (£32,718), Ferrari's 400E (£31,809), the Mercedes-Benz 500 SLC (price yet to be announced) and the Maserati Khamsin (£33,999)

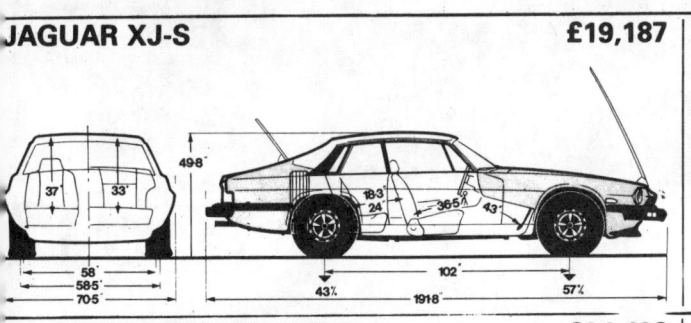

JAGUAR XJ-S — £19,187

Power, bhp/rpm	300/5400
Torque, lb ft/rpm	318/3900
Tyres	205/70 VR 15
Weight, cwt	34.3
Max speed, mph	145†
0-60 mph, sec	7.6
30-50 mph in kickdown, sec	2.8
Overall mpg	13.5
Touring mpg	16.1
Fuel grade, stars	4
Boot capacity, cu ft	8.4
Test Date	October 25, 1980

† Estimated

Now available only in automatic form, the XJ-S continues to provide an exceptional combination of performance, refinement and comfort at a price none of its rivals can beat. Economy improved with new digital fuel-injection, though still a thirsty car. Styling hasn't improved with familiarity, and cramped in the rear, but if you can afford the fuel bills the XJ-S remains one of the world's finest and most desirable cars.

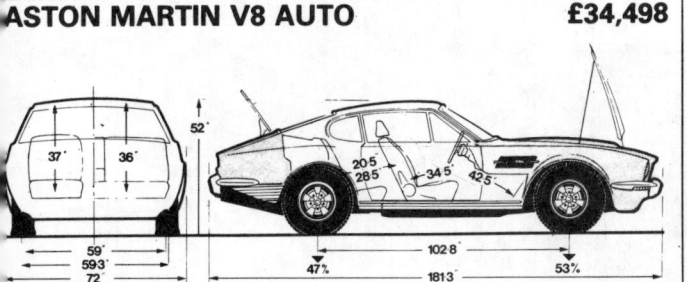

ASTON MARTIN V8 AUTO — £34,498

Power, bhp/rpm	Not available
Torque, lb ft/rpm	Not available
Tyres	225/70 VR 15
Weight, cwt	34.8
Max speed, mph	145†
0-60 mph, sec	7.5
30-50 mph in kickdown, sec	2.3
Overall mpg	10.7
Touring mpg	14.6
Fuel grade, stars	4
Boot capacity, cu ft	8.9
Test Date	October 21, 1978

†Estimated

A superb motor car, and now it is superbly made. Recent revisions have made vast improvement to interior, and overall quality is noticeably better. Otherwise as before: excellent handling, astounding performance, powerful and progressive brakes, ride reasonable at low speeds, good at high speed. Very expensive and rather thirsty, but if you can afford it, it's worth it. At the top of the supercar league.

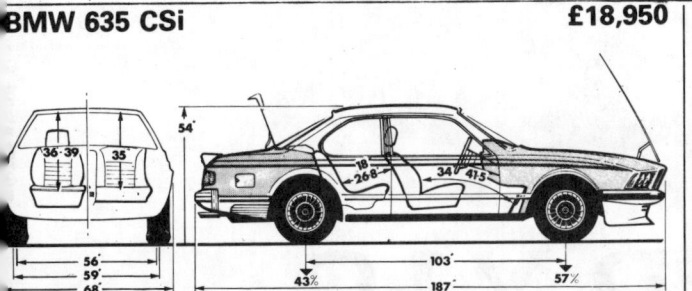

BMW 635 CSi — £18,950

Power, bhp/rpm	218/5200
Torque, lb ft/rpm	224/4000
Tyres	195/70 VR 14
Weight, cwt	29.7
Max speed mph	138†
0-60 mph, sec	7.6
30-50 mph in 4th, sec	6.0
Overall mpg	17.2
Touring mpg	—
Fuel grade, stars	4
Boot capacity, cu ft	12.5
Test Date	December 2, 1978

†Estimated

Top BMW coupé that with its large spoilers, five speed manual gearbox and stiffened suspension is designed as very much a driver's car. Less than ideal matching of the gearing to the engine characteristics spoils what is otherwise a fine machine, for it has excellent handling, roadholding and brakes, good performance (for price) and outstanding economy. Rear seat is cramped and the 635 is not the quest-test car of class.

FERRARI 308GT4 — £17,534

Power, bhp/rpm	255/7600
Torque, lb ft/rpm	210/5000
Tyres	205/70 VR 15
Weight, cwt	25.3
Max speed, mph	150†
0-60 mph, sec	6.4
30-50 mph in 4th, sec	5.1
Overall mpg	14.1
Touring mpg	18.7
Fuel grade, stars	4
Boot capacity, cu ft	5.0
Test Date	January 11, 1975

†Estimated

Mid-engine coupé powered by 255 bhp V8 giving outstanding performance and, in this company, fair economy. Mediocre gearchange. Nominally a 2 + 2 but tiny rear seats nor suitable for adults; boot is small. Roadholding exceptional, handling less precise and responsive than that of 246 Dino. Visibility good, heating and ventilation disappointing. Although production has ceased, the replacement Mondial has just become available.

MASERATI KYALAMI — £33,999

Power, bhp/rpm	270/6000
Torque, lb ft/rpm	289/3800
Tyres	205/70 VR 15
Weight, cwt	33.3
Max speed, mph	147†
0-60 mph, sec	7.6
30-50 mph in 4th, sec	6.2
Overall mpg	11.6
Touring mpg	14.7
Fuel grade, stars	4
Boot capacity, cu ft	8.9
Test Date	July 22, 1978

† Estimated

Notchback 2 plus 2 has De Tomaso-derived body, but the engineering is pure Maserati. Not as quick as some rivals, but more thirsty. Excellent high speed cruiser, handling good until limit is approached, but ride only fair. Engine less refined than those of many rivals, sounding strained at high revs; gearchange slow. Powerful progressive brakes. Interior finish leaves much to be desired. A disappointment — not in the same league as the rest.

PORSCHE 928 AUTO — £21,827

Power, bhp/rpm	240/5250
Torque, lb ft/rpm	280/3600
Tyres	215/60 VR 15
Weight, cwt	28.9
Max speed, mph	140†
0-60 mph, sec	7.6
30-50 mph in kickdown, sec	2.8
Overall mpg	18.5
Touring mpg	—
Fuel grade, stars	4
Boot capacity, cu ft	7.3
Test Date	Not published

† Estimated

Porsche's luxury sports car in its latest form is as quick as the Jaguar. Super-refined engine and low wind noise but potential refinement let down by excessive tyre roar. Superlative road holding, and excellent handling in all but the most extreme conditions, with mediocre ride. Excellent brakes. Beautifully made and lavishly equipped. Very spacious for two, but cramped rear seat. Economical for its class.

JAGUAR XJ-S

Enhancing its reputation for silky smooth performance and luxury

JAGUARS HAVE ALWAYS offered enthusiasts a certain flair, a lovely blend of excitement and charm first exemplified by the trend-setting flowing lines of the XK-120 introduced in 1948, later in the XK-E ("the greatest crumpet collector known to man," we cited in April 1964), right up to the current XJ6 sedan (tested in September 1980) and the XJ-S coupe. As we've noted before, though, a spiritual transition has occurred over the years, with successive Jaguar models offering more refinement at the expense of out-and-out performance.

But *aficionados* of the Coventry cat have reason for rejoicing, because the latest XJ-S displays improved performance with no tarnishing of its luxury GT aura. In its initial R&T road test (January 1976), the XJ-S certainly impressed us with its comfort and luxury, yet we lamented its move up-market away from the performance-oriented cats of the first and second generation. Today, the XJ-S continues to be a spiritual kin to the BMW 633CSi, the Mercedes-Benz 450 (now 380) SLC and, perhaps to a lesser extent, the Porsche 928. And if one grants this alignment of competitors, the Jaguar XJ-S fares quite well indeed.

Consider price, for instance. Whereas it would be elitist to say a car costing $30,000 is anything but expensive, note that the Jag's competitors cost even more. How about standard equipment? Cars of this ilk ordinarily have all the niceties. And so it is with the Jag: air conditioning, electric window lifts, AM/FM stereo/cassette, central locking; you name it, and it's probably standard equipment. What about the mechanicals? Cars of this class should have a well engineered independent suspension and a disc brake at each corner, and under the hood there should be more than adequate power for today's driving conditions. Here, the XJ-S looks particularly good, with elements such as inboard rear discs for less unsprung weight and the ultimate fillip in these days of fuel-conservative motoring—a lovely overhead-cam-per-bank vee engine with 12 (count them: twelve!) cylinders. As such, it's the only V-12 currently available in the U.S. market, and this evidently enhances what's already a certain exclusivity. And, after all, this is what cars in this almost-exotic range are all about: exclusivity.

What we have here, by the way, is not formally a 1981 XJ-S. According to Jaguar-Rover-Triumph, there is no 1981 model per se; rather, there is a supply of cars certified as 1980 models, and there's also a 1982 XJ-S waiting in the wings for mid-year introduction. No official word on what changes to expect, but we'd guess any modifications will be in the direction of yet more refinement and added standard equipment (cruise control, for instance) a la XJ-6. In particular, the word is there'll be no new bodywork and the V-12 will remain.

So, we've got every reason to expect that the XJ-S you see here is completely representative of the breed, at least through 1982. And what you see is a design that looks contemporary, despite its 1975 introduction, with prominent features being bulges here and there and a flying buttress rear-window treatment that's either especially distinctive or an automotive cliche, depending on whom you ask. Although there's an impression of litheness from without, once within the XJ-S there's no mistaking the fact that this is a fairly large automobile, with a wheelbase of 102.0 in., an overall length of 192.2 in. and a width of 70.6 in.

You sit low in the car, in an area that's most properly called a cockpit. The XJ-S has a relatively squat greenhouse and the hood looks (indeed, is) long. A whir of its starter brings to mind a Spitfire—the airplane, not the car—and you're aware of the 12-cylinder burble through a subdued bass note of the dual exhaust.

Like other aspects of the XJ-S, this engine has undergone refinement over the years, and it has become one of the smoothest and most responsive powerplants in the world. This latest version replaces a pair of oxidizing converters, air injection and exhaust-gas recirculation with a single 3-way converter linked via oxygen sensing to the Lucas/Bosch digital electronic fuel injection. Because of these changes in emission control and an increased compression ratio (7.8 to 9.0:1), horsepower increases from 244 bhp at 5250 to 262 at 5000; torque jumps from 269 lb-ft at 4500 to 289 at 4000. And the engine positively purrs under almost any condition. We say almost, because detonation made itself known now and again; like many modern high-performance powerplants, the Jag prefers unleaded high-test gasoline.

The 3-speed automatic transmission is the ubiquitous GM Turbo Hydra-matic 400, and it's the sort of mating that allows upshifts or downshifts essentially under the control of one's right foot. Which, come to think of it, is quite fortuitous because we continue to be unimpressed by the XJ-S's shift lever: Compared to the rest of the controls, it feels a trifle flimsy and its detent isn't especially amenable to manual shifting.

But no matter. Leave the lever in D and the car is effortlessly quick enough that its speedometer (a large handsome Smiths unit) profits from an attentive eye. Otherwise, as one staff member found, "You wonder why the rest of the traffic is moving so slowly, and then you find it's you moving so quickly." Or as another driver put it, "Here's a car that simply glides."

Glide with some manual shifting at the 6500-rpm redline, and you'll see 60 mph come up in 7.8 seconds, the quarter-mile in 15.9 sec at 90.5 mph. However, be prepared to pay for all this lead-foot activity at the fuel pump, because our day of track testing resulted in no more than 9.5 mpg. Omit this fill (as we always do in computing our data panel's mpg calculation) and our ordinary-use average worked out to 15.0 mpg. Still not what one would call fuel conservation of the highest sort, but then as one staff member noted, "Complaining about mpg in a car of this character is like living in a mansion and saying good help is so hard to find."

To cite another staff member: "This is a car that encourages me to go for a drive—simply for pleasure." Not that there aren't occasional irritations in said pleasure seeking, however. The same driver felt that the front seats offered insufficient lateral support combined with too much support for the lower back, and other staff members agreed. Nor is the rear seating all that roomy; like several other cars in this class, the XJ-S should be thought of most definitely as a 2+2. Ah, but we were unanimous in savoring the wonderful leather aroma of the XJ-S's seats and interior trim, and only one staff member lamented the loss of walnut veneer that graced many early Jaguars. And this Jag is certainly a quiet cat, rustling our noise level meter to only 64 dBA at 50 mph.

An aspect of the XJ-S that's changed since our last road test is its tires. This car was fitted with Pirelli P5s, the same 205/70VR-15 size as the Dunlop SP Sports of our previous XJ-S test but a much different tire in terms of performance. For example, we measured an improvement in slalom speed from a relatively

AT A GLANCE

	Jaguar XJ-S	BMW 633CSi	Porsche 928
List price	$30,000	$35,910	$38,850
Curb weight, lb	3890	3400	3370
Engine	V-12	inline-6	V-8
Transmission	3-sp A	4-sp M	3-sp A
0-60 mph, sec	7.8	8.4	8.1
Standing ¼ mi, sec	15.9	16.8	16.2
Speed at end of ¼ mi, mph	90.5	84.5	89.5
Stopping distance from 60 mph, ft	157	160	138
Interior noise at 50 mph, dBA	64	68	72
Lateral acceleration, g	0.726	0.754	0.811
Slalom speed, mph	56.6	55.5	59.7
Fuel economy, mpg	15.0	19.0	16.5

squealy 52.7 mph to a more commendable 56.6 mph. Our test driver sensed the car liked an early lead around the pylons with fairly sharp steering input. However, he and other staff members noted a failing in the Jag's power-assisted steering; namely, a quick turn of the wheel will beat the power assist, with the feeling that one is wrestling for an instant with steering of the unassisted variety. In general, though, the XJ-S's steering displayed a lightness in keeping with the car's luxury image, combined with a road feel bespeaking its considerable tire patches.

The ride/handling compromise is an especially good one. There's an initially soft response to minor road irregularities and very good fore/aft balance of springs and damping. Press on further, and the XJ-S responds with predictable understeer just that side of neutrality and exhibits an overall feeling of stability. Just the thing for getting quickly from here to there in confidence.

Its brakes were up to the rest of the XJ-S's performance capability, pulling the almost 2-ton car down from 60 mph in a distance of 157 ft; the 80–0 stops averaged 276 ft. There was some rear-lock sensitivity in these panic simulations, but nothing that couldn't be modulated through a fairly hard-feeling brake pedal that was nicely suited to modulation. This hard pedal, by the way, was primarily a panic-test phenomenon. In ordinary driving the Jag's brakes were well assisted; for example, they required a comfortable 24-lb pedal pressure for our 0.5g fade test, and exhibited no fade whatsoever during these six stops.

Last, we can observe that our own love/hate relationship with the Jaguar marque has often revolved around its quality control in assembly and mechanical durability in the longer term. And in this regard, our notebook views of this particular XJ-S displayed a certain schizophrenia. Comments ranged from "the trunk lid/fender fit is very poor," "the spray-on labeling of the instruments and controls isn't in keeping with a car in this class," and "the leather is lovely, but its stitching already shows signs of separating" to another view (maybe damning with faint praise) that summarized the car as "better than any Jag of recent memory." In truth, the car did appear rather better detailed and, unlike some others of the marque we've seen, it displayed nothing untoward mechanically during the period of our testing. We're told Jaguar is placing increased emphasis on quality control, and perhaps these efforts are bearing fruit.

So we're ready to rejoice along with other Jaguar *aficionados*: The pleasures and exclusivity of 12-cylinder motoring are still available, and this latest example of the Coventry cat enhances its reputation for silky smooth performance and luxury.

PRICE

List price, all POE	$30,000
Price as tested	$30,000

Price as tested includes standard equipment (air conditioning, AM/FM stereo/cassette, elect. window lifts, central locking system)

GENERAL

Curb weight, lb/kg	3890	1766
Test weight	4090	1857
Weight dist (with driver), f/r, %	55/45	
Wheelbase, in./mm	102.0	2591
Track, front/rear	58.6/58.6	1488/1488
Length	192.2	4882
Width	70.6	1793
Height	47.8	1214
Trunk space, cu ft/liters	13.6	385
Fuel capacity, U.S. gal./liters	24.0	91

ENGINE

Type	sohc V-12
Bore x stroke, in./mm	3.54 x 2.76 ... 90.0 x 70.0
Displacement, cu in./cc	326 ... 5343
Compression ratio	9.0:1
Bhp @ rpm, SAE net/kW	262/195 @ 5000
Torque @ rpm, lb-ft/Nm	289/392 @ 4000
Fuel injection	Lucas/Bosch electronic
Fuel requirement	unleaded, 91-oct

DRIVETRAIN

Transmission......automatic; torque converter with 3-sp planetary gearbox

Gear ratios:		
3rd (1.00)		3.07:1
2nd (1.48)		4.54:1
1st (2.48)		7.61:1
1st (2.48 x 2.40)		18.27:1
Final drive ratio		3.07:1

CHASSIS & BODY

Layout	front engine/rear drive
Body/frame	unit steel
Brake system	11.8-in. (300-mm) vented discs front, 10.9-in. (277-mm) discs rear; vacuum asst
Wheels	cast alloy, 15 x 6JK
Tires	Pirelli P5, 205/70VR-15
Steering type	rack & pinion, power assisted
Turns, lock-to-lock	3.3

Front suspension: unequal-length A-arms, coil springs, tube shocks, anti-roll bar/lower A-arms, fixed-length halfshafts, trailing arms, dual coil springs, dual tube shocks, anti-roll bar

CALCULATED DATA

Lb/bhp (test weight)	15.6
Mph/1000 rpm (3rd gear)	24.5
Engine revs/mi (60 mph)	2450
R&T steering index	1.20
Brake swept area, sq in./ton	219

ROAD TEST RESULTS

ACCELERATION

Time to distance, sec:
0-100 ft	3.4
0-500 ft	8.9
0-1320 ft (¼ mi)	15.9
Speed at end of ¼ mi, mph	90.5

Time to speed, sec:
0-30 mph	3.1
0-50 mph	6.1
0-60 mph	7.8
0-70 mph	10.0
0-80 mph	12.5
0-100 mph	19.5

SPEEDS IN GEARS

3rd gear (5700 rpm)	139
2nd (6500)	104
1st (6500)	63

FUEL ECONOMY

Normal driving, mpg	15.0

BRAKES

Minimum stopping distances, ft:
From 60 mph	157
From 80 mph	276
Control in panic stop	very good
Pedal effort for 0.5g stop, lb	24
Fade: percent increase in pedal effort to maintain 0.5g deceleration in 6 stops from 60 mph	nil
Overall brake rating	very good

HANDLING

Lateral accel, 100-ft radius, g	0.726
Speed thru 700-ft slalom, mph	56.6

INTERIOR NOISE

Constant 30 mph, dBA	60
50 mph	64
70 mph	72

SPEEDOMETER ERROR

30 mph indicated is actually	27.0
60 mph	54.0

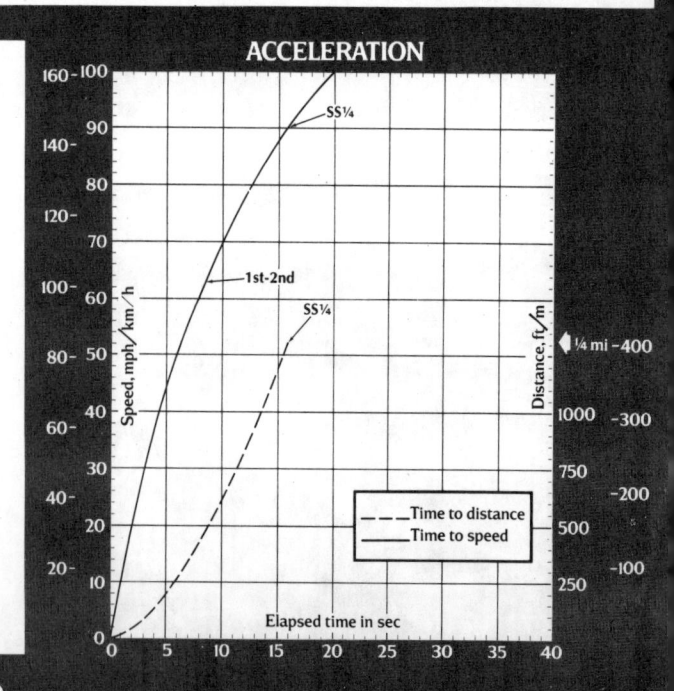

FIREBALL
The development of the decade?

The new Jaguar XJ-S with minor styling improvements.

When we look back, about ten years hence, will we regard the announcement of Jaguar's new fireball cylinder head as the development of the decade? I suspect so, it reveals what a tremendous amount has still to be learned about efficient combustion in the petrol engine.

Be sure that — right now — engineers the world over are examining what Jaguar have achieved, and wondering how the same lessons can be applied to their own products in the interests of getting more power out of less fuel.

It was in early 1976 that Harry Mundy, Jaguar's chief engineer on the power units side, went out to Switzerland to see the independent engineer and researcher, Michael May. A Volkswagen Passat engine had been adapted by May using his patented combustion chamber, to give far better performance, economy and emissions than the standard engine.

Co-operation with Michael May was tied up, and development proceeded fast, first on a single-cylinder mock-up engine, then on a slant-six obtained by milling away one side of a V12 cylinder block. Problems were encountered, which had to be solved, but the goal was promising: the ability to make the Jaguar

How the new Jaguar V12 HE (High Efficiency) May "Fireball" combustion chamber works. During the compression stroke, the air/fuel mixture is induced to swirl. Inset shows underside view of swirl pattern.

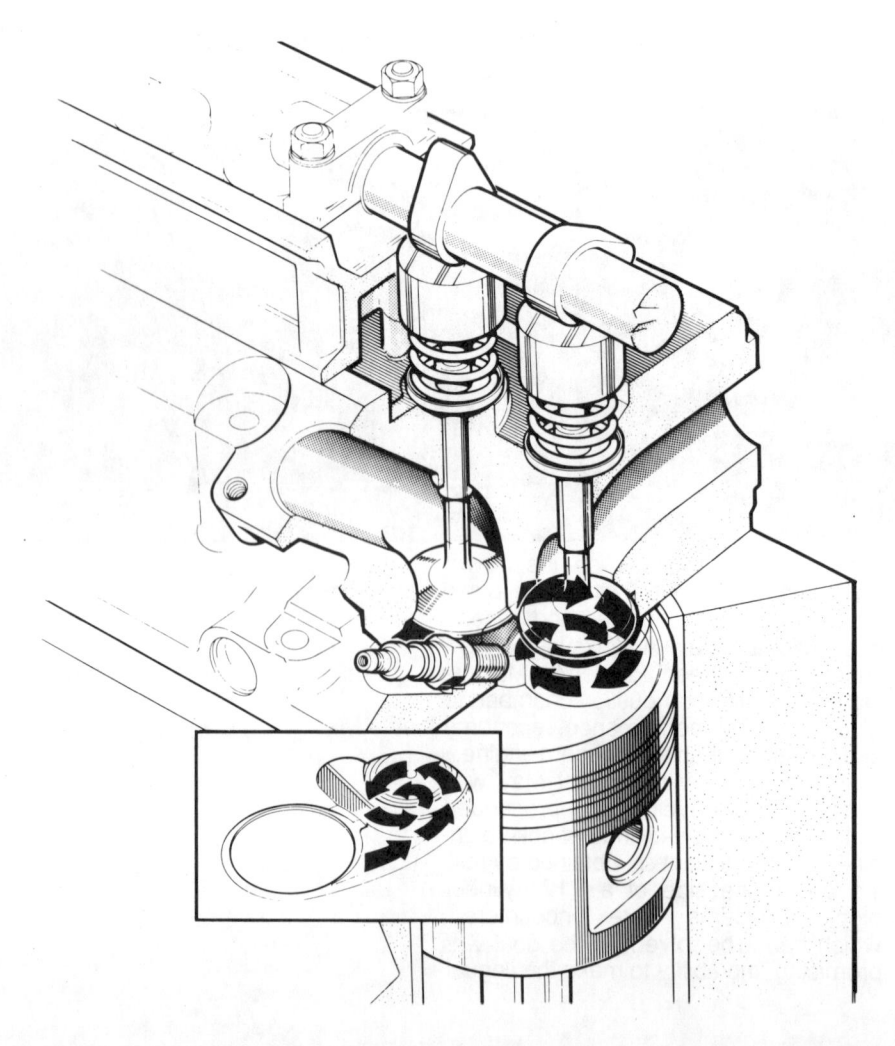

V12 engine give as good fuel econom[y] as cars with much smaller and less po[]werful engines.

The May combustion chamber, chris[]tened "Fireball", is a form of two-leve[l] spiral ducting. As the piston rises, durin[g] the compression stroke, the mixtur[e] charge is forced into rapid turbulence i[n] the main chamber. But careful shaping [of] the spiral tract ensures that there is a lo[w] turbulence concentration of mixture in th[e] region of the sparking plug.

To look at the cylinder head on its in[]ternal face, you wonder quite what th[e] fuss is about, as it is not apparent tha[t] there is anything very different from th[e] head of an ordinary engine. The resul[t] however, is that it makes the engine abl[e] to burn — very rapidly and very com[]pletely — exceptionally lean mixtures.

Because the burning is so well con[]trolled with this high swirl effect, very hig[h] compression ratios can be used with or[]dinary 97 octane fuel. Thus, the compres[]sion ratio of the new Jaguar engine ha[s] been pushed up to 12.5:1.

A high-burning rate, and very wea[k] mixture under high compression, calle[d] for special development of the ignitio[n] system. Lucas managed to meet the spe[]cial requirements by using a twin-coil sys[]tem. The secondary coil is used as [a] large inductor, to boost the spark voltage[] and it is mounted ahead of the radiator t[o] help keep it cool.

As before, the Lucas-Bosch fuel-injec[]tion system is retained, and only had t[o] be re-calibrated to give the weaker mi[x]tures which the new engine will eccep[t.] Automatic fuel cut-off acts as an econo[]my measure on the over-run.

This new version of the Jagua[r] 12-cylinder 5.3 litre engine is called th[e] HE (high efficiency), and it develops 29[0] bhp at 5500 rpm. It is the only engine fo[r] the Jaguar XJ-S, and is the power unit [of] the top models in the Jaguar XK an[d] Daimler ranges.

At the same time, the appearance o[f] the XJ-S has been improved, and the for[]mer massive bumpers developed for th[e] American market give way to sleeker[,] bright-finish bumpers. There are stylish[] new alloy wheels, and a double coach[]line along the sides emphasises th[e] length of the car.

Internally, there are further improve[]ments, with more use of leather for th[e] console, doors and side trim, while th[e] fascia panel is reshaped and made o[f] burr elm veneer. Air conditioning is stan[]dard, as is the latest Philips 990 radi[o] cassette unit which monitors an[d] chooses the strongest FM signal for an[y] chosen broadcast.

Although I have yet to experience it, [I] am delighted to record that the forme[r]

The new Jaguar Fireball HE engine in place.

wrongly-gated transmission selector, which I always believed to have been patently unsafe as well as inconvenient to use, has been replaced. There is, of course, no manual transmission version.

In the saloon range, the 6-cylinder DOHC engines with 3.4-litre and 4.2-litre capacities continue, and manual five-speed gearbox is now standard for both; automatic is optional.

In 12-cylinder form, the Jaguar saloon becomes the XJ-12 HE, and it has alloy wheels, electrically-operated sunroof and door mirrors, and headlamps wash/wipe. The boot floor is now covered in a new and practical mat, with carpet on one side and a tough, rubberised material on the other.

At the same time as these major improvements have been made, Jaguar emphasise the very real efforts they have been making over the past 12 months to improve quality, both in standards of assembly and in the components bought in from outside suppliers.

It all adds up to goods news for the buyer, and bad news for rivals such as Mercedes-Benz and BMW. Jaguar have certainly been living on their laurels too long, and it's more than time that they started trying to fight back to their former supremacy.

— *Stuart Bladon*

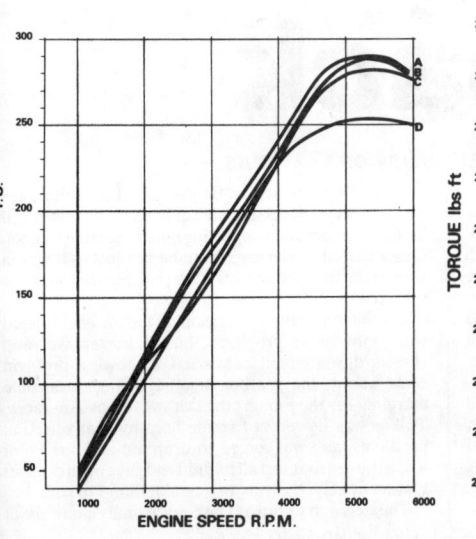

A 1981
SPECIFICATION
'MAY' CYLINDER HEAD
12.5:1 COMPRESSION RATIO

B 1980
SPECIFICATION
DIGITAL PETROL INJECTION
10.0:1 COMPRESSION RATIO

C 1976/79
SPECIFICATION
PETROL INJECTION
9.0:1 COMPRESSION RATIO

D 1973/75
SPECIFICATION
CARBURETTOR
9.0:1 COMPRESSION RATIO

Comparison of DIN power outputs of the Jaguar V12 engine over the years.

A 1981
SPECIFICATION
'MAY' CYLINDER HEAD
12.5:1 COMPRESSION RATIO

B 1980
SPECIFICATION
DIGITAL PETROL INJECTION
10.0:1 COMPRESSION RATIO

C 1976/79
SPECIFICATION
PETROL INJECTION
9.0:1 COMPRESSION RATIO

D 1973/75
SPECIFICATION
CARBURETTOR
9.0:1 COMPRESSION RATIO

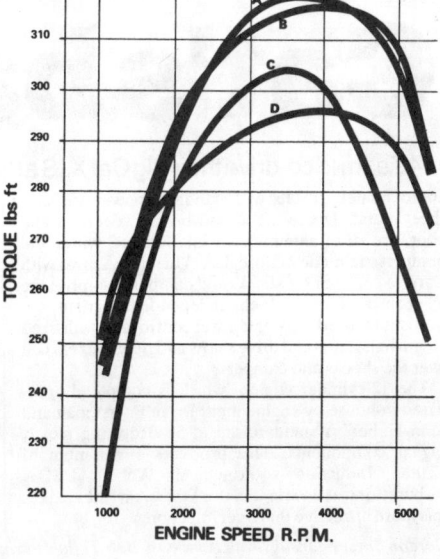

Comparison of torque ratings of the Jaguar V12 engine over the years.

Track Test

The immaculately prepared Group 44 Jaguar XJS at the recent Daytona 24 Hours where gearbox problems hindered the progress somewhat.

The Big Cat's back

GORDON SMILEY was invited by the Group 44 Team to co-drive their Big Cat XJS at the Daytona 24 Hours.

When the call came, I had nothing but positive feelings. Would I like to drive the Group 44 XJ-S Jaguar at Daytona for the 24 hours? Sure, I replied, sounds terrific. I had offers from other teams to drive a BMW M1 and a 512BB Ferrari but since I have been negotiating with Group 44 to do some other racing with them during 1982, I was pleased to accept their invitation. I had driven the 24 hours last year with the John Paul team but we dropped out early so I was looking forward to going the distance this time.

The Group 44 team has one of the finest reputations in the world of North American racing. Their thoroughness, professionalism and standards of preparation are more along the lines of Formula 1 than of a team competing with production based equipment. Team boss Bob Tullius, five times National Champion and two times Trans-Am Champion, is a man with a reputation for representing his sponsors like no one else in the business. As a result, the team is always adequately financed and equipped with the finest components available.

One of Tullius's other admirable qualities is his ability to secure, and keep, top flight personnel. Crew chief Lawton "Lanky" Foushee, is a "good ol' boy" from NASCAR racing but has been with Group 44 for 12 years. He and Tullius have assembled a fine staff.

The Daytona XJ-S is in fact the third of the "Big Cats" to be built by Group 44 for Jaguar-Rover-Triumph (the BL subsidiary in the US). XJ-S was in fact built for Tullius to drive in the 1981 Trans-Am series which he did with great success, winning more races than anyone else and finishing a close second in the series.

The car is of tube-frame construction, weighing approximately 2850lbs with standard appearance XJ-S sheet metal. Doors, hood and bonnet, deck lid and boot, are all of aluminium, while fenders, wings and bumpers are made of fibreglass. Unequal A arms with Carrera coil over the shocks and dampers comprise the front suspension, while the independent rear suspension is composed of a single lower control arm with fixed length half shafts and drive shafts and *dual* Carrera coil over the shocks and dampers.

The 12-cylinder engine, which is completely prepared in-house by engine men, Brian Fuerstenau and John Huber, is standardised at 5343cc using mostly Jaguar components. This produces a minimum of 525bhp. Induction is through six Weber 44 IDA carburettors, as a variant in the Trans-Am rules makes this more attractive than fuel injection.

Since Daytona was to be the car's first endurance outing, it was necessary to equip the car with a full lighting system and supporting electrical components. Other than this, the car was to be run just as it was in sprint race trim, again a credit to the preparation of the Group 44 team.

During the Trans-Am season, the team was afflicted with only minor problems, but as horsepower and chassis development continued a growing problem came about; the gearbox was beginning to become marginal for the job. At the last two Trans-Am races, Tullius was slowed at Laguna Seca by gear selection problems, and was forced to drop out at Sears Point when the gearbox lost all its fluid and gave up in disgust! Unfortunately, this was an omen of things to come!

The decision to run at Daytona actually came about

Gordon Smiley consults with team boss Bob Tullius during practice for Daytona.

through one of Tullius's long time sponsors, the Goodyear Tyre & Rubber Co. A somewhat unusual tyre war is heating up in the US, based around racing participation and Daytona was to be the first round of the battle. BF Goodrich had secured some Porsche 924 Turbos to run on their street radials. Firestone had equipped one of the two factory Mustang cars with their latest product and, of course, Goodyear wanted their NCT radials to be represented with an equal effort. So, the decision was made, the XJ-S would go to Daytona on street tyres!

This decision was a bit of a blessing in disguise for Tullius and the Group 44 team. Not only would they be serving the wishes of one of their major sponsors but the 24 hour race would give the team a chance to monitor the performance of the 12-cylinder engine in endurance trim. It would also give the crew a chance to practice their act over 24 hours and would provide the opportunity for the three projected drivers for the GTP car to work together.

I arrived in Daytona on Wednesday ready to begin practice. Group 44 team driver, Canadian Bill Adam (usually referred to as Jim Adams by the US commentators) was to arrive that evening. Bob took the car out first to get some initial impressions and came in to turn the car over to me saying everything felt very good. The tyres did not have the bite of race tyres but were generally good, even on the high speed banking. He said the engine felt good but the gear ratio was wrong. It was turning 7900rpm going into the third turn banking and the team only wanted to see about 7200 for the sake of reliability. He told me to lift off on the straight to keep the revs down.

My first impression driving out of the pits? What a nice piece of equipment I was in! The car in its pristine white finish, inside and out, gives you the feeling of a real race car that was going to respond to just about anything you asked it to do.

After acclimatising myself to car and circuit and completing the practice session, we returned to the garage area wtih a short list of alterations to make before the next session. One of the changes was to the gear ratio, not only to help preserve the engine but to slow us down! After calculating our speed, Lanky had to be helped into a chair when he discovered these crazy drivers were out running his nice car around on street tyres at 188.8mph! All of us were quite impressed and I finally realised what a futile job I had undertaken chasing the Big Cat around for those four Trans-Am races last year!

The next two days of practice continued uneventfully with Bob, Bill and myself, all settling in well to the car and the remarkable street tyres which were performing flawlessly under a lot of stress. Just before the final practice session we changed the gearbox, since second gear was not engaging smoothly and we did not want to take any chances. The team felt that running the car on street tyres would eliminate a lot of the strain on the gearbox especially when combined with the reduced pace of an endurance race.

Oh yes, the gear ratio. We changed ratios no fewer than five times during the two days in our efforts to reduce maximum rpm output of the engine. The team succeeded in doing this nicely but the mechanics, sadistic as they all are, left us with a car that would not slow down! We finally ended up with a maximum of 6900rpm on the straights, which the car would achieve twice a lap. This pleased the mechanics greatly, as it calculates out to 195mph, which did not please the drivers! The Goodyear engineers even looked at the three of us rather strangely whenever we went out! What a comforting feeling.

Race time arrived at 3.30 on Saturday and Bob brought the Big Cat round right in the thick of a 69 car field, doing everything he could to get clear of slower cars and staying out of the way of the fast guys. There were lots of both. Bill and I sat in the pits watching with admiration as the boss put together some very

Tullius in the TransAm version last September.

consistent laps and steadily moved up the field. That was until the 40mins mark, when Lanky's face, hidden behind the radio head set, suddenly turned white. Bob had just called in, the gearbox was gone!

Bob coasted in to a halt, got out and explained to Bill and I that it had gone suddenly. I checked my watch and to my utter amazement, just 25mins later, Bill drove out on the race track with a fresh gearbox installed! The crew brought another spare box out of the truck, just in case and the gearbox man went to work rebuilding the broken one.

Needless to say a deep concern started to settle on us all. Bob and I plotted ways to take it even easier on the gearbox. We were just going to have to rely on the brakes to stop us and when we got into a corner, ease the car into the right gear.

Bill came in after 1½ hours and turned it over to me. He said there wasn't a clear "gate" on second gear but the box seemed to be holding up all right. Out I went, using just my thumb and first finger on the lever to try and minimise the leverage and easing the clutch in and out and the throttle up and down. We knew we were not going to be the quickest car in the race, so we had to finish at all costs.

About 20mins into my segment, I started to pick up a vibration and I radioed this to the pits. About 20mins later I was called in and the crew changed the right rear tyre, the one that takes all the load on the banking at Daytona, with a replacement NCT radial that had some of the shoulder shaved off. Goodyear was trying to help maintain the temperature of the tyre across the tread surface to keep up with the tremendous speeds we were asking of it. This fixed the vibration and I continued for another one hour plus, when I again felt a possible tyre problem and came into the pits for a change, a refuel and to turn the car over to Bob.

Arriving in the pits, I discovered that a car had blown an engine and coated the circuit with oil. This accounted for the vibration which I thought might be a low tyre. Everything else looked fine and Bob was back out after a quick stop. 3mins later and he was back in — the gearbox had broken! Evidently the box gave up the ghost when I came off the banking for the last time into the pits and when Bob drove out all he could do was nurse the car back to the pits.

This time the change was made in an incredible 16mins, the close ratio sprint box being installed, including a complete flush of the lines to the cooler, as the team felt they might have contained some debris from the first failure, this accounting for failure number two.

I got back in about 4am and was very pleased with the solid feeling of the gearbox. It took me a few laps to get used to the different shift points on the track with the close ratio box but all in all everything was going well. Then I started having problems getting third gear. The lever would "hang up" going across the neutral gate.

About 20mins later I was coasting down the back straight with the gearbox stuck in neutral! I rolled almost to the middle of turns three and four and stopped on the infield grass. I called into the pits, told them where I was, what had happened, and asked for advice.

The safety truck wanted to tow me further away from the circuit to a safer location, so I got back into the car to radio the crew where we were going. Once I turned the radio back on, I could hear them calling me but I could not understand what they were saying. Finally, it struck me someone was saying, "Gordon, can you see me on the fence?" I turned round and looked into the spectator area to see my other crew chief, Jeff Eischen, standing on top of the fence with his radio set on!

After consulting with Jeff and Brian Fuerstenau, I somehow persuaded the gear lever into some gear, restarted and drove the car back into the pit, where the offending unit was changed once again. This time it was not an internal failure but a problem with the linkage itself. We lost about an hour with all this.

Following three more relatively uneventful segments, Bill came in about half way through his penultimate stint and said he could hear a noise in the rear of the car and that the gearbox was jumping out of third gear. It turned out that a bolt had broken in the rear suspension and since it would take a little while to replace it, half the crew set into changing the gearbox for the fourth time! Bolt and gearbox replaced, I went out for my final 1½ hours.

I was amazed to find that the car could still comfortably achieve the same 6900rpm as it had some 20 hours earlier and it still felt just as good at 190mph as it did all through the race. Every other component on the car performed flawlessly.

Bob finally brought the car across the finish line at the end of 24 hours, in the middle of the Porsche victory picture, completing 533 laps or 2046.72 miles. We were 21st overall, sixth in the GTX category (the car did not quite conform to the 1982 GTO category rules where we should have been) and third among the street tyres.

We estimated that we had lost about 80 laps with the gearbox problems which would have put us right in the thick of things for the street tyre honours, but then the name of the game in endurance racing is not to have the problems we did.

So, all in all, the Jaguar performed perfectly, apart from the gearbox affliction, the NCT Goodyear's were absolutely remarkable, the team functioned like a well-oiled machine and have many improvements planned for the next time out. Lastly, all three drivers managed to keep out of harm's way, so that the car looked just as beautiful at 3.30 Sunday as it did at 3.30 Saturday. Don't count out the Big Cat yet! ■

The clean lines of the Jaguar are clearly expressed on the Daytona infield.

Road Test

XJS-HE

State of the art

JOHN BOLSTER has driven most of the twelve cylinder cars produced in the last half-century and by comparison the Jaguar XJS-HE makes them all appear unrefined. It is even more of a Jaguar than any previous model.

For silence, smoothness and refinement the XJS-HE is beyond comparison making it an excellent long distance tourer.

Boot space is what one would expect of such a car.

One seldom needs to use more than half the throttle.

The internal finish is close to perfection.

For smoothness and silence, the 12-cylinder Jaguar has always been incomparable. It made all other 12-cylinder cars — and I have driven nearly all those produced in the last half-century — feel comparatively unrefined. As for eights, straight or otherwise, they just didn't compete.

These virtues were combined with so much performance that, unless it was normally kept in reserve, the driver's licence did not remain with him for long. Blessed with an excellent compromise between ride and handling, this Jaguar was one of the world's great cars. In many ways it approached perfection, yet it failed because it had a fuel consumption which was at first uncomfortably heavy and later, under worsening inflation, simply unacceptable.

The Jaguar engineers were only too well aware of the problem and the use of fuel injection, with a 10:1 compression ratio, saved petrol and made even more power available. Yet this was only a start and the brilliant techniques of the Swiss engineer (and racing driver), Michael May, were adopted. Known as the 'Fireball' combustion chamber, this makes use of turbulence to a greater extent than in any previous application. The split level combustion chamber, of bath-tub shape, has an inlet valve located in a shallow collecting zone and an exhaust valve set higher up, close to the sparking plug. A swirl-inducing, ramped channel connects these areas.

High efficiency

It is these swirl characteristics that make the burning of extraordinary weak mixtures possible, using an extremely powerful twin-coil electronic ignition system and the Lucas digital injection design. An even higher compression ratio of 12.5:1 is now employed. This HE (high efficiency) engine will pull a 'longer' gear than its predecessor, 2.88:1 instead of 3.07:1, yet another contribution to fuel economy. The official fuel consumption figures prove the remarkable results that have been achieved and they are here shown, with former mpg in brackets 56mph, 27.1mpg (21.9mpg); 75mph, 22.5mpg (18.6mpg); urban cycle, 15.6mpg (12.7mpg). For a luxury car weighing 33.6cwt these figures are certainly acceptable and for a machine with a maximum around 155mph they are phenomenally economical.

The transmission is the GM400 3-speed automatic, with hydraulic torque converter. The brakes are servo-assisted, dual-circuit discs, ventilated in front. Front suspension is by semi-trailing wishbones and coil springs with an anti-roll bar; the Adwest power-assisted rack and pinion steering has a Saginaw rotary vane pump. At the rear, there are lower transverse wishbones, the driveshafts acting as the upper, lateral links, with trailing arms and tin concentric coil spring and damper units for each wheel. Light alloy wheels with 6½ in JK rims carry 215/70 VR 15 tyres.

A 2-door coupé giving excellent comfort for the front occupants, the XJ-S HE cannot be regarded as a full 4-seater for long journeys. At the present stage of the art, a car of this weight would not be expected to corner at racing speeds, but it handles extremely well for such a luxurious car and the steering gives the driver every confidence, even in violent gales.

Perhaps the most delightful feature of this Jaguar is the wonderful reserve of power. One seldom uses more than half throttle and it is all too easy to wander along at 120mph when 70mph was intended, unless the cruise control is employed. An air conditioning system is installed and it is effective if not quite up to Rolls-Royce standards.

I have previously praised the silence, smoothness, and refinement of the car, allied with its extraordinary performance and riding comfort. Perhaps no car is less tiring when driven on a long journey. Yet I must criticise the electric starter, which is no quieter than that of any ordinary car and may be embarrassing in the early hours of the morning. I have owned many cars that started more quietly, notably a 'bullnose' Morris Cowley.

It is difficult to put into words the perfection of the elm burr veneer fascia and door fillets, and the generous use of Connolly hide gives the interior the ultimate look of luxury. Perhaps all that one can say is that the XJ-S HE is even more of a Jaguar than any previous model. Driven continuously at 140mph it still cannot be called economical, but at more practical speeds it uses no more petrol than some cars of about half its size. ∎

Unless the cruise control is employed it is easy to wander on at 120mph rather than 70mph.

The familiar lines of the XJS-HE are retained although the rear view is the one we are most used to seeing.

JAGUAR XJ-S HE

£18,950

Specification

Cylinders, capacity	V-12, 5345cc
Bore & stroke	90 x 70mm
Valve gear	Single overhead camshaft per bank
Compression ratio	12.5:1
Fuel system	Lucas-Bosch Digital Electronic Injection
Power/rpm	299 bhp (DIN) at 5500rpm
Torque/rpm	318 lbf ft (DIN) at 3000rpm
Gear ratios	1.0, 1.5, 2.5:1 x torque converter
Final drive	Salisbury limited-slip 2.88:1
Steering	Power-assisted rack and pinion
Brakes	Servo-assisted split dual circuit discs, ventilated front
Wheels	Light-alloy 6½JK, fitted 215/70 VR 15 tyres
Suspension (F)	Semi-trailing wishbones and coil springs with anti-roll bar
Suspension (R)	Lower wishbones, fixed length driveshafts, radius arms and dual concentric coil springs

Dimensions

Wheelbase	102ins
Track (F/R)	58.6/58.0ins
Length	186.75ins
Width	70.6ins
Weight	33.6cwt
Boot	10.9cuft

Performance

Maximum speed (approx.)	155mph
0-30mph	3.1 secs
0-50mph	6.0 secs
0-60mph	7.3 secs
0-80mph	11.2 secs
0-100mph	16.6 secs
0-120mph	26.3 secs

Fuel

Urban/56/75mph	15.6, 27.1, 22.5mpg

> "Our job, our mission, is simply to build magnificent motorcars, cars of the highest possible quality in every way." —Jaguar Chairman John Egan

Jaguar XJ-S

DRIVING IMPRESSION EUROPE

by Jim McCraw

PHOTOGRAPHY BY MATT KEEFE

Jaguar picked a great way to celebrate its 50th anniversary. The 1982 XJ-S is a luxury GT that sports the most powerful, efficient, and sophisticated V-12 engine Jaguar has ever built. Its wrappings are of the highest quality, and it is the most completely packaged S since the car's introduction in 1975.

Though the new XJ-S has been revitalized on the outside with new bumpers, badges, wheels, and tires, and has the most luxurious interior treatment in the car's history, it is the engine that has brought it back to life. It boasts a 10% improvement in fuel economy, along with the highest horsepower rating (262 hp at 5000 rpm) and the highest compression ratio (11.5:1) available in any car sold in America.

The 5.3-liter (326cid) V-12 engine, dubbed the H.E. engine (for High Efficiency) by the Coventry factory, is equipped with new cylinder heads containing the patented May Fireball combustion chambers developed by Swiss engineer Michael May. Though several European manufacturers have participated with May in development programs for a variety of engines, Jaguar is the first to bring the May chamber into volume production—primarily as a way to get around the U.S. gas-guzzler tax and stringent Common Market pollution standards.

The May Fireball chamber design uses a split-level valve arrangement. The intake valve is positioned in the head in a small pocket, close to the mating surface. The exhaust valve is positioned in its own pocket, with the pocket volume much larger and the exhaust valve set deeper in the head, farther away from the mating surface. The two pockets are connected by a swirl-inducing channel. On the engine's compression stroke, the rising piston pushes the air/fuel mixture out of the intake pocket, through the channel, and into the exhaust pocket, imparting a swirling motion to the mixture as it travels, and compacting the charge into the spark plug area. When the plug fires the lean, swirling mixture, flame initiation and propagation are such that the mixture is burned evenly, rapidly, and completely at high compression levels. (The European XJ-S H.E. engine will use even higher 12.5:1 compression.)

Jaguar began work on the May head in 1976 with single-cylinder test engines, refining chamber shapes and channel configurations, and using mixtures as lean as 23:1. After single-cylinder programs produced good lab results, Jaguar engineers put together a 2.6-liter inline six engine (exactly half of a V-12) and installed it. Final V-12 development began in 1979. Once the final combustion chamber shape was determined, very little else had to be changed—good news for the small, cash-strapped Jaguar works. The dished pistons of the old V-12 were scrapped in favor of flat-top pistons, and the top ring lands of the new pistons were moved closer to the crowns to assist in cutting hydrocarbon emissions. The high compression and lean mixture inherent in the May design demanded a dead reliable 7000-rpm ignition system as well. The H.E. engine will use a twin-coil, 8-amp Lucas magnetic pickup system, with the secondary coil mounted ahead of the radiator for cooling, in conjunction with small, tapered-seat spark plugs. The Lucas/Bosch digital electronic fuel injection was recalibrated for the H.E. engine's needs, but was otherwise unchanged. It's a closed-loop feedback system that's tied into the V-12 engine's complete dual exhaust system containing *six* catalytic converter units on each side.

With an additional 18 hp and 21 lb.-ft. of torque (now 290 at 3000 rpm) over the '81 V-12 engine, plus two full points of compression (11.5 vs. 9.5:1), the Jaguar engineers had an inherently clean and powerful engine with significantly improved fuel economy. They took one extra step with the new XJ-S gearing, going from 3.07:1 to 2.88:1, for an additional increment in fuel economy and an even higher maximum

The XJ-S the critics have been demanding for years

speed (27 mph per thousand rpm, versus 24.8 mph per thousand with the 3.07 gearing). The European version is rated at 155 mph, though no rating was given by the U.S. subsidiary for the cars that will be sold here. But the U.S. subsidiary, Jaguar Rover Triumph, *did* tell us that the new S will be rated by the EPA at 15 mpg city, 21 highway.

Jaguar being Jaguar, it would have been inappropriate to issue a new XJ-S version without a fresh injection of Jaguar luxury accoutrements. These start with about twice as much Connolly leather as the '81 version had. Connolly hides are now used for seats, door panels, armrests, rear side panels, and sun visors, and the steering wheel rim is leather-covered as well. And, to complement the leather, real elm burl (on the dash panel, glovebox door, center switch panel, and door trim panels) has been reintroduced in the Jaguar sports car for the first time since the 1957 XK-140.

New standard equipment items added to the XJ-S for 1982 include twin power mirrors, AM/FM/4-speaker cassette system with seek/scan and station memory, intermittent wipers, door-edge warning/courtesy lights, interior courtesy light with 15-sec. delay, and pile-carpeted trunk compartment.

On the outside, the formerly all-black wraparound bumpers are trimmed with wraparound chrome caps front and rear; the rear end badging has been changed slightly; and the cat's-head logo now adorns the leading edge of the hood and the centers of the all-new alloy wheels. The five-spoke, five-slot wheels are wider by a half inch (15 x 6.5 in.), with deep-set lug holes and domed centers. U.S. cars will have these wheels wrapped by Pirelli P5 steel-belted radial tires in 215/70VR15 size, the P5 being selected for its combination of soft ride characteristics, low rolling resistance, and grip.

The term that Jaguar Cars, Ltd. chairman John Egan used over and over again in his description of Jaguar's reorganization in 1980 and its subsequent programs—including the new XJ-S—was "magnificent motorcars." "Our job, our mission, is simply to build magnificent motorcars, cars of the highest possible quality in every way." And the U.K. versions that we drove were exactly that. With 12.5:1 compression, a rating of 299 hp (DIN), fewer concessions to emissions, and other changes including Dunlop D7 tires, the right-hand-drive XJ-S cars we drove in the English Midlands were spectacularly smooth and incredibly fast in spite of their 3900-lb. weight.

The first thing that struck our fancy about the new XJ-S was its new interior, which, with all that burl elm and additional leather, looked more traditionally British than ever. The seats are a compromise between all-out luxury and sportiness, but can't be considered "driving seats" in the new argot. They move fore and aft and the backs recline, and that's it. No adjustable bolsters, no lumbar adjustment, no tilt feature. Yet somehow the seats are supremely comfortable without all the knobs, handles and cranks. The elm trim, added at the request of U.S. dealers and customers (make no mistake, the U.S. is far and away the largest Jaguar market) adds luxury and warmth to an already sumptuous interior, and was fitted and finished beautifully.

The biggest problem we had in enjoying the XJ-S to the fullest was the lack of knee room and foot room, especially foot room. The XJ-S is a 102-in. wheelbase coupe stuffed with a V-12 engine and a GM TurboHydramatic 400 automatic; this creates a gigantic tunnel and leaves very little space to store an unused left foot while driving. Aside from that tiny cavil, the

CONTINUED ON PAGE 152

SPECIFICATIONS

GENERAL
- Vehicle type: Front-engine, rear drive, 4-pass., 2-door coupe
- Base price: $32,100
- Options on test car: None
- Price as tested: $32,100

ENGINE
- Type: V-12, aluminum alloy block and heads, water cooled, 7 main bearings
- Bore & stroke: 3.54 x 2.75 in.
- Displacement: 326 cu. in. (5345 cc)
- Compression ratio: 11.5:1
- Fuel system: Electronic fuel injection
- Recommended fuel: Premium unleaded
- Emission control: Federal
- Valve gear: Twin overhead cams
- Horsepower (SAE net): 262 at 5000 rpm
- Torque (lb.-ft., SAE net): 290 at 3000 rpm
- Power-to-weight ratio: 14.8 lb/hp

DRIVETRAIN
- Transmission: 3-speed automatic
- Final drive ratio: 2.88:1

DIMENSIONS
- Wheelbase: 102.0 in.
- Track, F/R: 58.6/59.2 in.
- Length: 191.3 in.
- Width: 70.6 in.
- Height: 49.6 in.
- Ground clearance: 5.5 in.
- Curb weight: 3900 lb.
- Weight distribution, F/R: 56/44

CAPACITIES
- Fuel: 24 gals.

SUSPENSION
- Front: Independent, double wishbones, coil springs, hydraulic shocks, stabilizer bar
- Rear: Independent, upper & lower links, radius arms, coil springs, hydraulic shocks, stabilizer bar

STEERING
- Type: Rack and pinion, power assist
- Turns lock-to-lock: 3.0
- Turning circle, curb-to-curb: 38.4 ft.

BRAKES
- Front: 11.2-in. discs, ventilated, power assist
- Rear: 10.4-in. discs, power assist

WHEELS AND TIRES
- Wheel size: 15 x 6.5 in.
- Wheel type: Aluminum alloy
- Tire make and size: Pirelli P5 215/70VR15
- Tire type: Steel-belted radial
- Recommended pressure (psi), F/R: 24/28 psi

AutoTEST Jaguar XJS HE
Fireball efficiency

Jaguar XJS H.E.
Jaguar's May-patents high-turbulence, ultra-high-compression, lean-burn version of the aluminium-alloy-cased V12 introduced last July fitted to the high-performance, high-refinement flagship coupé of the XJ range. Still (regrettably) available only in automatic form; former under-gearing is largely overcome with 2.88 instead of 3.07 final drive and one-width-size wider Dunlop D7 tyres on new 6.5 in. light alloy rims. Styling and equipment are improved outside and in. XJS lauched September 1975 with injection form of 5.3-litre V12 first seen in July 1972 XJ 12.

PRODUCED BY:
Jaguar Cars Ltd., Browns Lane, Allesley, Coventry, Warwickshire CV5 9DR

TO THE EXPERIENCED yet still enthusiastic driver, there is a tremendous attraction in unfussy efficiency and power. Jaguar's XJS has always been particularly fascinating as the supreme example world wide of the embodiment of quietness, smoothness, efficiency and power, the last two in the fullest senses, and not just performance. Its efficiency was always magnificent in performance remembering its weight, and in ride combined with cornering and faithful handling void of any treachery — and good enough in fuel consumption bearing in mind its 1¾ tons and latterday restriction to automatic transmission.

The adoption by Jaguar's engine design team led by Harry Mundy of the May "Fireball" high turbulence, high compression lean burn principle to the now classic 90×70mm V12, is the main reason for an extraordinary leap forward in fuel efficiency, achieved not with any loss in power but a slight increase, making this XJS overall the fastest we have tested, in either automatic or manual form, in spite of a 12 per cent improvement in overall fuel consumption. Experience with this test car and our long term test XJ12 suggest that there are no serious disadvantages to this 12½-to-1 compression 97-octane engine, in spite of the doubts muttered by some rivals perhaps envious of Jaguar's world lead into production with this type of cylinder head, and it seems significant that others ranging from Porsche to Ford are now following up the principle (not necessarily exactly on Michael May lines) hard.

Performance
Smoothly dominating

No other car made has so near perfectly mannered a power unit. Turn the key first thing on a cold morning, and starting is almost instantaneous, yet as on all good modern fuel injection systems, not followed by the fussy high idling speed of a carburettor system in the same circumstances. Driveaway is correspondingly totally untroubled. Initial heater warm up, signalled by the starting of the first gentle warming blow of the automatic air conditioning system, is remarkably rapid for the size of engine, and invites, road and traffic permitting, the engineeringly conscientious driver's preliminary firmer squeezing of the as ever superbly progressive Jaguar accelerator; the unconscientious who have floored the throttle pedal from the start are rewarded better than they deserve with only the slightest shade less than the usual thrust in the back.

The surge of power always available may be equalled by one or two much rarer and more expensive automatic transmission cars, but none accompany it with such smoothness. The only cir-

— so that traction with stall-converter standing starts is remarkably good, with only three yards of wheelspin after take-off. The MIRA twin horizontal straights are a mile long, easily allowing the Jaguar to reach 130 mph in just over 3sec more than half a minute. The figures speak for themselves; 60 reached in 6.5sec, 80 in 10.5, the quarter-mile in 14.9 at just under 100 mph, 120 in 25.4 and so on.

For once, we were truly lucky with testing weather at MIRA, with between 3 and 5mph of wind. The maximum speed runs abroad were not so blessed, suffering from 10 to 20mph conditions, yet the car's straight stability made 150 mph one way and 157 mph the other wonderfully and reassuringly relaxed. With the higher final drive, the XJS

Left: Even when cornered hard, the XJS does not roll much by most roadgoing standards; note typical Jaguar XJ slight understeer

cumstance in which the Jaguar V12 disappoints in this respect is on tickover, whose puzzling lumpiness is felt rather than heard in the way it shakes the car. One cannot help wondering whether something like the special extra injection idling system claimed by Mercedes to aid idle efficiency and smoothness on their V8 might help.

The only other small criticism arises when one first uses the full acceleration after a lot of slow traffic running and cold starts, which can cause two momentary minor irritations. Misfiring due to plug fouling can take a mile or so of hard driving to clear. If this does not happen, but if there has been a great deal of short slow journey work before finally flooring the pedal, one can encounter a brief but at the first time alarming burst of knocking at around 4,500 rpm, accompanied by some pale blue exhaust smoke. Jaguar say that this is due to small amounts of combustion deposit built up during the town running causing detonation as they are burnt off, which produces the smoke. They also assure us that the condition is not in the least bad for engine life, as proved by endurance running tests in which the engine has been made to detonate continuously in this fashion. It must be stressed that the phenomenon is brief and transient; once it has happened, it cannot be repeated by immediately slowing and then accelerating again, which makes the maker's explanation entirely credible. We feel however that some mention should be made in the handbook, as a reassurance to owners.

Remarkably remembering its size, the XJS H.E. comes into its own as much when driven quickly along winding country roads as it does on motorways and twin track highways. Its other qualities of virtually perfect handling, excellent grip and braking (dealt with later) are of course an essential complement to give and take driving — but the way in which it so urgently yet smoothly restores one to the often astonishing speed which it achieved before the corner is exhilarating. The totally competent performance of the XJS and its all round refinement bring out the best in any keen driver, forcing him or her to drive better. The performance provides no excuse for ever becoming impatient with slower traffic, since as with other very fast cars, it is so much easier to dispose of the other car on the road without irritating him unnecessarily.

The General Motors transmission is as before, mostly pretty smooth, making virtually jerkless up and down changes when one is driving gently. It retains this smoothness on full throttle kickdown, perhaps helped by the small (around a second) delay in response, only becoming jerky if one uses the selector to engage second from say 80 mph when braking for a corner, which is accompanied by a squeak from the back tyres. The selector still lacks the ideal refinement of movement found in the Mercedes gate type one, but at least it now has correctly arranged stops, so that movement is not restricted between D and 2, but does not allow unintentional movement into neutral or, in the hands of the very clumsy, reverse. There is a safety restriction which prevents first being engaged at too high a speed if selected by a thoughtless driver. Similarly, the box changes up when accelerating with 1 selected at the 6,500 rpm red line; second is not limited in this way, the red line corresponding to 115 mph. Maximum kick-down speeds are 38 mph from 2 to 1, and 88 from top to 2.

The performance measurements at MIRA were as always with every XJS we have tested, impressive. There proved to be no point in over-riding the transmission's 64 (1st to 2nd) and 100 mph (2nd to top) automatic change-up points, which correspond respectively to an indicated 5,900 and 5,700 rpm. Jaguar's classically correct double-wishbone-geometry independent rear suspension works virtually as well as a de Dion (or live) axle would in keeping the driving wheels free of camber change on acceleration — in contrast to cars with semi-trailing arm rear ends H.E. is geared overall at 26.88 mph per 1,000 rpm, so that it turns out to be only very slightly under-geared.

Economy
The 20 mpg XJS?

No, not if you drive it as it begs to be driven when mood and road permit, fast — but yes, even the 21 mpg XJS if you pay only flippant regard to the British 70 mph overall speed limit (which this car's great margin of dynamic safety makes such an entirely unnecessary and irritating measure) and provided you match 80 mph restraint with not more than half-throttle acceleration. Such restraint is still more than enough to push the Jaguar ahead of the great majority of other cars from the lights. We several times drove it

Above: That famous shape, still somewhat controversial. Five-spoke aluminium alloy wheels and the side stripe are main H.E. distinguishing marks. Rear quarter is undeniable block to over-shoulder vision.

A real bonnetful, which does not lend itself to attention from anyone less than a really knowledgeable specialist. Bonnet itself is pleasingly counterbalanced, but surprisingly lacks bonnet lamp

95

Front seats, deliberately kept thin to save space inside, are exceptionally comfortable thanks to correct shaping of details like lumbar areas

Far right: Back seating is carefully shaped too, but for full size adults demands compromises in legroom from those in front. Inertia reel rear belts are extra. Note sensible pockets for oddments

160 mpg speedometer (on left) has twiddle-zero trip; 7,000 rpm revcounter has 6,500 rpm red line. Centre vertical instruments (from left) are temperature, oil pressure, fuel, battery volts. Stalks are left-hand-drive layout, with signalling (with switch of optional cruise control set) on left and wipe/wash on right. Horns are sounded by pressing any part of centre piece of steering wheel spoke. Hidden switches on each side of telescopically adjustable steering column are lighting on left, ignition on right. Centre panel switches on each side of clock (from left) are hazard, rear window heater, map lamp and interior lamps. On each side of superb Philips MCC stereo radio-cum-cassette player (standard on XJS) are automatic temperature control (left) and distribution (right)

thus, incredulous that 5.3 litres of supposedly friction-prone V12 capable of 56 bhp per litre (good for a refined road-going engine of this size) propelling not far short of two tons laden could be so comparatively frugal — and it was always so, the consumption varying between 20.1 and a best interval of 21.2 mpg.

Our 16.0 mpg overall test consumption includes of course a high proportion of hard to very hard driving, when the car can return as little as 12 to 14 mpg. It is unlikely that many owners will find the car that thirsty, unless they are particularly severe with the right foot; the average for our non-testing interval consumptions was between 16 and 19 mpg. This improves the range of the comparatively easily filling 20 gallon tank with its single central filler (a minor relief at petrol stations after the double tanks of the XJ saloons) to between 250 to 300 plus miles. Oil consumption worked out at a quite reasonable 1,300 miles per pint.

Noise
Still very quiet, but . . .

Mechanically, the XJS H.E. is the quietest super-car made, which adds an extra dimension to the pleasure of driving it. Ambling gently through town, one is hardly conscious of the engine at all. There is the usual small but noticeable transmission whine in first gear, made so by the quietness of the engine. The same is true of the power unit at a cruising 70 to 90 mph; it is only just noticeable. When accelerating hard, the voice of those twelve cylinders is heard, still less than on any competitive car, yet it is pleasantly obvious none the less. Experience with this XJS and our similarly-engined long term test XJ12 makes us suspect that the May-head V12 may have slightly more obvious combustion noise than before; the difference if there is one is undeniably so small that it would require a back-to-back test with the previous model to be sure.

As on the saloon contemporary, there is a price paid for the undeniably better steering response of the Dunlop D7 tyre in its Jaguar form. Compared with the previous Dunlop Super Sport (on the 24 January 1981 Road Test XJS), the H.E. is less impressive in its reduction of road noise, even if to be fair one has to point out that it is still distinctly better than the average in this respect. This applies both to road roar over coarse surfaces, and to bump-thump.

The test car suffered very slightly from perceptible gear whine from the rear axle gears, on drive but not over-run, at around 90 mph. There proved to be some aerodynamic noise at higher speeds from the right hand screen pillar, but not from the left one; the main source of such noise when going fast is however the air conditioning system, to an extent which we do not recall from previous XJS test cars — switch to the air-conditioning-off/intake-shut position and the noise disappears.

When running gently, the occasional whirr from the air conditioning system's servo motors is a mild distraction only as before. The higher-geared steering (compared to the saloon) seems more prone to hiss when manoeuvring, although still not at all bad in that respect. Overall, in spite of our criticism of the added tyre-transmitted noise, it must be stressed that the XJS H.E. is without doubt marvellously quiet still, and much more so than any car in its class — it shames most big saloons apart from its sister Jaguars too.

Road behaviour
Exemplary

If an *Autocar* Road Test was simply confined to criticism, there would be very, very little one could contribute under this heading. We wonder that Jaguar don't put the higher geared steering (2.9 turns for a quite tolerable-for-the-size-of-car 37ft 2in. mean turning circle diameter) on the saloons. It has the same excellent accuracy, but its better response is delightful, as it its better feel; combined with the greater reluctance to roll, it makes the XJS a remarkably handy car to throw about when the need arises. The car has superb straight stability at all speeds, and notices side wind less than most, which adds to its comfort on a long fast trip. Yet this stability is not the product of too much understeer. The car can only be provoked into tail-out attitude with power on wet surfaces, or at lower speeds through tighter corners; lifting off in a bend taken very fast, partly of course because such provocation is reduced by the automatic transmission, merely tightens the line slightly, reducing understeer. The car's road manners are utterly reliable.

Full throttle traction on a good surface has already been mentioned. At the other end of the scale, in snow and ice, the copy-book wheel location, unhelped by the limited slip differential which does not work when there is too little grip to generate enough diff-locking clutch pressure, is still responsible for surprisingly good traction in the worst conditions provided of course that one is gentle with the accelerator — something which the always beautifully sweetly progressive Jaguar throttle linkage makes particularly easy.

Ride is still pretty impressive, verging on the firm side but not too much so, although those better-responding and even better-gripping Dunlops are responsible for an extra sharpness over correspondingly sharp-edged bumps. As in the saloon, one has to treat "sleeping policemen" — those speed-restricting humps found on some urban roads — with greater respect than the absorbing abilities of the suspension dictates, because the somewhat restricted ground clearance allows the front to ground after the hump is taken as quickly as many good-riding cars permit. For the first time, we noticed a tendency for the car to produce an odd lateral rocking motion over indifferently surfaced main roads, which jogged one's head sideways in a way we do not recall on previous XJS models.

Brakes are perfectly weighted, and most refreshingly for nowadays, biased just right when two-up for truly balanced ultimate braking of over 1g at 16 lb pedal effort — neither too light nor too heavy. Fade resistance in our deliberately performance-related and therefore severe test was excellent. Seen in the dusk of a long test day, the red glow of both front and rear discs after the tenth ½g stop from 98 mph, was remarkable — the more so as the rise in pedal effort had not suggested such heat. The Jaguar is incidentally the only car we can recall in which our fade test heats the back brakes to red heat — an indication of how much less brake cooling there is when they are mounted inboard.

Behind the wheel
Very comfortable

The seats for both driver and front passenger don't look special, and they lack the stepped lumbar support adjustment of the saloons, but they work exceptionally well, so that none of our testers had any complaint about them. There is enough rearward adjustment to suit drivers of just over 6ft, and Jaguar's familiar provision of a telescopically adjustable steering column makes suiting a wide range of driver sizes easier than in many other cars, (although as before one wishes that the stalks came with it). Headroom is good, with about 1½in. to spare for our taller testers (6ft again).

Visibility is good straight ahead, and directly aft, but spoiled to the side by the clutter of A-post and the small window guide channel, particularly when looking at the right hand door mirror, and over the left shoulder by the intrusio of the no-doubt aerodynamically commendable sail-fairing behind the near-side rear quarter.

Controls are straightforward, simple and obvious, even to a newcomer to the car. The use of a left-hand-drive indicator stalk is less irritating on an automatic than on a manual box car, since the left hand is not so busy; sensibly, Jaguar stick to a steering wheel spoke horn control, which is ideal as it can be worked from any point across the spokes. The combined automatic air conditioning cum heating system works as well as ever, which is to say very well, particularly now that there is some sort of minimal adjustment of upper and lower differential temperatures, even if it is not as ideal as Rolls-Royce's more independent system.

Living with the XJS H.E.

As with most cars of this class, the XJS makes a very aimiable mistress. Little things please greatly; the way in which the stalk controls move, without any undue friction: the provision of delay switch-off for the courtesy lights: the combined puddle and door-open lights: the to-most-eyes very attractive elm burr wood now discreetly gracing dash and doors: the high quality leather upholstery (cloth trim is optional for those who do not care for leather): the generally excellent finish, which suggests that one can believe in Jaguar's well-publicised quality purge of last year: the sensibly sized and well proportioned lockable glove compartment on the passenger's side with its flip-up vanity mirror.

The Philips AC 994 MCC stereo radio with its ability to re-tune automatically on to the most powerful FM station as you travel is mostly very satisfactory, although we would criticise the way in which it sometimes cuts out momentarily due to interference — most irritating in the middle of broadcast conversation, as one can miss a vital word or two; otherwise its quality of performance and reproduction is very high. It pays of course to read the instructions on how to use it, as it isn't at first the simplest of sets to work.

Headlamps work very well, with very good range, spread and intensity. Remote control door mirrors are welcome, but one wishes that Jaguar would follow Ford and some others in fitting a mildly convex glass to the nearside; the flat glass used gives a hopelessly narrow field of view in a mirror whose only use is to show the fullest width rather than give good identification.

The boot is deep and short, but not too much so, allowing good space for most needs in this sort of car. It is neat the way in which the handbrake on the right-hand side of the driver drops down out of the way once released, whether or not it is engaged. Rear seating is typical of a coupé, requiring slight stooping for the tall (6ft) occupant, and some sacrifice of legroom for the same-size seat occupant.

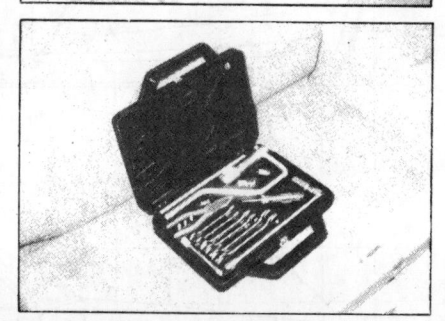

The XJS range

Since, in spite of our repeated plea for a manual overdrive gearbox, ideally the shelved Mundy design, with which the full virtue of the engine might again be exploited, there is nothing but an automatic XJS H.E., one cannot speak of an XJS range. At £19,708, it is the most expensive Jaguar (only some Daimler-trim XJ saloons are higher priced) — although still less costly than any real competitor; for the person who needs something not far off the same performance and economy plus saloon space and visibility, there is of course the XJ 12 H.E. at £18,938.

Remote control electric mirror works well, but left hand one would be much more useful with convex gass for better field of view

Optional extra cruise control is worked from this switch in end of signalling stalk

Boot is on short side but makes up for this with its depth. Spare wheel and battery live behind cover at front; superb toolkit is not standard at present

HOW THE JAGUAR XJS H.E. PERFORMS

Figures taken at 15,400 miles by our own staff at the Motor Industry Research Association proving ground at Nuneaton and abroad.

All Autocar test results are subject to world copyright and may not be reproduced in whole or part without the Editor's written permission.

TEST CONDITIONS:
- Wind: 3-5 mph (10-20 mph for maximum speed runs)
- Temperature: –4 deg C (25 deg F)
- Barometer: 30.0 in. Hg (1016 mbar)
- Humidity: 80 per cent
- Surface: dry asphalt and concrete
- Test distance: 2,678 miles

MAXIMUM SPEEDS

Gear	mph	kph	rpm
Top (mean)	153	246	5,700
(best)	157	253	5,850
2nd	118	190	6,500
1st	70	113	6,500

ACCELERATION

FROM REST

True mph	Time (sec)	Speedo mph
30	2.8	31
40	4.0	41
50	5.2	51
60	6.5	61
70	8.3	73
80	10.5	85
90	12.8	96
100	15.7	107
110	19.6	118
120	25.4	130
130	33.2	137

Standing ¼-mile: 14.9 sec, 98 mph
Standing km: 26.9 sec, 123 mph

IN EACH GEAR

mph	Top	2nd	1st
0-20	—	—	1.7
10-30	—	—	2.0
20-40	—	—	2.5
30-50	—	—	2.5
40-60	—	4.0	2.5
50-70	—	4.2	3.2
60-80	—	4.6	—
70-90	—	4.8	—
80-100	—	4.9	—
90-110	8.6	6.6	—
100-120	10.4	—	—
110-130	12.7	—	—

FUEL CONSUMPTION

Overall mpg: 16.0 (17.7 litres/100km)
3.52 mpl

Constant speed (manufacturers' figures)

mph	mpg	mpl	mph	mpg	mpl
30	27.0	5.94	70	21.0	4.62
40	26.4	5.81	80	19.2	4.22
50	25.3	5.57	90	17.3	3.81
60	22.9	5.04	100	14.4	3.17

Autocar formula: Hard 14.4 mpg, Driving Average 17.6 mpg and conditions Gentle 20.8 mpg
Grade of fuel: Premium, 4-star (97 RM)
Fuel tank: 20.0 Imp. galls (91 litres)
Mileage recorder reads: 3.8 per cent long

Official fuel consumption figures
(ECE laboratory test conditions; not necessarily related to Autocar figures)
- Urban cycle: 15.6 mpg
- Steady 56 mph: 27.1 mpg
- Steady 75 mph: 22.5 mpg

OIL CONSUMPTION
(SAE 20W/50) 1,300 miles/pint

BRAKING

Fade (from 98 mph in neutral)
(Pedal load for 0.5g stops in lb)

	start/end		start/end
1	30-24	6	40-44
2	32-40	7	40-45
3	36-42	8	44-48
4	36-42	9	44-46
5	40-44	10	50-44

Response (from 30 mph in neutral)

Load	g	Distance
20 lb	0.26	116 ft
30 lb	0.55	55 ft
40 lb	0.72	42 ft
50 lb	0.95	31 ft
60 lb	1.05	29 ft
Handbrake	0.32	94 ft

Max. gradient: 1 in 3

WEIGHT
- Kerb, 34.1 cwt/3,824 lb/1,735 kg
- (Distribution F/R, 56.2/43.8)
- Test, 37.8 cwt/4,234 lb/1,921 kg
- Max. payload, 720 lb/327 kg

DIMENSIONS

- OVERALL LENGTH 186.75"/4743
- OVERALL WIDTH 70.6"/1793
- Turning circles: Between kerbs L, 37ft. 0in., R 37ft. 4in.
- Boot capacity: 15 cu. ft.
- OVERALL HEIGHT 50"/1270
- GROUND CLEARANCE 5.5"/140
- WHEELBASE 102"/2591
- FRONT TRACK 58.6"/1488
- REAR TRACK 58"/1473
- SCALE 1:35
- OVERALL DIMENSIONS in/mm

PRICES

Basic	£15,819.00
Special Car Tax	£1,318.25
VAT	£2,570.59
Total (in GB)	**£19,707.84**
Seat Belts	Standard
Licence	£80.00
Delivery charge (London)	£138.00
Number plates	£20.00
Total on the Road (exc. insurance)	**£19,945.84**

EXTRAS (inc. VAT)
- *Cruise control £323.78
- *Lap and diagonal rear seat belts £49.44

*Fitted to test car

TOTAL AS TESTED ON THE ROAD	**£20,319.06**
Insurance	Group 9/OA

SERVICE & PARTS

Change	Interval 7,500	15,000	30,000
Engine oil	Yes	Yes	Yes
Oil filter	Yes	Yes	Yes
Gearbox oil	No	No	Yes
Spark plugs	No	Yes	Yes
Air cleaner	No	Yes	Yes
Total cost	£88.10	£140.76	£168.21

(Assuming labour at £15.50/hour inc. VAT)

PARTS COST (including VAT)

Brake pads (2 wheels) – front	£17.14
Brake pads (2 wheels) – rear	£17.14
Exhaust, complete	£440.24
Tyre – each (typical)	£142.99
Windscreen (tinted, laminated)	£109.25
Headlamp unit	£59.80
Front wing	£166.75
Rear bumper	£285.14

WARRANTY
12 months' unlimited mileage with optional Supercover extension for second and third years

SPECIFICATION

ENGINE
- Head/block: Front, rear drive
- Head/block: Al. alloy
- Cylinders: 12, in 60 deg. vee/wet liners
- Main bearings: 7
- Cooling: Water
- Fan: Viscous and electric
- Bore, mm (in.): 90.0 (3.54)
- Stroke, mm (in.): 70.0 (2.76)
- Capacity, cc (in.³): 5,345 (326.0)
- Valve gear: Ohc
- Camshaft drive: Chain
- Compression ratio: 12.5-to-1
- Ignition: Breakerless
- Injection: Lucas digital electronic
- Max power: 299 bhp (DIN) at 5,500 rpm
- Max torque: 318 lb ft at 3,000 rpm

TRANSMISSION
- Type: GM 400 3-speed

Gear	Ratio	mph/1000rpm
Top	1.00-2.40	26.88
2nd	1.48-3.55	18.16
1st	2.48-5.95	10.84

- Final drive gear: Hypoid bevel with Salisbury Power-lok limited slip diff.
- Ratio: 2.88-to-1

SUSPENSION
- Front – location: Ind. double anti-dive wishbones
- – springs: Coil
- – dampers: Telescopic
- – anti-roll bar: Standard
- Rear – location: Ind. drive shaft, radius arms, bottom links
- – springs: Twin coil
- – dampers: Twin telescopic
- – anti-roll bar: Standard

STEERING
- Type: Adwest rack and pinion
- Power assistance: Standard
- Wheel diameter: 15.7 in.
- Turns lock to lock: 2.9

BRAKES
- Circuits: Twin, split front/rear
- Front: 11.18 in. dia. disc
- Rear: 10.38 in. dia drum
- Servo: Vacuum
- Handbrake: RH side lever, rear discs

WHEELS
- Type: Al. alloy
- Rim width: 6½ in.
- Tyres – make: Dunlop SP Sport Super D7
- – type: Steel radial tubeless
- – size: 215/70VR15 in.
- – pressures: F 32, R 30 psi (normal driving)

EQUIPMENT
- Battery: 12V 68Ah
- Alternator: 67A
- Headlamps: 110/220W
- Reversing lamp: Standard
- Electric fuses: 12
- Screen wipers: 2-speed plus intermittent
- Screen washer: Electric
- Interior heater: Air blending auto. temp. control
- Air conditioning: Standard
- Interior trim: Leather seats, pvc headlining
- Floor covering: Carpet
- Jack: Screw cantilever
- Jacking points: Two each side, under sills
- Windscreen: Laminated, Triplex X/XX
- Underbody protection: Bitumastic

HOW THE JAGUAR XJS H.E. COMPARES

Jaguar XJS H.E. (A) — £19,708
Front engine, rear drive
Capacity 5,343 c.c.
Power 299 bhp (DIN) at 5,500 rpm
Weight 3,824 lb/1,735 kg
Autotest 24 April 1982

Aston Martin V8 (A) — £39,999
Front engine, rear drive
Capacity 5,340 c.c.
Power 304 bhp (DIN) at 5,000 rpm
Weight 3,970 lb/1,800 kg
Autotest 14 October 1978

BMW 635 CSi (A) — £19,329
Front engine, rear drive
Capacity 3,453 c.c.
Power 218 bhp (DIN) at 5,200 rpm
Weight 3,447 lb/1,564 kg
Autotest 6 January 1979 (Four-speed manual)

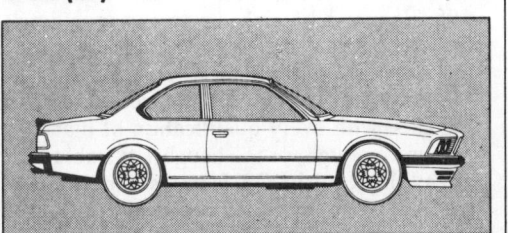

Ferrari 400i (A) — £35,300
Front engine, rear drive
Capacity 4,823 c.c.
Power 310 bhp (DIN) at 6,400 rpm
Weight 4,034 lb/1,830 kg
Autotest 4 October 1975 (365GT4 2+2 manual)

Maserati Kyalami (A) — £29,900
Front engine, rear drive
Capacity 4,930 c.c.
Power 280 bhp (DIN) at 5,600 rpm
Weight 3,836 lb/1,740 kg
Autotest 8 July 1978 (4,136 c.c. manual)

Porsche 928S (A) — £25,251
Front engine, rear drive
Capacity 4,664 c.c.
Power 300 bhp (DIN) at 5,900 rpm
Weight 3,390 lb/1,537 kg
Autotest 5 April 1980 (Five-speed manual)

MPH & MPG

Maximum speed (mph)
Jaguar XJS H.E. (A)	153
Porsche 928S*	152
Ferrari 400i (365)*	150
Maserati Kyalami (4.1)*	147
Aston Martin V8 (A)	146
BMW 635 CSi*	140

Acceleration 0-60 (sec)
Jaguar XJS H.E. (A)	6.5
Porsche 928S*	6.7
Ferrari 400i (365)*	7.1
Aston Martin V8 (A)	7.2
Maserati Kyalami (4.1)*	7.6
BMW 635 CSi*	8.5

Overall mpg
BMW 635 CSi*	17.5
Porsche 928S*	17.5
Jaguar XJS H.E.	16.0
Maserati Kyalami (4.1)*	15.3
Aston Martin V8 (A)	13.0
Ferrari 400i (365)*	11.0

Figures for manual version tested

Anyone looking hard enough at the XJS H.E. to be considering buying a car like it must presumably confine his or her view to Grand Touring fixed head cars with automatic transmission and 2+2 accommodation. That restricts the choice effectively to only a handful. There isn't an automatic Quattro (135 mph, 7.3 sec and 19.1 mpg manual), and we have yet to test the Mercedes 500 SEC (£28,700). Judging by the 500 SE saloon automatic which we have tried (145 mph, 7.5 sec, 15.2 mpg), the slightly better shaped but not much lighter coupé would not beat the Jaguar, even if it approached it. For the rest, they are listed as rivals here; we have done an automatic Aston Martin, but only manual gearbox versions of the Maserati and Porsche, and of the older 365 GT4 2+2 Ferrari. Our Maserati Kyalami test of July 1978 was of a manual gearbox car with the smaller of the two engine sizes now available (4.1 and 4.7 litres). Knowing how much speed and acceleration is stolen from any car by the inefficiencies and losses of torque converters and automatic epicyclic gearboxes, it seems perfectly safe to say as we did in the sub-headline of our previous XJS Road Test, that the XJS is still the world's fastest production automatic, adding to this that in H.E. form it is now by no means the thirstiest.

ON THE ROAD

It is still true to say as we did last time that apart from quietness, there is little to choose between the Jaguar, Aston Martin and Maserati.

The Porsche in this its most powerful form can turn into something of a handful on its limit, a product we suspect of its deliberately high polar moment of inertia (the rear gearbox giving it more "dumbell" effect once the back end starts to slide) and the rear suspension itself, which can be provoked into breaking away a little more easily than is well-bred if one decelerates abruptly in a bend taken fast. It must be stressed however that it is a much easier car to drive tail-out than the 911 series.

The BMW disqualifies itself from this company with its still too-easily provoked tail – the result of semi-trailing arm camber change.

In spite of its size and weight, the Jaguar is superb in its grip and ultimate behaviour, totally without treachery, with good feel in its nicely geared steering, excellent response, and all this combined with a pretty acceptable if not saloon-gentle ride.

SIZE & SPACE

Legroom front/rear (in.)
(seats fully back)
Maserati Kyalami	43/35
BMW 635 CSi	42½/32
Aston Martin V8	43/31
Jaguar XJS H.E.	42/30
Ferrari 400i	41/29
Porsche 928	41/27

It is unusual to find an Italian supercar at the head of an interior legroom list, the reason here being that the Kyalami is nearer the full size saloon than the others; space in the back demands less than the others do in compromise of the comfort of a tall front occupant.

Although ideally the Jaguar should emulate Porsche in coupling its stalk controls with the steering wheel when one moves the wheel backwards or forwards, its telescopic steering column is a far better way of accommodating the needs of a wider range of drivers than other systems.

The Jaguar also has the benefit of arguably the best combination of air conditioning and heater with automatic control.

VERDICT

For unalloyed, somewhat exhibitionist fun, the Aston Martin is still a marvellous motor car, with its great power and performance. It is also the equal of anything front-engined in its road behaviour, which is impeccable and entirely trustworthy. For a different sort of excitement, the BMW is attractive, provided that if one fancies oneself as someone who is going to drive the car fast, one also is prepared for occasionally unexpected applications of opposite lock. Driven far enough from its limit, the 635 is a very pleasant car, if not in the same performance class as the others. The man who wants something different will not be dissuaded from the Ferrari or Maserati, which are both excellent in their own ways, if not true all-rounders. The 928S is also not perfect, having slightly unreliable ultimate manners, but it is a remarkable and mostly very civilised machine nevertheless. For all round satisfaction – a tiger of a car when need be, the most refined, relatively speaking, of Grand Touring 2+2s when driven less hard, let alone gently, and always a gentleman in character – the Jaguar is even more so the paragon, especially now that its fuel economy is so much better than before. It is also more motor car for the money than the others, in every way.

JAGUAR XJ-S

Can HE mean High Excitement?

ROAD & TRACK ROAD TEST

IT'S UNTHINKABLE NOT to take your full pleasure from a car like this. And of course we're in the car-celebration business, so we may as well admit right off that we enjoyed the 1982 Jaguar XJ-S right up to and including the legal limit. We may even have inadvertently edged over the line a time or two, just out of sheer ebullience.

This, after all, is one of the true state-of-the-art Grand Touring cars, with every qualification of that lofty breed, including pedigree enough for the haughtiest stable. The XJ-S offers every one of the requisite GT features: solid comfort, quiet speed, firm control, power when needed, and individual, recognizable styling—adding up to a taut but delicate elegance. And of course significant GTs must sport significant amenities, so it also has all those special conveniences to lighten long touring miles.

The most important feature, however, in the 1983 version of the Jaguar XJ-S lies under that traditionally long, hydraulically assisted hood. The familiar 5343-cc V-12 incorporates an unusual combustion chamber both to generate higher energy efficiency and to lower its torque peak into a more usable rpm range. The energy goal was a 10 percent restraint on an engine well known for its thirst.

The revamped V-12 incorporates a patented "Fireball" combustion chamber developed by Michael May, a Swiss engineer. The purpose of the design is to provide the advantages of a stratified-charge system without its complexity and expense. The

AT A GLANCE	Jaguar XJ-S	BMW 633CSi	Mercedes-Benz 380SEC
Curb weight, lb	3890	3350	3665
Engine	V-12	inline-6	V-8
Transmission	3-sp A	5-sp M	4-sp A
0–60 mph, sec	8.2	8.1	10.4
Standing ¼ mi, sec	16.3	16.4	17.9
Speed at end of ¼ mi, mph	89.0	85.5	77.0
Stopping distance from 60 mph, ft	157	na	149
Interior noise at 50 mph, dBA	65	65	65
Lateral acceleration, g	0.726	na	0.732
Slalom speed, mph	56.6	na	56.1
Fuel economy, mpg	13.5	est 19.0	19.0

combustion chamber is shaped a little like a bathtub, and on the compression stroke it produces a concentrated air/fuel mixture with low turbulence in the area of the sparkplug and a lean, swirling mixture in the rest of the chamber. The end result is to allow a very high compression ratio of 11.5:1 to be used, and also to bring about a decrease in emissions.

Of course this lean and highly compressed mixture needs special ignition, and Jaguar engineers have developed a high performance system using an electronic distributor, an amplifier unit and twin coils; a combination that is able to provide the necessary 12 sparks per revolution and control them up to 7000 rpm.

The engine still has the same power output; 262 bhp at 5000 rpm and a torque peak at 290 lb-ft, but the latter now comes at 3000 rpm instead of the previous 4000. In addition to the convenience of earlier torque availability, the new torque peak made it possible for Jaguar to lower the rear axle ratio numerically from 3.07:1 to 2.88.1 in the interest of improving fuel economy without losing performance. And successfully, too; at least as far as the EPA folks are concerned. Last year's 13/19 city/highway rating rose to a somewhat more presentable 14/22, and we suspect that most drivers' averages would round out to about 14 or 15 mpg in normal driving.

Ours, however, did not. We averaged 13.5 mpg, and of course we will blithely blame it on the nature of the car itself. It is simply such a pleasure to deep-foot this machine, especially during one's early relationship with it, that the XJ-S easily routs your resolutions to hoard fuel. Further, a staff consensus was that the car's natural operating range lies up there in the 80- to 90-mph decile—in recognition of its revised engine, axle ratio and other characteristics, and that it would really fall into its stride in that speed range. We did feel a certain transitory regret over our fuel profligacy but, as one staffer noted, one sweet night of wretched excess is worth years of dull moderation.

The Lucas fuel injection system is digitally controlled, and has a separate automatic injector for cold starts. We had no opportunity to test the latter *in extremis*, but never experienced even momentary hesitation when lighting off the engine, hot or cold. It always started instantly, and never failed to elicit comments regarding its alacrity and silence from any new audience. It may be well to note here, though, that the sound-deadening of this car is especially good, and that what goes on outside the passenger compartment is much diminished by the time it reaches the inside.

We did detect a slightly uneven idle now and then, as though a cylinder wasn't firing cleanly and regularly, a fault some of us were willing to lay at the door of the U.S. emission-control fathers. On the other hand, the Editor Emeritus has driven the same cars in England, set up to their specifications, and says they also had this same characteristic.

Another odd symptom, particularly noticeable in a car of such fully controllable power and action, was brief indecisiveness or "hunting" for the right rpm when the car decelerated. The engine revolutions would drop to about 800 rpm and then return quickly to 1000 as the injectors came in, a transition that always caught our attention.

As always, we must raise the issue of dependability, and it is good to report that Jaguar has taken some direct action in this direction by establishing a 24-month/36,000-mile warranty to cover all models.

Among automobiles of this stature, we are always bemused by the manufacturer's statement, "Optional Equipment Available: None." This is nothing less than a bold announcement that everything that could (or should) be of interest to this particularly well-off and sophisticated class of buyers already has been anticipated and provided. Nothing more is needed. And in the case of the Jaguar XJ-S, that is just about the case: The car is fitted out very well indeed.

The list of features is impressively long, but we will skip all but a few of the more impressive specialties. In the first place,

PHOTOS BY JEFFREY R. ZWART

101

the interior is all leather and wood: The latter is contoured burl elm under an almost too-perfect finish. The sound system is one of those do-everything rigs with tape player, scanning, full-circumference faders, and memory buttons that can handle two stations apiece. The rear speakers are set back into caves in the leather. The doors have interconnected electric locks, and operating either lock operates the other, inside or outside—a surprisingly useful convenience. And there is a nice safety touch: Any heavy impact to the car actuates a fuel shut-off switch by the driver's left knee. Fuel flow can be restored by pressing a button on top of the switch.

All this is not to say, however, that we feel the interior of the Jaguar has attained perfection. Several drivers noted that a little more padding at the lower front edge of the seat would keep them from sliding forward during hard deceleration. And most of us didn't find the vertical, sliding-scale instruments particularly preferable to the more conventional round type.

On the road the car is tremendously impressive with lots of smooth power on tap and seemingly always more to come. It has the long, even power-flow of a turbine, punctuated by pleasant little surges of acceleration with each range change. The transmission is GM's 400-80 3-speed automatic, and it fits very acceptably and positively into the British power system. We did find it a little reluctant to downshift, but the engine's great torque turned this into a minor problem.

The 1982 XJ-S is actually somewhat slower in straight-line acceleration than its predecessor, recording a quarter-mile figure of 16.3 seconds (up from 15.9), and a 0–100 time of 21.3 sec, compared with 19.5. By today's standards of performance, however, either car is so fast that it makes the comparison somewhat academic. This Jaguar will never embarrass you when it comes time to call a halt to all this enthusiasm. The all-disc braking system will pull this nearly 2-ton machine down from 60 mph in 157 ft and from 80 mph in 276 ft. We encountered minor rear-lock sensitivity but the pedal was easy to modulate. And the pedal pressure never wavered from 24 lb during our fade test.

To sum it up, then, the Jaguar XJ-S HE (the latter standing at the rear in 2-in. chrome letters to proclaim the High Efficiency of Mr May's little bathtubs) is a car that can lay legitimate claim to full credentials as one of the world's very finest 2+2 Grand Touring automobiles. It has the requisite supple but controlled ride, the superb handling and braking, and the ability to instill that extra feeling of confidence and safety. It also has that wonderfully smooth flow of power, and a slightly heavy feel to the whole machine that gradually lightens up as the speed increases. It is, without question, an excellent example of the breed.

PRICE

List price, all POE $32,100
Price as tested includes standard equipment (air conditioning, leather interior, AM/FM stereo/cassette, elect. window lifts, elect. mirrors, central locking system, cruise control)

GENERAL

Curb weight, lb/kg	3890	1766
Test weight	4090	1857
Weight dist (with driver), f/r, %		55/45
Wheelbase, in./mm	102.0	2591
Track, front/rear	58.6/59.2	1488/1504
Length	191.3	4859
Width	70.6	1793
Height	49.6	1260
Trunk space, cu ft/liters	13.6	385
Fuel capacity, U.S. gal./liters	24.0	91

ENGINE

Type		sohc V-12
Bore x stroke, in./mm	3.54 x 2.76	90.0 x 70.0
Displacement, cu in./cc	326	5343
Compression ratio		11.5:1
Bhp @ rpm, SAE net/kW		262/195 @ 5000
Torque @ rpm, lb-ft/Nm		290/393 @ 3000
Fuel injection		Lucas electronic
Fuel requirement		unleaded, 91-oct

DRIVETRAIN

Transmission .... automatic; torque converter with 3-sp planetary gearbox

Gear ratios: 3rd (1.00)	2.88:1
2nd (1.48)	4.26:1
1st (2.48)	7.14:1
1st (2.48 x 2.40)	17.14:1
Final drive ratio	2.88:1

CHASSIS & BODY

Layout	front engine/rear drive
Body/frame	unit steel
Brake system	11.8-in. (300-mm) vented discs front, 10.9-in. (277-mm) discs rear; vacuum assisted
Wheels	cast alloy, 15 x 6½JK
Tires	Pirelli P5, 215/70VR-15
Steering type	rack & pinion, power assisted
Turns, lock-to-lock	3.3

Suspension, front/rear: unequal-length A-arms, coil springs, tube shocks, anti-roll bar/lower A-arms, fixed-length halfshafts, trailing arms, dual coil springs, dual tube shocks, anti-roll bar

CALCULATED DATA

Lb/bhp (test weight)	15.6
Mph/1000 rpm (3rd gear)	26.3
Engine revs/mi (60 mph)	2280
R&T steering index	1.20
Brake swept area, sq in./ton	219

ROAD TEST RESULTS

ACCELERATION

Time to distance, sec:
0-100 ft	3.4
0-500 ft	9.0
0-1320 ft (¼ mi)	16.3
Speed at end of ¼ mi, mph	89.0

Time to speed, sec:
0-30 mph	3.4
0-50 mph	6.3
0-60 mph	8.2
0-70 mph	10.5
0-80 mph	13.5
0-100 mph	21.3

SPEEDS IN GEARS

3rd gear (5300 rpm)	140
2nd (6500)	112
1st (6500)	66

FUEL ECONOMY

Normal driving, mpg 13.5

BRAKES

Minimum stopping distances, ft:
From 60 mph	157
From 80 mph	276
Control in panic stop	very good
Pedal effort for 0.5g stop, lb	24

Fade: percent increase in pedal effort to maintain 0.5g deceleration in 6 stops from 60 mph .... nil
Overall brake rating .... very good

HANDLING

Lateral accel, 100-ft radius, g 0.726
Speed thru 700-ft slalom, mph 56.6

INTERIOR NOISE

Constant 30 mph, dBA	60
50 mph	65
70 mph	72

SPEEDOMETER ERROR

30 mph indicated is actually	29.0
60 mph	56.0

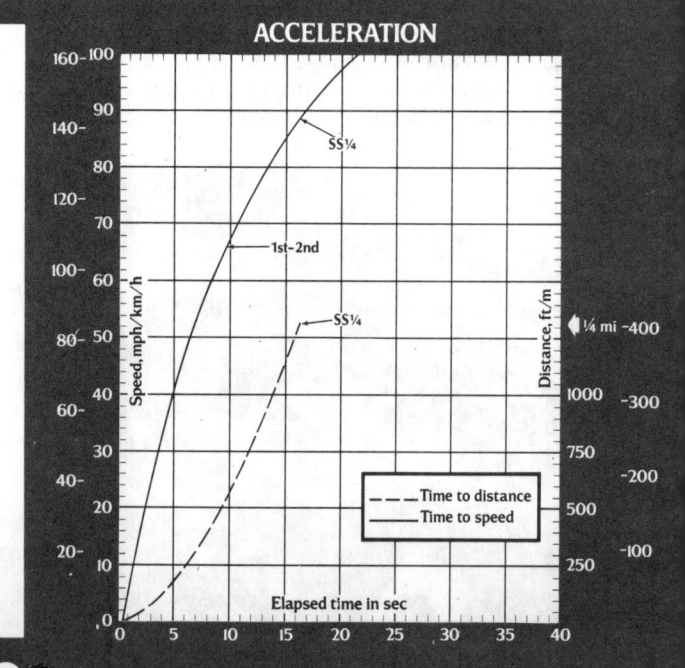

TUNING TOPICS

Race modifying Jaguar's XJ-S for European endurance events

WHEN I heard that Jaguar was to go racing in European Touring Car Championship events again my first reaction was, "Oh no! We've seen all this before. Why should it work now, when it was such a disaster in 1977?" Then I started getting phone calls from Germany. At eight o'clock in the morning it's difficult to argue with an irate trans-European caller who wants to know, "if Jaguar can race that XJ-S as a touring car, why cannot Porsche bring the 911S, or BMW the M1?" There were questions about the numbers produced and the size of the rear seats, all good touring car stuff in which the "other guy" is always trying to run a rampant cheater, while the injured entrant speaking to you has never run even a gram underweight or with an engine so much as a c.c. oversize. Up to this point, where the calls from Europe grew more persistent and the stories in the weeklies ever more specific, I had hoped nobody would be interested in the Jaguar until it had proved itself. Then, just a few months after the project was first seriously voiced by Tom Walkinshaw Racing (TWR) to sponsors Motul, the French oil concern, the car appeared at Monza for the March opening round of the 1982 European Touring Car Championship. It was immediately obvious that Tom Walkinshaw, who was sharing the driving with "Chuck" Nicholson, had tackled the problems of racing a Jaguar with a professional acumen that was totally absent in the original over-complex Broadspeed XJ Coupé project. Instead of a mass of engines strewn over the paddock after a year long gestation period, this young TWR / Motul project — which has blessing and co-operation from Jaguar Cars, but is reportedly *not* a direct factory-financed project — was surviving hours of Monza practice with honour. It had troubles of course, but it was fastest for much of qualifying time and was not tempted into a pole position battle when a BMW 528i tipped it to quickest time by $^2/_{10}$ths for a four-hour race. Listening to the accounts of fellow journalists and rivals — one of whom described the car as; "simply gorgeous-looking, but no way a touring car!" — I felt compelled to go down to Kidlington and see the black beast for myself. Especially as it had comfortably led the opposition (9 sec. lead after 18 laps) until a chicane scuffle, not an initial mechanical failure, subsequently led to retirement. A gearbox oil cooler line was so badly damaged that retirement followed after 45 laps, roughly one hour of racing time.

Kidlington, just a few miles outside the Oxford ring road, has been home for TWR and the increasing web of other activities that have seen Walkinshaw move from full time racer to preparation specialist, retailer and racer. To gauge the worth of the cliché "Canny Scott," you need look no further than Tom. Amiable until contradicted or crossed, Tom is a muscular Scottish gentleman who realised that formula car racing might be for those of less intelligence when the variety of machinery he drove in Formulae Atlantic and 5000 broke beneath him. One memorable outing literally leaving him with just the steering wheel to brandish at Brands Hatch, questioners asking "what happened?"

Ford, BMW, Mazda were the manufacturers who employed his talents as a driver. Ford got a rapid RS2000 and Capri pilot but were not keen enough to stay Tom's preference for European long distance events. BMW were the obvious home and he drove all manner of cars for Munich / Alpina, ironically defeating Jaguar on their last 1977 UK appearance at Silverstone, when the green Alpina BMW CSL held off Andy Rouse in the Jaguar until a crash decided the matter minutes from the end. So Tom certainly saw how Alpina BMW did things in endurance racing: from air jacks to careful attention that ensured, "not a nut and bolt goes on the car that isna working, and *working hard*, for a living".

For Mazda, Walkinshaw became not only a driver capable of winning one of touring car racing's most valued prizes — the 24 hours at Spa-Francorchamps (1981) with journalist Pierre Dieudonné — but also of bringing back two British touring car Championships *via* the RX-7 and the talents of Win Percy.

Walkinshaw is still involved with Mazda as the wealth of fascinating artistic impressions around the office testify, but now it is "more on the rallying side and development for that sport". There is a Walkinshaw-owned Mazda dealership at nearby Dorchester and the man also has financial ownership of DART, the Silverstone-based sole distributors of Dunlop racing tyres, also an Alpina-BMW franchise for the UK and the responsibility of running the factory Rover 3500 team in Britain. It was this last link that led to the Jaguar project and continental confusion, because everyone automatically assumed that TWR would run the Rover in Europe this season and opposition was readied on that basis.

Tom takes up the story of how the racing Jaguar project began. "Racing saloon cars are an important part of our business, so when a new set of regulations were issued we automatically evaluated them, just as we look at everything in this field." Those new regulations were those for the international Group A, basically catering for touring cars produced in numbers of over 5,000 per annum. Tom continued, "we knew there was a good chance of the Belgians adopting Group A, besides the recommendation that it form the basis of the European Touring Car series instead of Group 2."

Of course the regulations kept changing, not so dramatically as those for Group C endurance racing, or the sudden lurches in policy that seem to afflict Grand Prix, but enough to keep TWR revising their estimations of potential winners. Eventually they decided to simplify their search for a competitive Group A project, "early in September last year we drew up a spec of what *we* wanted in an ideal Group A car," recalled Walkinshaw with a deprecatory grin.

"The priorities were that the suspension be of the wishbone type, to accommodate every inch of rubber allowed by the regulations [a MacPherson strut hogs that vital space of course — J.W.]. Secondly we wanted a car with fairly large diameter wheels, because you are allowed to go up on diameter 2″, or down for that matter. If we went up 2″ on a car with 14″ wheels that would obviously allow a lot of extra space for brakes, which would be free in Group A any way. Looking at the engine regulations we could see that they allowed a fair bit of internal modification but, and this is a bit puzzling to me, the induction and exhaust were pretty restricted. So we needed a car with fuel injection because at least with injection we could force the fuel in and not have to rely on the pulsing action of an efficient exhaust manifold to take it all out again quickly, which is very much the case with a carburated engine.

"We then looked at what the opposition was likely to be for our 'ideal' car; in fact what cars could be used whether by ourselves or the opposition. By this stage our investigation had purpose because BL Motorsport had asked to see what the prospects were of running the Rover in Europe during 1982.

"We thought the possibilities included Mercedes, who have a 5-litre V8, the well-proved BMW, Turbo Ford Mustang and Chevrolet Camaro." Obviously TWR were only interested in winning outright, so the tendency was to look at the bigger capacity cars. This tendency to favour cubic capacity, plus a factory preference now apparently leaning toward rallying, ruled Mazda out of the reckoning.

Examining the Rover in detail and with experience that has seen the BL hatchback five-door end Ford's reign with the Capri in British events, Walkinshaw reluctantly concluded that it wasn't the right car for the job. It had MacPherson struts: "we *could* cram in 10" overall width without modifying the body, but it really would be a tight fit," felt Walkinshaw, the present Rover V8s are only carburated and, though Tom did not mention it to us, it may have been that the lack of disc-braked back axle was a factor too. Certainly Tom was not keen on the Rover's cast iron exhaust manifolds for, "these are very important on a V8 and I knew that it would be very hard to get the fuel consumption we wanted, which is roughly the same as BMW. This is vital, so as not to incur the extra pit stops, which need a 2-3s lap speed advantage to offset the extra fuel stop." There is a definite equation here between speed and consumption, and the listener ends up thinking that a lightly modified big engine giving over 5½ m.p.g. and roughly 400 b.h.p. is the right way to go.

As soon as the equation between speed over four hours (and one 24 hour race, for Spa has always been important to TWR) and m.p.g. materialised, the Rover with 3½-litres working hard began to look as though, "it would sort out the early season European races. Then, when the big cars began to get their act together we would be in trouble. So why not have a big car in the first place?" Tom asked himself. When the Jaguar name was mentioned, the tendency was to instantly dismiss it, remembering what had gone on is Europe in 1977.

That the XJ-S was homologated at all was owing to the Australians, who wanted to run it in their Bathurst endurance race. BL had homologated the XJ-S in Group 1 and there it had lain, neglected by Europeans but a force to be reckoned with in modified American form (courtesy Bob Tulius) for Trans Am. When the FIA decided to implement Group A in 1982 they automatically transferred a lot of cars from Group 1 into the new Group A. The XJ-S was amongst them.

Now Tom sat down with long standing friends at Jaguar. Informal meetings and social gatherings formed the backcloth against which the Jaguar return to saloon car racing went from passing jest to a deadly serious project. TWR had one of the 1977 Broadspeed coupés out of the Syon House collection. It was painfully obvious that the big coupé, originally a brilliant conception, had become "a committee car, a four-wheeled equivalent of a racing Camel," in one observer's words. The old coupé was really heavy (nearly 1,700 kg.) despite an amazing specification that is even now not freely commented upon. The weight limit for the biggest capacity class in 1982 Group A would be 1,400 kg. Two vital principles were now established: this racing XJ-S would weigh in at the limit and it would *not* be a complicated car. All its systems were surveyed and the lessons of the past painfully absorbed: no more fans cooling extra radiators, or pumps lubricating pumps. This Jaguar would outwardly be a steel shell XJ-S benefitting from modifications rather than re-engineering. As Tom says, "if Jaguar do things the way they do, and it's been a very successful way for the XJ-S is acknowledged to be one of the best cars in the World, then who am I to tell them they're wrong, designing my own Jag racer?"

Only in one important area would the basic principles of the road car be abandoned: the inboard rear brakes would have to go. On the TWR racer you will find 13" vented units by AP on all four corners, using four piston calipers. The ensemble is the result of the British company's response to the increasing speeds offered by turbocharged Grand Prix cars.

By late October 1981 Walkinshaw had convinced himself that a racing Jaguar could be a serious challenger under the new Group A rules. Now he had to convince others. "I knew from the start that Jaguar were out," Tom commented quietly, "they were just filled up with new model work and it had always been plain to me that there would be no finance in the direct sense."

Part of that plain speaking included a session for Walkinshaw with Jaguar Cars Chairman John Egan. To his eternal credit, Egan was not frightened by the company's previous experience and the promise of technical cooperation in the form of the fashionable "Think Tank" process was agreed. In plain English that means that if TWR have a problem they can go and ask Jaguar how they recommend solving it, rather than trying to fabricate a pure racing solution without consultation.

To illustrate such a process I would cite the case of cooling ducts, fans, and auxiliary radiators. TWR were able to talk directly with the aerodynamicists who developed the car originally and come away knowing where the high and low pressure areas were once the car was in motion. Utilising such knowledge virtually all cooling requirements could be met by some of the cleverest and best integrated touring car ducting I have ever seen, rather than using fans to create an airstream. Ford boast about air management on their Escort, but when you look at how TWR make moving air all but stand on its head to reach the neatly tucked away oil coolers for engine and transmission, the term takes on a whole new meaning. I did ask what the drag factor (Cd) was, but I didn't seriously expect an answer apart from "very good", which one would expect from a professional team. Preferable to a glip lie anyway . . .

There are a number of subsidiary trade

THE interior of the Walkinshaw Racing XJ-S lacks any trim other than the Recaro seat for the driver; apart from a message attached to the steering wheel urging drivers to "be kind to your pussy", it is purely functional.

sponsors upon the Jaguar's black flanks, but the name of Motul is the result of an October 1981 approach by TWR. No figures were relayed to us of course — "a lot", and a wistful grin from Walkinshaw was the reply — but the French suppliers of the Synthetic 300 V race oil agreed to let Tom have four months before deciding whether the Jaguar project could be fielded in Europe. It would only be raced, as agreed with Jaguar and Motul, *if* it proved competitive in testing. Perhaps the surprise of the project, and another lesson learned from the past, has been Tom's basic approach of racing rather than testing to progress the car. He reasons, "while you can control the pace at which you are running there will be things that don't break, but that will break as soon as you have to run at the pace of others on a race track." Also integral to the decision to go ahead were Dunlop, for Tom needed the technical support from Birmingham.

Turning the Jaguar into prototype metal TWR worked to a brief which included:—

1, The car must be on the weight limit.
2, It must be easy to service.
3, *Nothing* outside the essentials to be included, applies particularly to cooling and lubrication systems.
4, No buy British Policy. What's best for the job goes on (e.g. BBS Mahle wheels; Bilsteins, Recaro seat etc.).
5, Car to be maintained on a "lifing" aeroplane basis. Access to parts that need to be routinely serviced, replaced (choice of three differential ratios for example) or repaired, under racing or between races, must be good.
6, A minimum of 6 m.p.g.
7, Engine to be developed for m.p.g. and endurance rather than power, but no less than 380 b.h.p.

There are only some 20 people at TWR Kidlington so this would be no massive manpower effort. From theory into metal racing XJ-S would occupy November 1981 to early February 1982. Three men were assigned to maintain and assemble the car under chief mechanic Kevin Lee and team manager Paul Davies. The chassis would be the responsibility of long time TWR employee Eddie Hinckley. Tom completed all the test driving. Perhaps the most important role was that of New Zealander Alan Scott, for he had the responsibility of making Jaguar's magnificent V12 into a reliable and comparatively abstemious performer. Based on what had happened in 1977 one would have dismissed his chances of achieving much.

The tendency toward oil surge, inevitable in such a large wet sump unit, has been reduced by an internal sump baffling system that Walkinshaw wryly describes as, "capable of coping with a Torrey Canyon disaster!" Total capacity of the system is around 18 pints and very high flow rates are maintained to provide a steady 70 p.s.i. reading. The system is far less complex than one might have imagined, using a smaller sump pick up point than might have been expected, looking successfully toward minimising recovery in pressure should surge become beyond control. An extra breather was added to the system after the initial test run of February 6/7. It had been found that the high oil level being run within was allowing a very slow return feed during the longer corners of British Leyland's Gaydon proving ground.

From an extra power viewpoint TWR were almost overcome by the largesse of Jaguar engineering in standard form. The single overhead camshaft per bank motor is rated at 299 b.h.p. in standard 90 × 70 mm. (5,343 c.c.) trim, comfortably in excess of the 240 b.h.p. BMW expected to extract from their competition Group A version of the Munich 2,788 c.c. six. Thus it is not surprising to hear, especially bearing in mind the spirit of the regulations (which are biased in favour of modification rather than replacement engineering in 1982-Group A) that a very high percentage of standard parts are retained. The valve gear camshaft profiles took some time to establish in conjunction with the hybrid production Lucas-Bosch electronic injection system. Mechanical injection from Lucas was avoided as the company have now abandoned the production of such systems (the Cosworth DFV's future requirements have been carefully established and catered for by a "once-and-for-all production" leaving the team little alternative).

The production pistons are replaced but the May principles are retained in as much as the compression is a highish 12:1. The bottom end of the production vehicle was described to me as, "very strong and certainly man enough to cope with the power and r.p.m. we anticipated". At Monza in March the V12 required only 6,500 r.p.m. to build a convincing 12s lead in the opening laps, but the r.p.m. limit will normally be 7,000. "It's safe to 8,000/8,500 revs", says Tom candidly, "but then you wind up walking the 500 b.h.p. and 3 m.p.g. path, and we had already established that was not what we wanted to win European events". We did not voice the thought, but for 20 lappers in Britain next year, just such exciting b.h.p. figures might be entirely suitable. At Monza consumption averaged 5.5 to 5.9 m.p.g. Further work with Lucas on part throttle economy and absent progressive metering should realise 6.5 m.p.g.

Tom handed over a power and torque set of readings for the first engine they constructed. At peak values the torque was up from a production rating of 316 lb. ft. at 4,000 r.p.m. to 387 lb. ft. at 6,500 r.p.m. Between 5,000 and 7,000 r.p.m. over 300 b.h.p. was available, culminating in 403 b.h.p. as the best reading.

From its February testing debut in the privacy of Gaydon, to the March centrepiece of attention in the Monza paddock, the team have biased development on practical engineering rather than theory. That first weekend at Gaydon saw a 100 mile shakedown and was continued the following weekend at the same venue. There were three days at Goodwood — "when we found out it's cold in February and did a little bit of tyre sorting as well as getting some friendly police to chase off a certain press photographer!" By March 2nd they had progressed to Silverstone briefly before meeting a March 4th obligation to test at Zolder, Belgium, with Bilstein gas dampers. Ferodo made sure their pads were coping manfully with the extraordinary demands of a large touring car in full flight. They say Jaguar enthusiasts in the surrounding Belgian flatlands are still humming a 12 cylinder rhapsody at the memory of that private test day. All testing was conducted on Weber downdraught carburetters while the injection system was adopted for the first race, dispensing with the prototype's cutaway bonnet.

March 21st was the racing debut, but Tom estimates they fitted in something like 6 hours of practice on the preceding two days of that Monza meeting. "We had to ensure we were ready for a 4 hour race, checking out pit stops with the air jacks and centre-nut wheels. We reckon 30 secs. or so for 115-litres (25.33 Imperial gallons) of fuel and changing all four wheels and tyres. So far as the tyres go, I'd just say they are round and black, wet or dry. We do expect to change them at our pitstops and we do have some qualifying rubber ready for use, but hope that it will never be required." Winning races appeals more than split second pole positions at present.

"We'll take the flak", is how Tom sees the first half of the season with a car that is only months old in a race development for a new series of regulations. A second car will only be completed when they are happy with the specification and performance of the first. When that second car arrives, the drivers are presently predicted as Porsche dealer Peter Lovett and Piérre Dieudonné.

Judging by the gleam in Walkinshaw's far-sighted grey eyes he is looking for that happy state of affairs to arrive around June when the Championship emerges from comparatively slow circuits that are hardest on a big car onto the spaces of Brno's magnificent public road layout. Late June for Austria's Oesterreichring; Nürburgring on July 4 and the reduced, but still challenging Spa-Francorchamps track from July 31st to August 1st. All except Spa are 4 hour races. Those with a taste for the Nürburgring might care to note that Jaguar's previous best result of the 1976-77 XJ Coupé schedule was at the 'Ring, when a hastily repaired machine managed second, driven by Derek Bell/Andy Rouse.

Looking beyond this largely experimental opening season in Europe there is an obvious future for the XJ in the British Touring Car Championship of 1983. Then the rules are expected to be as for the European series (Group A) whereas the present rules are an untidy hybrid of Group 1 and have a 3½-litre restriction. The latter capacity regulation should disappear too.

Perhaps we all expected too much from Broadspeed and Jaguar last time, an expectation eagerly fanned by the initial BL PR support. This time, at Tom's request (and that of Jaguar Cars, I suspect), there's not a PR man to be seen, just a small team of extremely tough racers putting together a car capable of bringing back the European Touring Car Championship to a Jaguar driver for the first time since German sportsman Peter Nöcker took the title in a 3.8 saloon. That was in 1963, the first year that the European series was held.

Since then only Mini Cooper S derivatives (1964: Warwick Banks and 1968, when John Handley and John Rhodes won a division apiece) and a 1965 title win for John Whitmore's factory Lotus Cortina have brought Britain European victories. Companies like Alfa Romeo (1970) Ford Cologne (Weslake V6 Capris, 1971/72) and Audi (1980) are more typical European winners. The big winners? BMW. After their class struggles of the sixties, developing a turbo 2002 to beat the Porsche 911 before that was outlawed as a touring car, BMW emerged with the winged CSL in 1973 and won every title until 1980, when Audi interrupted by winning the Makes title. Last year the 635 CSi brought BMW back a European Championship title, but with a production requirement of 5,000 p.a. BMW are temporarily forced to rely on the 528i, which could obviously do with a 535i cousin if it is to even start in the same overall horsepower / speed class as the Jaguar. — J.W.

Technical notes

I was not able to detail the racing XJ-S fully at the time of writing but the following appeared salient when walking around the dissembled car in late March, following its Monza debut.

Engine: as per text, aluminium wet sump V12 of 5.3-litres with electronic ignition (CDI system), racing Serck front radiator, oil cooler, electric fan for water radiator. Currently rated at 400 b.h.p. with 7,000 r.p.m. limit. Time to change engine, approximately 1 to 1½ hours.

Transmission: Manual 4-speed Jaguar gearbox with one set of homologated racing ratios available. Racing single plate clutch, limited slip differential. Oil cooling for gearbox and differential. Time to change gearbox, 1 hour. Time to change differential (choice of three ratios) half hour. Geared for 155 m.p.h. at Monza.

Body: Steel two door with internal alloy roll cage and light tubular engine bay stays. Weight reduced from 1981 Geneva Show catalogue figure of 1,750 kg. to 1,400 kg. Built-in air jacks. Quick release systems for fluid lines, electrical connections and front wings. Premier 120-litre petrol tank.

Cockpit: Untrimmed bare steel with single Recaro race seat. Fabricated centre console with switchgear for: ingnition and fuel (4); differential and oil cooler; lights; fan booster for demist. Willans harness, racing steering wheel (4-spoke).

THE fabricated centre console.

Instrumentation: 0-150 p.s.i. oil; 50-150°C oil; Jones 0-9,000 r.p.m. tachometer; 40-120°C water; 0-30-70 p.s.i. fuel pressure.

Steering: Power-assisted, std. rack, extra castor and camber geometry.

Suspension: Alternative lightweight titanium coil springs available; Bilstein adjustable ride height gas dampers; hybrid production and race fabrication of double front wishbones and rear double coil spring per side / lower wishbone system around Jaguar principles.

Brakes: AP vented competition discs, 13 x 1⅝" thick; four piston calipers.

UK appearances: Donington, May 2nd; Silverstone, September 12th.

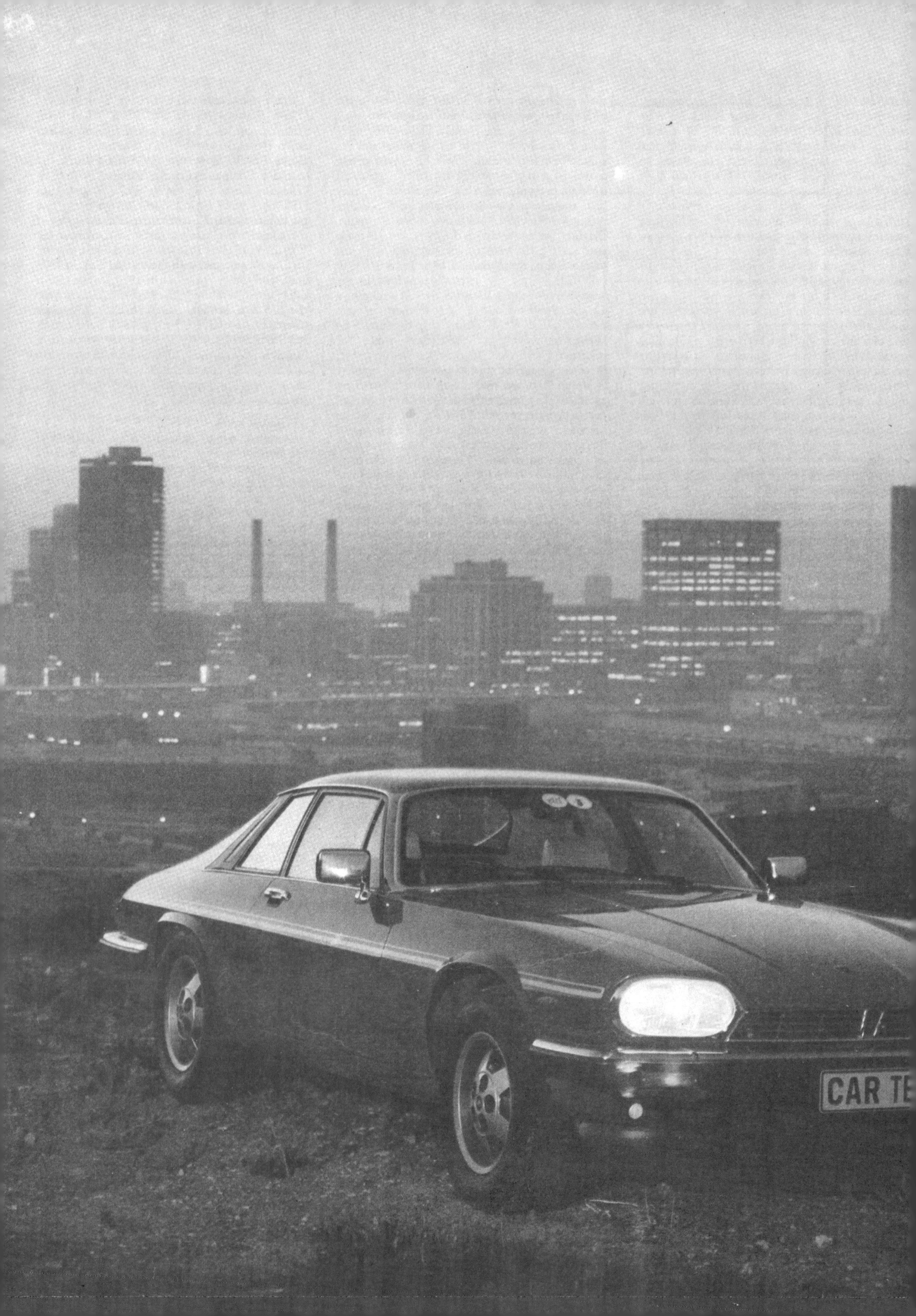
CAR TE

CAR TEST
JAGUAR XJS-HE AUTOMATIC, BY LEYLAND

The new "Fireball" V12 engine creates an electrifying speed car— which costs as much as a medium-sized house!

It's low, lean and wicked-looking. It's a Jaguar, and well-accredited legend makes it one of the world's fastest production cars.

There are a handful of Jaguar XJS-V12 coupés in South Africa. Many people may not even have seen one. But take our word for it: nearly everything they say about this modern British supercar is true!

Its looks are dramatic. A bigger car than one would expect, nearly five metres long and close to two metres wide. But it stands low — only just more than waist-high to an adult — you can stand next to it, and comfortably rest an arm on the roof. It turns the scales at a massive 1,7 tons, and its configuration is a generous 2+2, two-door coupé.

YEARS OF CHANGE

The XJS is the latest of the "Big Cats" from Jaguar Cars, of Coventry, and it has evolved steadily since it was launched nearly a decade ago. The car itself has not changed much in form: it has gained the rather unwieldy-looking impact-absorbing bumpers required in some export markets, and its furnishing and trim have been improved.

But the principal change has been mechanical. The original V12, triple-carburettor motor was low-stressed and thirsty: we drove it during a visit to Britain in 1976, and found it rather unimpressive under urban road conditions.

KEY FIGURES	
Maximum speed	217,7 km/h
1 km sprint	29,1 seconds
Terminal speed	177,0 km/h
Fuel tank capacity	91 litres
Litres/100 km at 80 (ECE/EPA)	10,07
Optimum fuel range at 80	904 km
*Fuel Index	13,09
Engine revs per km	1 385
National list price	R64 300
(*Consumption at 80, plus 30%)	

Since then there have been three changes — fuel injection came first, then a more-refined digital fuel injection system, and finally the spectacular May "Fireball' cylinder head which raised com-

Below: The latest V12 engine incorporates the new May "Fireball" cylinder head principle — designated by the "HE" in the car's title.

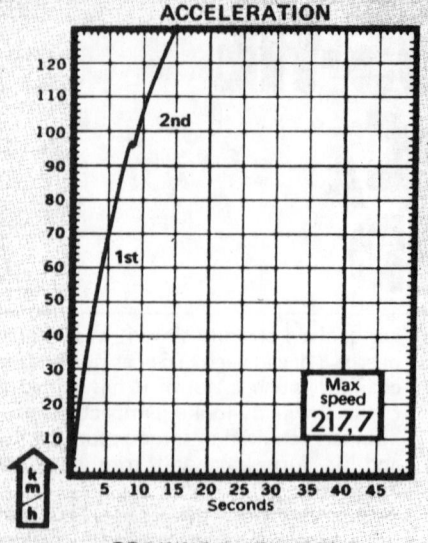

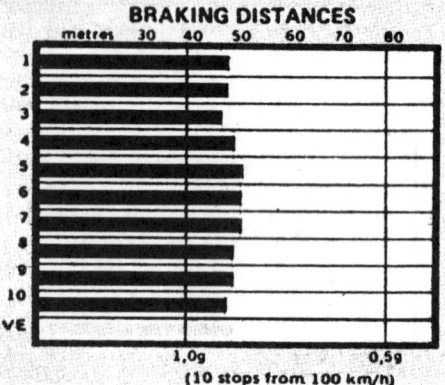

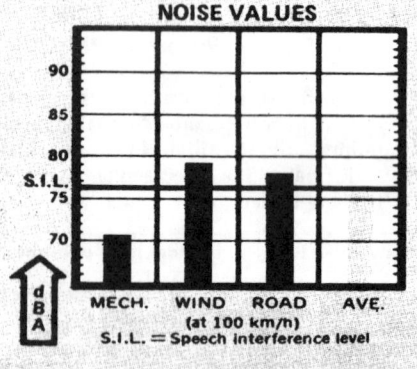

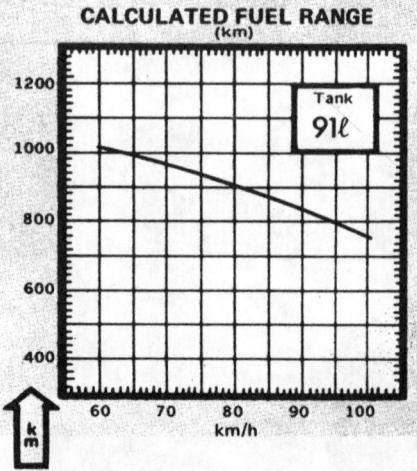

test — Jaguar XJS V12 (Model HE)

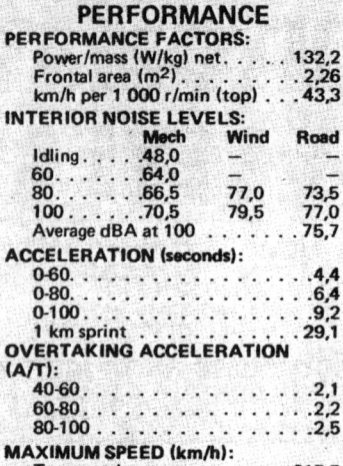

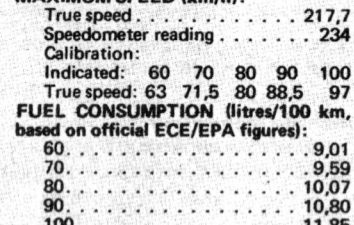

PERFORMANCE

PERFORMANCE FACTORS:
- Power/mass (W/kg) net 132,2
- Frontal area (m²) 2,26
- km/h per 1 000 r/min (top) . . . 43,3

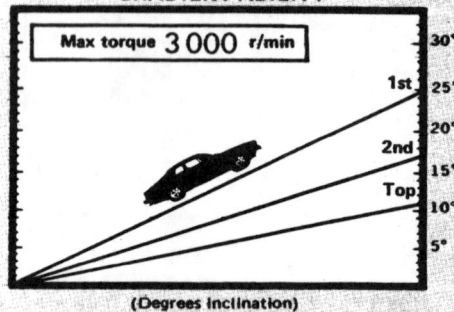

INTERIOR NOISE LEVELS:

	Mech	Wind	Road
Idling	48,0	—	—
60	64,0		
80	66,5	77,0	73,5
100	70,5	79,5	77,0
Average dBA at 100			75,7

ACCELERATION (seconds):
- 0-60 4,4
- 0-80 6,4
- 0-100 9,2
- 1 km sprint 29,1

OVERTAKING ACCELERATION (A/T):
- 40-60 2,1
- 60-80 2,2
- 80-100 2,5

MAXIMUM SPEED (km/h):
- True speed 217,7
- Speedometer reading 234

Calibration:

Indicated:	60	70	80	90	100
True speed:	63	71,5	80	88,5	97

FUEL CONSUMPTION (litres/100 km, based on official ECE/EPA figures):
- 60 9,01
- 70 9,59
- 80 10,07
- 90 10,80
- 100 11,85

BRAKING TEST:
From 100 km/h
- Best stop 3,4
- Worst stop 3,7
- Average 3,58

GRADIENTS IN GEARS:
- Low gear . . . . 1 in 2,2
- 2nd gear . . . . 1 in 3,3
- Top gear . . . . 1 in 5,2

GEARED SPEEDS (km/h):
- Low gear 95,3
- 2nd gear 158,9
- Top gear 238,3
(Calculated at engine power peak — 5 500 r/min)

TEST CONDITIONS:
- Altitude . . . . at sea level
- Weather . . . . fine and warm
- Fuel used . . . . 98 octane
- Test car's odometer . . . . 9 246 km

WARRANTY:
12 months.

TEST CAR FROM:
Leyland SA, Blackheath, Cape.

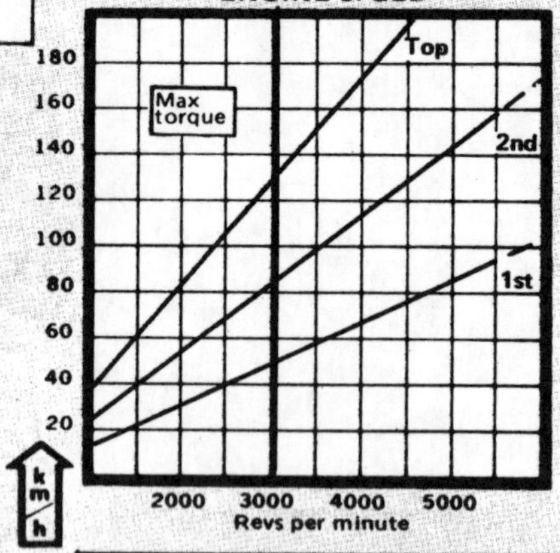

IMPERIAL DATA

ACCELERATION (seconds):
- 0-60 m-p-h 8,7

MAXIMUM SPEED (m-p-h):
- True speed 135,3

FUEL ECONOMY (m-p-g, ECE/EPA):
- 50 m-p-h 27,9
- 60 m-p-h 25,0

CRUISING AT 100

- Mech. noise level . . . . . . 70,5 dBA
- 0-100 through gears . . . . . 9,2 seconds
- Litres/100 km at 100 (ECE/EPA) . 11,85
- Optimum fuel range at 100 . . . 768 km
- Braking from 100 . . . 3,58 seconds
- Maximum gradient (top) . . . 1 in 6,5
- Speedometer error . . . . . . variable
- Speedo at true 100 103
- Tachometer error . . . . . . negligible
- Odometer error . . . . . . not measured
- Engine r/min at 100 . . . . . . 2 310

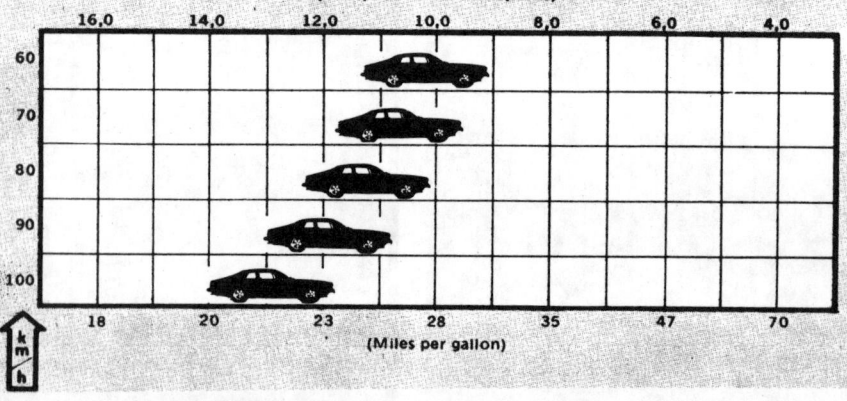

pression ratio and efficiency, giving solid gains in torque characteristics and a big jump in fuel economy, both specific and overall.

'FIREBALL' ENGINE

This latest engine development adds the letters "HE" (for "High Efficiency") to the designation of the current model, and makes it a phenomenon among performance cars. The new cylinder head, featuring a dual combustion chamber system on something like a figure-8 pattern, was developed by Swiss consulting engineer Michael May, to give optimum swirl of incoming mixture, high combustion efficiency, and allow very high compression ratios (as high as 13,0 to 1 on standard 97-octane fuel).

The Jaguar Cars Division of BL was one of the first manufacturers to show an interest in the May "Fireball" head, as it was named, and a special version was developed for the Jaguar V12 aluminium alloy engine, together with a special Lucas dual-coil electronic ignition system to complement the Lucas digital fuel injection. To meet the requirements of the 12,5 to 1 compression ratio used, compact tapered-seat spark plugs are fitted.

LONGER TRANSMISSION

The HE "Fireball" development was not confined to the engine: the automatic transmission (which has become standard on this V12 model) was changed as well. The earlier Borg-Warner Model 12 automatic gearbox made way for the latest GM Turbo-Hydramatic 400. At the same time, the final drive ratio was changed from 3,07 to 1, to a long-legged 2,88 to 1, to get full value from the improved torque characteristics of the Fireball engine.

This makes the best use of the engine, with magnificent overall gearing: 43,3 km/h per 1 000 revs in top, so that the car whispers along at cruising speeds, yet with abundant pull in reserve: we found it would climb a one-in-five gradient in top! A limited-slip diff and low-profile high-speed radials transmit the power to the road.

Wider (6,5 J) alloy wheels are fitted to the new model, and extra-powerful Cibié quartz-halogen headlights have become standard.

WOOD AND LEATHER

The interior is very much Jaguar, with Connolly leather upholstery and trim, and magnificent burred elm veneer wood panelling on fascia and doors. The front seats are low, spacious and supremely-comfortable, while the rear seat is fairly generous by plus-2 standards: an average-sized adult manages a tight fit, with his head against the roof and knees tucked around the front seat backrest. The luggage trunk, on the other hand, is roomy and deep, with the spare wheel set vertically inside and the battery alongside it.

There is a full inventory of equipment: air-conditioning, power windows, power mirrors, central locking, stereo radio/tape deck, power steering, and the rest.

Surprisingly, instrumentation is functional and uncomplicated, with a 260-km/h full-dial speedometer, rev-counter red-lined at 6 500, and four strip-type gauges at centre covering minor functions. An analog clock is set at centre in the fascia.

ROAD PERFORMANCE

In the modern line-up, we cannot remember an automatic which comes anywhere near this big Jag.

It really strains at the leash when it is torque-started against the footbrake — the driver has to hold hard! — whisking away from rest with only a touch of wheelspin (thanks to the diff lock). And with its super-long legs, it sprints all the way from zero to 100 in 1st!

This amazing small-twelve engine is quite willing to go all the way to 6 500 revs, as it powers 1,7 tons of car to 80 in 6,4 seconds, and 100 in 9,2 seconds.

Maximum speed is academic, but we never take works claims on trust, and we ran the XJS at full bore both ways on an 8-km straight to record an average of 217,7 km/h — or 135,3 m-p-h, at a full 5 300 revs. That was with air-conditioning off, incidentally — with the cooler working, top speed was 213,6 (132,8 in m-p-h).

This fifth-wheel reading was confirmed by the very accurate rev-counter (allowing for slip loss) while the speedometer overread by a fairly mild 7,5 per cent at this hefty top speed — it registered 234 and 230, respectively.

FUEL ECONOMY

We have combined two offical sets of figures — ECE (Economic Community of Europe) and EPA (Environmental Protection Agency in the US) — to present tables and graphs of the XJS-HE fuel consumption at steady speeds.

The picture is quite impressive for a car of this mass and performance capability: about 10,0 litres/100 km at 80, and inside 12,0 litres/100 km at a steady 100. The ECE gives it an "urban cycle" figure of 15,6 m-p-g — which is just over 18 litres/100 km. Our Fuel Index suggests that it should give something like 20 m-p-g overall — 13 litres/100 km. As the XJS has a 91-litre tank, range potential is impressive by big-car standards.

SOUND AND BRAKING

To the best of our recollection, this is the quietest car we have ever tested. With its engine ticking over at 2 300 revs, it registers only 70,5 decibels of sound at a true 100 km/h — which is virtually inaudible. Road noise is also magnificently low, and wind noise is the biggest offender — yet still inside the 80 decibels level.

Jaguar was a pioneer of disc brakes in modern car production, and has remained among the leaders in brake technology ever since. The XJS has a well-balanced all-disc system, with ventilated discs at front to maximise cooling, and it stops firmly and fast every time — even from very high speeds. It averaged 3,58 seconds in our 10-stop sequence from a true 100, with only a very mild tendency to lock front wheels.

TEST SUMMARY

You could buy a comfortable house for the South African price of the XJS (which includes that hefty bite of import duty) — but this is not a car to be assessed by ordinary monetary standards. It's the kind of conveyance that most of us can only dream about, and it will always remain exclusive and utterly extravagant.

It's a magnificent machine — and a connoisseur collector's item, from the day it is built. •

SPECIFICATIONS

ENGINE:
- Cylinders . . . . V12 (Model HE)
- Fuel supply . Lucas digital electronic fuel injection
- Bore/stroke . . . . . . 90,0/70,0 mm
- Cubic capacity . . . . 5 345 cm^3
- Compression ratio . . . . . . 12,5 to 1
- Valve gear . . . o-h-v, two single o-h-c
- Ignition . . . twin coils and electronic distributor
- Main bearings . . . . seven
- Fuel requirement . . 98-octane Coast, 93-octane Reef
- Cooling . . . water; visco-coupled and electric fans

ENGINE OUTPUT:
- Max. power I.S.O. (kW) 223
- Power peak (r/min) 5 500
- Max. usable r/min. 6 500
- Max. torque (N.m) 433
- Torque peak (r/min) 3 000

TRANSMISSION:
- Forward speeds . . . . . 3, automatic (GM 400)
- Selector . . . . console
- Low gear . . . . 2,50 to 1
- 2nd gear . . . . 1,50 to 1
- Top gear . . . . direct
- Reverse gear . . . . 2,00 to 1
- Final drive . . . . 2,88 to 1
- Drive wheels . . . . . rear, limited slip

WHEELS AND TYRES:
- Road wheels . . . . alloy sports
- Rim width 6,5J
- Tyres . . . . . 215/70 VR 15 radials

BRAKES:
- Front . . . . 284 mm ventilated discs
- Rear . . . . 264 mm discs
- Pressure regulation . . . dual circuits
- Boosting . . . . vacuum servo

STEERING:
- Type . . . . rack and pinion, power assisted
- Lock to lock . . . . 2,9 turns
- Turning circle . . . . 11,6 m

SUSPENSION:
- Front . . . . independent
- Type . . . . . coils and wishbones, anti-roll bar
- Rear . . . . independent
- Type . . . . . . twin coil spring units, dual-jointed half-axles

CAPACITIES:
- Seating . . . . 2 + 2
- Fuel tank . . . . 91 litres
- Luggage trunk . . . . . 325 dm^3 net

Tail-piece

Stylish Eventer estate conversion is a definite improvement over the original XJS silhouette; increase in weight over coupé is minimal

Magnificent V12 engine makes the XJS the only estate to top 150mph

Load platform is over six feet in length and claimed capacity is as big as a Rover's. Fuel tank and spare wheel now under boot floor

Rarity: Jaguar XJS Eventer estate

Lynx's elegant estate conversion adds space to the XJS's 150mph pace. We drive the ultimate in high-speed hatchbacks

"An extremely up-market Scimitar" is the disconcertingly modest way in which Lynx pulicist Derek Green describes the Hastings-based specialist converters' latest project, an elegant shooting-brake version of Jaguar's mammoth XJS coupé, to be known as the XJS Eventer. It's an understated description on just about every conceivable count: after all, who else has an estate car that can top 150mph? Which other load carrier has a twelve-cylinder engine or, for that matter, which other current supercar can accommodate all the skis, poles, boots and countless changes of wardrobe required for a fortnight's high-class winter sports in St Moritz?

It is to Lynx Engineering's great credit that they saw in Jaguar's magnificently fast but notoriously cramped big coupé not just a superb open-topped sports car (their convertible conversion, available since August 1980 and tested by *What Car?* in Sept 80) but also the potential for a relatively spacious grand touring estate, much along the lines of the Lancia HPE but offering unparalleled levels of performance and refinement. The XJS estate is not, however, just an exercise to produce the quickest cargo carrier in Europe: it is part of Lynx's total commitment to Jaguars, a commitment which has provided painstaking restorations of classic C, D and E-Type racing machinery as well as the production of the highly authentic D-Type replicas which have led unkinder commentators to regard Lynx as somewhat irreverent interpreters of the Jaguar tradition.

It could indeed be considered mildly heretical calmly to slice the roof off a brand new £20,000 motor car of the finest breeding and radically to redraw the lines laid down by some of Jaguar's best designers; incautious, too, to tamper with the XJS's very concept as a fabulously fast tourer for two.

Yet deep down, Lynx were convinced that the XJS was missing out somewhere: while everyone admired its engineering, its performance and its outstanding refinement at speed, it was really very impractical for the bulk of buyers. Despite an overall length of over sixteen feet and a weight of almost two tons it offered only cramped 2+2 coupé accommodation, with neither the open-air advantages of a soft-top nor the compensating space of a grand tourer. "Where do you put your golf clubs in an XJS" is perhaps the most crucial indictment of its uncompromising coupé concept.

Conscious that this impracticality had prevented many XJS admirers from turning into XJS buyers, Lynx sought to tap this seam of discontent by making the Eventer conversion as attractive and as convenient as humanly possible. Accordingly, there is a practical, low and flat loading floor which, with the rear seats folded, is an unbelievable six feet long and 46 cubic feet (more than a big Rover) in volume; the rear seats fold individually, and the whole rear is trimmed in the best hides and carpet for maximum luxury.

There are many who believe that any change to the Jaguar's bulky exterior styling must be an improvement: what is more certain is that, just as with the successful XJS Convertible conversion, it is only with the removal of the coupe's severe 'flying buttress' rear window surround that the XJS reveals the real elegance of its Malcolm Sayer inspired lines. From the very outset Lynx refused to have their hands tied by committing themselves to existing glass shapes available off the shelf for other makes; they were determined, too, to improve upon the original car and would never have accepted such a grossly ugly end product as Avon's estate conversion of the Jaguar saloon.

Few would dispute, then, that just as the convertible is even more a Jaguar than the XJS, so is the Eventer estate which, with its smooth, flowing roofline and slender side pillars, has a lighter, more modern look than its parent. It is perhaps incidental, though amusing, to record that once the glass shapes had been decided it was discovered that there was after all an off-the-shelf rear window that fitted the outline perfectly – that from the Citroen Ami 8, one of the most hideous designs ever to reach production.

An incidental benefit of the new roofline is an increase in headroom – always at a premium in the top Jaguar – but by far the biggest change noticed by occupants is the dramatically improved brightness and space of the interior. Gone is the cramed, claustrophobic feel of the coupé: the all-round view is very much better, reversing is made easier, and the rear view mirror no longer gives a restricting tunnel vision effect.

Rear passengers have better headroom, too, though the back axle housing unfortunately prevents any increase in the meagre legroom the standard car offers.

Existing XJS owners will naturally wonder what has happened to the huge twenty-gallon fuel tank, normally occupying the space between the back seat and boot bulkheads: it now fills, together with the now-horizontal spare wheel which it neatly surrounds, the whole of the space under the load floor, right up to bumper level. Only 1½ gallons fuel capacity has been lost in the operation.

In city traffic, as we have said, the XJS Eventer is a much more manageable beast than the standard car thanks to its much-improved vision: on the open road it is an equal complement to say that it is every bit as outstanding as the original product. There's still the phenomenal, silken potency of the superb V12 that turns even the most modest 50mph overtaking manoeuvre into an unconscious surge into three-figure speeds and beyond, so silent and deceptively fast is the car; there's the same eerie absence of engine, road and wind noise, the latter despite the new roofline, and there is the identical feeling of decadent superiority and isolation from other traffic and the world around.

Slightly stiffer rear springs give the Eventer a fractionally firmer ride – no bad thing for the XJS – and the only other clue to the conversion is a slight rear axle whine, audible perhaps only because everything else is so uncannily silent. Lynx say that an improved metal sandwich rear floor panel will cut this incongruous sound out completely.

Handling, too, proves to be just as on the standard car, with no obvious loss of body rigidity but perhaps slightly less lightness to the steering. There is evidence, too, that the new tail has slightly better aerodynamics than the coupé's cutaway shape, Lynx's demonstrator's high-speed 16.6 mpg average being less immodest than the standard HE's.

With its rare, if not wholly unique, combination of speed, space and style, Lynx see the Eventer appealing not so much to the existing XJS driver (who may well have come to terms with his car's impracticality) but to owners of up-market estate cars such as Mercedes and Range Rover – the county set who have a big and prestigious estate without needing the full carrying capacity, yet for whom a Porsche 924 is too small, an Opel Monza too pedestrian and a top Capri unspeakably common.

The cost of the conversion – undertaken, of course, to the very highest standards of British coachbuilding – is £6950 plus VAT and takes around six weeks from first incision to final completion. A dozen or so have already been ordered, largely by word of mouth following the prototype's appearance at the Jaguar open day in August, and Lynx see no reason why they could not carry out 40 conversions in the first full year. Considered as a £28,000 new car the Eventer appears an intimidating proposition – but, to return to our original question, what else can offer so much of Jaguar's famed grace, space and pace, even at that elevated price level?

RoadTest

THE TIMELY arrival of two new and more frugal XJ-S models — one a convertible and both powered by a new generation all-alloy 3.6-litre straight six — not only opens up a fresh chapter for Coventry's Big Cat but gives Jaguar the ammunition to make a deeper dent in the lucrative luxury car market. Capitalising on an upturn in sales of late, the auto-only V12 XJ-S HE lives on and remains remarkable value for those who can afford to run it. Less powerful but lighter, and equipped with five-speed manual transmissions, the six-cylinder cars, though not that much cheaper, step more assertively into BMW/Mercedes/Audi territory.

The hinge-pin of the XJ-S revival is under its bonnet. Although the XJ-S was launched after the XJ saloon, its new AJ6 engine will go on to power the XJ40, the current saloon's replacement. The performance of the fuel-injected 24-valve unit (which effectively supersedes the venerable XK "six") is therefore of vital interest as the future success of Jaguar's mainline model range depends on it. For those who want the AJ6 development story, it was told in *Motor* w/e October 15, but we review the salient points here. With its chain-driven twin camshafts, shim adjustment for the directly-actuated tappets and straight-six configuration, the AJ6 is an engine in the classic Jaguar mould not far removed from the old XK. Where the new 3,590cc unit differs considerably, however, is in its head design. Made, like the block, from weight-saving aluminium alloy, the head features four valves per cylinder, the inlet pair feeding the cylinders with a mixture precisely metered by the same Lucas digital electronic fuel injection used on the V12. Also lifted from the larger engine is the seven main bearing crankshaft for which is claimed immense torsional strength. Like the injection the ignition is electronic and, on a compression ratio of 9.6:1, the engine develops 225 bhp (DIN) at 5,300 rpm and 240 lb ft of torque at 4,000 rpm. This compares with 299 bhp (DIN) at 5,500 rpm and 318 lb ft of torque at 3,000 rpm for the V12.

Completing the new drive train is the Getrag five-speed gearbox, chosen by Jaguar as being the best unit available and toting long-striding fourth and fifth gear ratios in the interests of cruising refinement and economy. Otherwise, there's little to distinguish the six-cylinder XJ-S 3.6 Coupé and Cabriolet from the V12 HE under the skin, the only changes to the suspension being slightly longer travel and softer rate springs at the front to compensate for the lighter engine. Visually, the Cabriolet is clearly a different animal with its abbreviated roofline and more obvious boot, but the 3.6 Coupé we test here differs from its V12-engined counterpart only in the adoption of different style alloy wheels and the accentuated bonnet bulge needed to accommodate the physically large AJ6 engine. Equipment levels are maintained at a high standard: items include air-conditioning, electric windows and door mirrors, a stereo radio/cassette player and central locking. Unlike the Cabriolet, which shares the HE's "star" type alloy wheels and full leather upholstery, the 3.6 Coupé has ambla interior trim with leather seat facings.

At £19,248, however, the junior XJ-S coupé is priced closer to the £21,752 HE than many industry observers were expecting. The way Jaguar tell it, the small differential shouldn't come as any surprise since the manual 3.6 Coupé can do everything its automatic V12-engined counterpart can (both are acredited with 145 mph top speeds and 7½ second 0-60 mph times) except consume petrol at the same rate. On that reckoning, the HE must be the 3.6 Coupé's closest rival but don't doubt that the competition from other stables is anything less than capable and keenly priced. The best of the bunch include Audi's 4-w-d Quattro (£17,722), BMW's 635 CSi (£24,995), Ferrari's Mondial Quattrovalvole (£30,710), the Mercedes-Benz 380 SEC (£28,560) and the Porsche 928S (£30,679).

The Jaguar fires-up promptly and reliably from cold and, under the guidance of its digital Lucas injection, pulls cleanly and tractably from the word go. In these respects, the AJ6 performs as well as any Jaguar engine in our experience, but be prepared for a few surprises when you begin to crack on. There's no question that the engine pulls strongly from low revs and with real vigour from about 3,500 rpm but the *quality* of its exertions will come as something of a shock to anyone familiar with the silky urgency of Jaguar's V12 or the gentlemanly refinement of the old XK unit. If we can trust our test car's engine as being representative — in fact it had *two* during its time with us, the first one suffering a chipped valve and consequent loss of power at the test track — then we must register some disappointment. Nor is it just a matter of holding back the superlatives: all our testers agreed that de-

JAGUAR

spite sounding impressively crisp and urgent without being in any way loud, the engine is afflicted by an underlying roughness present throughout the rev range which isn't only heard but felt through the toeboard. What's worse, it's prominent in the middle part of the rev band most frequently used in normal motoring. Both BMW and Mercedes make smoother six-cylinder engines. Until the demise of the XK unit, so will Jaguar.

On the other hand, the AJ6 engine does *feel* potent. Despite the rather low 5,800 rpm red-line, mid-to-upper range punch is such that constant cog-swapping isn't a prerequisite for brisk progress. There again, throttle response isn't as clean as it might be and the injection's overrun fuel cut-off sometimes promotes jerkiness when throttling back, a condition exacerbated by a degree of snatch in the transmission. In short, the AJ6-engine XJ-S has sharp claws but entirely lacks the seductive purr of the V12 HE.

Neither is it as fast when the chips are down. Jaguar claim a top speed of 145 mph for the 3.6 Coupé but we could manage only 136.8 mph (in fifth) round Millbrook's high-speed bowl on an almost perfectly still day — a long way short of the claim and, indeed, the 152.4 mph achieved by the V12 HE (299 bhp) round the same circuit, but about what we'd expect on 225 bhp. It puts the Jaguar near the back of the field in terms of outright speed, however, since of our selected rivals only the automatic Mercedes 380 SEC (132.6 mph) and possibly the Audi Quattro (an estimated 135 mph) are slower.

And while capable of shading its automatic V12-engined stablemate in the sprint from a standstill to 60 mph (7.2 against 7.5 sec), the 3.6 Coupé would be left at the by all the manual rivals with which we've compared it in this test: the lights V8 Ferrari Mondial QV disposes of the yardstick increment in a rousing 6.4 sec, the Audi Quattro and automatic Porsche 928S take 6.5 sec apiece, the BMW 635 CSi 6.9 sec and the automatic Mercedes 380 SEC a comparatively sluggish 9.4 sec. In its ultra long-striding fifth gear (nearly 30 mph/1,000 rpm), the important 50-70 mph increment is covered in an unexceptional 11.5 sec, but this drops to a far more spirited 7.4 sec in fourth, and if we take low speed flexibility in this gear as the criterion, the Jaguar's 30-50 mph time of 6.9 sec is bettered only by the Ferrari's 5.6 sec in the comparison table. Overall, though, the 3.6 Coupé is by no means the quickest of the quick. Then again, neither is it the most powerful or expensive: only the Quattro is cheaper.

Despite the significant and commendable gains in economy achieved by Jaguar's engineers with the Fireball headed version of its 5.3-litre V12 powerhouse, *socially respectable* fuel economy is the *raison d'etre* of the AJ6-engined XJ-S. But again it disappoints. The 3.6 Coupé's overall consumption of 18.9 mpg is nothing special by absolute standards and represents only a marginal improvement over the 16.3 mpg returned by the XJ-S HE we tested in 1981 and most of our selected manual rivals are thriftier, the BMW (22.5 mpg) by a big margin. Given a gentle right foot and considered use of the long fifth gear, most owners should be able to match or exceed our estimated touring consumption of 23 mpg, on which basis the large 20-gallon tank allows a practical range of more than 400 miles. Four star petrol is recommended.

The five-speed Getrag gearbox shifts with positive, meaty precision, even if it doesn't win any prizes for lightness. Beyond an annoying tendency to baulk into first and reverse (especially when cold), its short-throw action holds no perils for the driver, with the movement into fifth (on a dogleg to the right and forward) being as easy as the movement from second to third. The intermediate maxima of 36, 62, 92 and 128 mph at 5,800 rpm aren't exceptionaly tall but are well spaced: given the engine's broad power band, you're seldom caught in the "wrong" gear. At 70 mph in fifth (28.9 mph/1,000 rpm), the engine is turning over at a lazy 2,422 rpm. The heavyish clutch needs to be pushed fully to the floor to disengage the drive but has a late and rather abrupt take-up point which makes smooth gearchanges more demanding than they should be.

Although more realistically weighted than it used to be, XJ-S steering remains largely devoid of feel, effort at the steering wheel rim staying constant even when the front wheels are beginning to lose adhesion on a slippery surface. This is at odds with the steering's crisp responses and sensible gearing which help create a sense of wieldiness that belies the car's size and weight. As with the V12 HE, the 3.6 Coupé's handling is exceptional for its balance, surefootedness and unruffled poise on bumpy surfaces. But while the influence of modern Pirelli P5 215/70 VR 15 tyres and the spring/damper revisions designed to compensate for the 175 lb lighter AJ6 engine has improved the car's turn-in characteristics and lessened its tendency to heave and float over undulations taken at speed, tautness is still compromised for comfort in the final analysis. It remains a lack of damping control more than anything else that sets the cornering limit on twisty country roads. That limit is impressively high: the Pirelli rubber grips very well wet or dry and a natural tendency towards mild understeer changes to progressive oversteer in maximum effort corners. Apart from slightly more small bump harshness — insignificant by normal standards — the 3.6 Coupé matches the V12 HE for overall ride comfort which means that it rides very well indeed. It soaks up lumpy surfaces with supple aplomb and no trace of sogginess, and the high

XJS–3.6C

The advent of the six-cylinder AJ6 engine marks a new beginning for Jaguar's XJ-S. But does the new power unit mean diminished standards for the Coventry supercat?

MOTOR ROAD TEST No 10/84
JAGUAR XJ-S 3.6C

The Jaguar's facia remains a pleasing mixture of modern and traditional, though the instruments have never been attractive

Physically large AJ6 engine only just fits under Jag's bonnet: it's potent but lacks smoothness.

Left: though sumptuously trimmed, the rear of the XJ-S is no place for adults to travel any distance. Above: the leather-faced front seats aren't much to look at but proved comfortable and supportive

standard of comfort is enhanced by suspension which works very quietly. As always, the brakes are firm, progressive and tremendously powerful, capable of hauling the heavy XJ-S to rest from three-figure speeds without drama or fade, wet or dry.

The quality of the AJ6 engine's vocalisations apart, this XJ-S upholds the tradition for refinement set by its V12 stablemate. Our noise measurements show that while it's not as serenely quiet in any department, it's not *that* far behind and in terms of wind noise and road roar there's little to choose between the two cars: both set standards it would be hard to better.

A new engine can do nothing to change the fact that the XJ-S is one of the least space efficient packages money can buy. A far from compact 15ft 6in long, it can really only seat two adults in comfort over any distance (the occasional rear seats are OK for small children). The interior has changed remarkably little over the years bar a constantly improving standard of trim and equipment. The current cabin boasts the best finish of all with none of the cheap-looking chrome-plastic embellishments foisted upon the early cars but, instead, a good-looking and feeling mixture of leather, plastic and polished wood. Its basics remain much the same with ample room in the front, a well thought-out driving position and supportive, well shaped seats. The instrumentation earns low marks for styling and presentation but is well positioned, readable and comprehensive. The standard air conditioning is impressive in as much as it is possible to dial-up a temperature for the interior but poor in its ability to provide proper bi-level heating/ventilation, the flow of air from the ram flow vents at either end of the facia being insufficient to dispel stuffiness with the heater on.

The development of a virtuoso solo act into a broader market three-car range should ensure the survival of an endangered breed and for that we must be grateful. The 3.6 Coupé and Cabrio are the future of the XJ-S and their AJ6 engine is the future of Jaguar itself. The preceding comments relating to that engine's mechanical refinement should give Jaguar cause for concern. It simply isn't good enough — not only by the standards of the V12 XJ-S which sell for little over £2,000 more than the 3.6, but by the standards of arch rivals BMW and Mercedes. As things stand, the junior XJ-S is a much lesser thing than the promise of its famous name. What improvements further development might bring we can only ponder.

PERFORMANCE

WEATHER CONDITIONS
Wind	0-5 mph
Temperature	39°F/4°C
Barometer	29.1 in Hg/985 mbar
Surface	Dry tarmacadam

MAXIMUM SPEEDS
	mph	kph
Banked circuit	136.8	220.1
Best ¼ mile	138.3	222.5
Terminal Speeds:		
at ¼ mile	92	148
at kilometre	114	183
Speeds in gears (at 5,800 rpm):		
1st	36	58
2nd	62	100
3rd	92	148
4th	128	206

ACCELERATION FROM REST
mph	sec	kph	sec
0-30	2.6	0-40	2.2
0-40	4.1	0-60	3.9
0-50	5.5	0-80	5.4
0-60	7.2	0-100	7.7
0-70	9.6	0-120	10.8
0-80	12.2	0-140	14.3
0-90	15.2	0-160	19.2
0-100	19.6	0-180	25.6
0-110	24.4	0-200	34.8
0-120	31.3		
Stand'g ¼	15.6	Stand'g km	28.4

ACCELERATION IN TOP
mph	sec	kph	sec
20-40	11.4	40-60	7.2
30-50	10.7	60-80	6.4
40-60	10.7	80-100	6.9
50-70	11.5	100-120	7.7
60-80	12.6	120-140	8.7
70-90	14.1	140-160	9.8
80-100	15.2		

ACCELERATION IN 4TH
mph	sec	kph	sec
20-40	7.2	40-60	4.4
30-50	6.9	60-80	4.2
40-60	7.1	80-100	4.7
50-70	7.4	100-120	4.5
60-80	7.3	120-140	4.4
70-90	7.1	140-160	5.0
80-100	7.8	160-180	6.5
90-110	9.1		
100-120	11.6		

FUEL CONSUMPTION
Overall	18.9 mpg / 14.9 litres/100 km
Govt tests	15.7 mpg (urban) / 36.1 mpg (56 mph) / 29.4 mpg (75 mph)
Fuel grade	97 octane / 4 star rating
Tank capacity	20 galls / 91 litres
Max range*	440 miles / 708 km
Test distance	1,387 miles / 2,232 km

*An estimated fuel consumption computed from the theoretical consumption at a steady speed midway between 30 mph and the car's maximum, less a 5 per cent allowance for acceleration.

NOISE
	dBA	Motor rating*
30 mph	59	7.5
50 mph	63	9.5
70 mph	69	15
Maximum†	74	21

*A rating where 1=30 dBA and 100=96 dBA, and where double the number means double the loudness
†Peak noise level under full-throttle acceleration in 2nd.

SPEEDOMETER (mph)
Speedo 30 40 50 60 70 80 90 100 110 120
True mph 29.5 39 49 59 68 78 87 97 106 115
Distance recorder: 0.2/fast per cent —

WEIGHT
	cwt	kg
Unladen weight*	32.5	1,651
Weight as tested	36.2	1,839

*with fuel for approx 50 miles

Performance tests carried out by Motor's staff at the Motor Industry Research Association proving ground, Lindley.

Test Data: World Copyright reserved. No reproduction in whole or part without written permission.

GENERAL SPECIFICATION

ENGINE
Cylinders	Six, in-line
Capacity	3,590cc (218.9 cu in)
Bore/stroke	91/92mm (3.58/3.62in)
Cooling	Water
Block	Aluminium alloy
Head	Aluminium alloy
Valves	Dohc (4 valves per cylinder)
Cam drive	Duplex chain
Compression	9.6:1
Fuel system	Lucas digital fuel injection
Ignition	Lucas contactless electronic system
Bearings	7 main
Max power	225 bhp (DIN) at 5,300 rpm
Max torque	240 lb ft (DIN) at 4,000 rpm

TRANSMISSION
Type	Getrag 5-speed manual
Clutch dia	9.5 in
Actuation	Hydraulic
Internal ratios and mph/1,000 rpm	
Top	0.76:1/28.9
4th	1.00:1/22.0
3rd	1.39:1/15.8
2nd	2.06:1/10.7
1st	3.57:1/6.2
Rev	3.46:1
Final drive	3.54:1

BODY/CHASSIS
Construction	Unitary, all steel
Protection	Multi-stage phosphate spray and dip; anti-corrosive electroprimer, "adhesion promoter", thermoplastic acrylic paint; oil sanding, oven treatment; wax injection into box sections; anti-chip mill coating; full underbody sealant

SUSPENSION
Front	Independent by semi-trailing wishbones and coil springs, with anti-drive geometry; anti-roll bar
Rear	Independent by lower wishbones with fixed length driveshafts acting as upper links; twin concentric spring and damper units for each wheel

STEERING
Type	Rack and pinion
Assistance	Yes

BRAKES
Front	Ventilated discs, 11.2in dia
Rear	Discs, 10.4in dia
Park	On rear
Servo	Yes
Circuit	Dual, split front/rear
Rear valve	Yes
Adjustment	Automatic

WHEELS/TYRES
Type	Alloy, 6JK × 15
Tyres	215/70VR15
Pressures	32/30 psi F/R (normal) / 32/32 psi F/R (full load/high speed)

ELECTRICAL
Battery	12V 68Ah
Earth	Negative
Generator	Alternator, 75 Amp
Fuses	22
Headlights type	Halogen H4
dip	110 W total
main	120 W total

Make: Jaguar. **Model:** XJ-S 3.6
Maker: Jaguar Cars Ltd, Browns Lane, Allesley, Coventry CV5 9DR.
Price: £15,450.00 plus £1,287.50 Car Tax and £2,510.62 VAT equals £19,248.12 total.

Other possible rivals include the Porsche 944 Lux (£15,309), TVR's Tasmin 350i (£16,400) and Jaguar's own XJ-S V12 auto (£21,752)

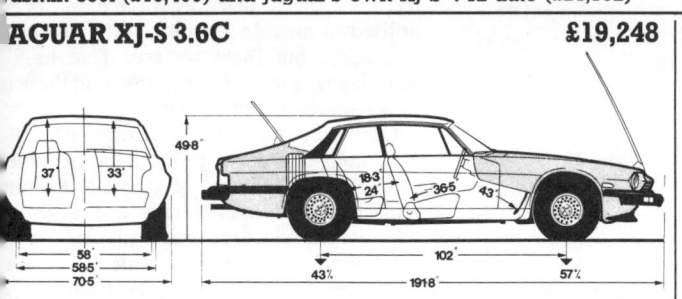

JAGUAR XJ-S 3.6C — £19,248

Power, bhp/rpm	225/5,300
Torque, lb ft/rpm	240/4,000
Tyres	215/70 VR 15
Weight, cwt	32.5
Max speed, mph	136.8
0-60 mph, sec	7.2
30-50 mph in 4th, sec	6.9
Overall mpg	18.9
Touring mpg	—
Fuel grade, stars	4
Boot capacity, cu ft	10.9
Test Date	March 3, 1984

Jaguar's new smaller-engined six-cylinder XJ-S excels in most areas but is badly disappointing in the most important one of all. Its 3.6-litre AJ6 engine, while a lusty performer, is neither as smooth nor as economical as it should be to be competitive in this class. Otherwise it's good news: a positive gearchange, a beautifully balanced chassis with good grip and a superb ride, fine finish and appointments. Still a desirable, if indulgent 2+2, but could be so much better.

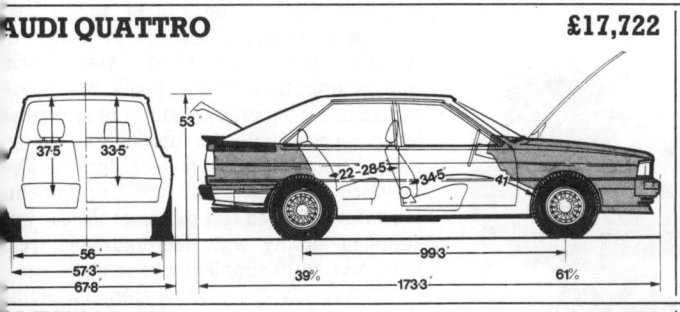

AUDI QUATTRO — £17,722

Power, bhp/rpm	200/5,500
Torque, lb ft/rpm	210/3,500
Tyres	205/60 VR 15
Weight, cwt	24.8
Max speed, mph	135e
0-60 mph, sec	6.5
30-50 mph in 4th, sec	8.2
Overall mpg	19.9
Touring mpg	—
Fuel grade, stars	4
Boot capacity, cu ft	7.6
Test Date	March 21, 1981

Now that it's looking as if its unique format — ultra-high performance and four-wheel drive — *has* started a new trend, the Quattro is more than ever a milestone in car design. It combines phenomenal roadholding and traction with performance, refinement, economy, comfort and accommodation in a way that has no equal, against which its weaknesses — poor ratios and slow shift, unprogressive heating, sparse instruments — are minor failings. Soon to be available with anti-lock braking.

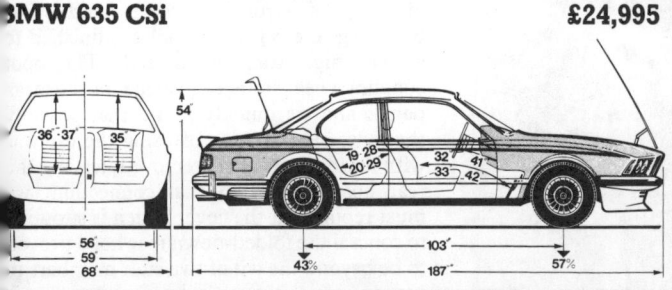

BMW 635 CSi — £24,995

Power, bhp/rpm	218/5,200
Torque, lb ft/rpm	229/4,000
Tyres	220/55 VR 390
Weight, cwt	28.0
Max speed, mph	137.1
0-60 mph, sec	6.9
30-50 mph in 4th, sec	7.2
Overall mpg	22.5
Touring mpg	—
Fuel grade, stars	4
Boot capacity, cu ft	12.5
Test Date	July 10, 1982

BMW's sleek UK flagship, the 635CSi, is rapid and refined in its latest form, yet sets new standards of economy in the supercoupé class. Handling and dry road grip are excellent, too, though in the wet traction is poor and the tail needs watching. Comfortable ride, superb gearchange, anti-lock brakes are further plus points, but rear accommodation, as in Jag, is cramped. Tested as a manual but also available with dual-mode, four-speed automatic.

FERRARI MONDIAL QUATTROVALVOLE — £30,710

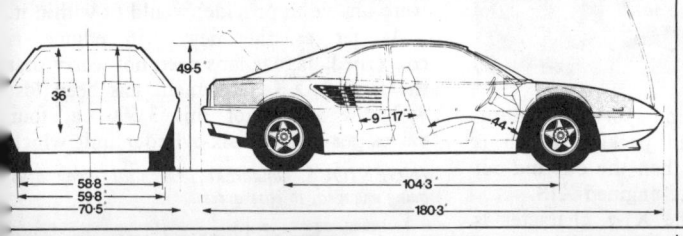

Power, bhp/rpm	240/7,000
Torque, lb ft/rpm	192/5,000
Tyres	240/55 VR 390
Weight, cwt	28.5
Max speed, mph	146.1
0-60 mph, sec	6.4
30-50 mph in 4th, sec	5.6
Overall mpg	18.6
Touring mpg	—
Fuel grade, stars	4
Boot capacity, cu ft	not measured
Test Date	October 30, 1982

Ferrari's 400i takes on Jaguar's big-engined XJ-S in the V12 supercar stakes. In the manual-only Mondial, *Quattrovalvole* power delivers supercar performance with respectable economy, while other virtues include safe, balanced handling, superb brakes, a comfortable ride and driving position, smooth engine and satisfying gearchange. Low gearing leads to highish noise levels, however, and the heating and ventilation leave something to be desired.

MERCEDES-BENZ 380 SEC — £28,560

Power, bhp/rpm	218/5,500
Torque, lb ft/rpm	225/4,000
Tyres	205/70 VR 14
Weight, cwt	31.8
Max speed, mph	132.6
0-60 mph, sec	6.4
30-50 mph in kickdown, sec	3.1
Overall mpg	15.2
Touring mpg	—
Fuel grade, stars	4
Boot capacity, cu ft	not measured
Test Date (SEL)	September 27, 1980

In typical Mercedes fashion, the 380 SEC is superbly engineered and impeccably finished. Good performance and handling are matched by excellent refinement, a very comfortable ride, and full four-seater accommodation. Only available with a four-speed automatic transmission which has exceptionally high overall gearing but is still responsive and smooth. Surprisingly economical for its size and performance, thanks principally to the Mercedes-Benz "Energy Concept"

PORSCHE 928 S — £30,679

Power, bhp/rpm	310/5,900
Torque, lb ft/rpm	295/4,100
Tyres	225/50 VR 16
Weight, cwt	30.3
Max speed, mph	149.1
0-60 mph, sec	6.5
30-50 mph in kickdown, sec	2.2
Overall mpg	17.1
Touring mpg	—
Fuel grade, stars	4
Boot capacity, cu ft	7.3
Test Date	December 3, 1983

In its latest form Porsche's splendid 928S is stunningly quick and respectably economical with its new four-speed automatic transmission. Handling and road-holding superb, and potent brakes now have anti-lock system as standard. Very spacious and comfortable cabin for two (but a cramped 2 + 2), excellent driving position and instrumentation, but ride and refinement not in the Jaguar class. Manual option would be more in keeping with its sporting character.

Jaguar XJ-SC 3.6

The Big Cat goes convertible

IT was the summer of 1981 which heralded the arrival of the new broom at Jaguar, John Egan blowing the cobwebs off the range with the introduction of the HE series of V12-engined cars, developments which finally enabled the basically splendid Coventry products to gain the market acceptance they truly deserved. This was achieved largely through attention to painstaking detail and quality control, aspects of the production process which had been sadly neglected in the late 1970s. The recent Jaguar sales success story needs no further embellishing within the pages of MOTOR SPORT, for we eulogised over the XJ-S HE qualities three years ago. Now it has been our pleasure to sample what must be one of the most attractive machines in the range, the XJ-SC 3.6 — basically an XJ-S cabriolet fitted with the latest AJ6 six-cylinder engine and mated to a delightful five-speed Getrag manual box. The result is a magnificent blend of almost regal, boulevard splendour and sports car agility: arguably, an all-round package which is even more appealing than the out-and-out brute power of the V12-engined XJS.

In broad terms, the XJ-S character is unchanged, but the cabriolet model is being built in limited quantities to customer order only, marking the return of open-air Jaguar motoring for the first time in ten years. The last convertible Jaguar was the V12 convertible E-type, something of an unpredictable handful in anything less than bone dry road conditions, so the XJ-SC is light years ahead in terms of refinement.

The specialist body building and coach trimming skills of Park Sheet Metal Ltd. in Coventry and Aston Martin Tickford's new Bedworth, Warwickshire, factory have been drawn together to produce a taut, rattle-free end product. Accommodating only two passengers, the XJ-SC still outwardly looks like a two-plus-two notwithstanding the fact that the passenger compartment appears a little smaller than the fixed-head version. Behind the comfortable, leather-faced front seats are twin lockable storage boxes above which is a luggage platform fitted with a forward retaining rail. Frankly, I tend to feel it's a shame that this space hasn't been utilised to provide a couple of "occasional" rear seats, but there we are. That hasn't been Jaguar's intention and that's all there is to it — a shame, in my view.

Obviously, the major modification to the XJ-S body in order to produce the Cabriolet model is the removal of the roof and its replacement with cant rails and a centre bar which incorporates tubular steel strengthening. In addition, added structural reinforcement is provided by means of a stiffened transmission tunnel and a rear cruciform member beneath the car. All XJ-SCs come equipped with twin interlocking targa roof panels which are locked into position by a pair of neat circular levers recessed into the interior roof lining and a removable rear "half-hardtop" is also included in the standard equipment. However, for road test purposes the car was supplied simply with the fold-down rear hood section to supplement the targa top.

Impressively quiet, both from the point of view of structural rattles and wind buffeting, the XJ-SC cabriolet is finished to a very high standard indeed. The boot contains a neat storage envelope for the targa panels and we quickly found that possibly the most pleasant way of using the car was with those panels alone removed, leaving the rear hood raised. In that connection, we must report that the cover which is supplied to conceal the folded-down rear hood proved to be beyond the wit of two staff members to secure: it was, in our view, tailored to far too stringent dimensions and nothing less than a perfectly folded hood (which we, clearly, were unable to provide) would fit within it.

As far as the new AJ6 engine is concerned, Jaguar fans need not worry that the famous XK refinement has been lost with the advent of this 3,590 cc, four valves-per-cylinder six-cylinder unit which

INGENIOUS handbrake lever falls away after being engaged, to ease access.

installed beneath the bonnet canted over an angle of 15-degrees. When one opens the front-hinged bonnet, two things strike you. Firstly, how compact the AJ6 unit looks when compared with the 5.3-litre V12 which we've become used to admiring in the XJ-S engine bay, and secondly, just how long are the inlet tracts. Their 17 in. length provides a worthwhile ram effect and contributes to the impressive torque characteristics (240 lb/ft at 4,000 rpm) of this twin overhead camshaft, 91 × 92 mm unit. Running with a compression ratio of 9.6:1, the AJ6 engine develops a healthy 225 b.h.p. at 5,300 rpm which compares very favourably with the 295 bhp at 5,500 r.p.m. produced by the big 5.3-litre V12 unit.

Ignoring the discreet badges on the boot-lid, and the fact that there is only a single fuel filler cap on the nearside rear wing, there is little to betray the new engine's presence beneath the bonnet when one slips in behind the leather-rimmed steering wheel. The matching "white on black" circular speedometer and rev. counter are unchanged from the regular XJ-S and, indeed, there seems no really good reason to call for any changes in this area. Even when one fires up the six-cylinder engine, there is no perceptible torsional vibration of any sort to indicate that this is not an XK unit — in fact, at tick-over speeds, it would be all too easy to imagine you are dealing with a V12.

The notchy, precise movement of the Getrag five-speed gearbox positively invites energetic handling and only after the engine passes about 5,000 r.p.m. does the silky smooth throb give way to a rasping, slightly urgent growl to betray the fact that this is only the six-cylinder engine. With a kerb weight of 3,660 lb., the XJ-SC is just over 100 lb. lighter than its V12-engined stablemate and actually manages to stay with the 5.3-litre engined car on acceleration

UNCHANGED below the waist, the XJ-SC retains the Coupé's elegance.

from 0-60 m.p.h., before tailing away as the V12 really gets into its stride. Fuel economy and ease of high speed cruising are aided by an extremely high top gear providing 28.4 m.p.h. per 1,000 r.p.m. in fifth gear which means that the unit is turning at just over 2,400 r.p.m. at the 70 m.p.h. legal limit.

During the course of our spell with the car the fuel consumption worked out at a very reasonable 20.8 m.p.g. over a programme which included some very high speed motoring mixed with some "pottering" through the Suffolk country lanes, using only third and fourth whilst on the move. The Getrag gearbox never demonstrated any reluctance or stiffness whatsoever and, in our view, a manual transmission undoubtedly enhances the XJ-S range's overall appeal.

The servo assisted four wheel disc brake set-up performed well thoughout the test,

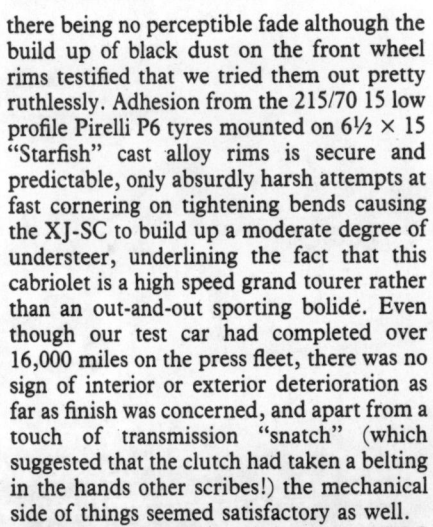

NEW AJ6 ENGINE makes its first appearance in XJ-S but will soon power XJ40 saloon.

there being no perceptible fade although the build up of black dust on the front wheel rims testified that we tried them out pretty ruthlessly. Adhesion from the 215/70 15 low profile Pirelli P6 tyres mounted on 6½ × 15 "Starfish" cast alloy rims is secure and predictable, only absurdly harsh attempts at fast cornering on tightening bends causing the XJ-SC to build up a moderate degree of understeer, underlining the fact that this cabriolet is a high speed grand tourer rather than an out-and-out sporting bolide. Even though our test car had completed over 16,000 miles on the press fleet, there was no sign of interior or exterior deterioration as far as finish was concerned, and apart from a touch of transmission "snatch" (which suggested that the clutch had taken a belting in the hands other scribes!) the mechanical side of things seemed satisfactory as well.

The long, low lines of the Jaguar XJ-S range are now to everybody's taste — in fact, few luxury coupes have prompted such debate in this respect over the past decade. I have never felt the coupé to be as awkward as some critics who, I suspect, are confusing "individuality" with "awkward styling". However, I feel that the cabriolet's lines *are* more attractive than the basic coupé's and will be very interested to see how many of these beautifully finished, bespoke motor cars the Jaguar company will be required to build for its discerning customers. All too often, we feel, it is becoming fashionable to look beyond our shores for true (or apparent) excellence in high performance car design. It is refreshing to know that machines such as the XJ-SC are being made closer to home, in Coventry, and that the Jaguar Company's star is once again in the commercial ascendancy. On the strength of this machine, it certainly deserves to be.

A.H.

GO CAT GO!

Tyrrell Grand Prix driver Martin Brundle races a factory Jaguar XJS and uses a TWR-modified XJS on the street. So how does track success influence roadgoing Big Cats?

SHOULD RACE results influence your next car purchase?

Group A racing cars — which resemble reasonably closely the cars which you or I could buy in any new car showroom — were introduced for the 1982 European Touring Car Championship.

"Homologation" is the key word — the word which has been well worn over the years. If a manufacturer wants to race a certain specification on his car then that same specification must be available through the showroom, and a minimum number must have been produced and available. When these criteria are met that part — a bootlid spoiler which improves aerodynamics and, therefore, straightline speed and cornering, for example — is "homologated" and can be raced.

The regulations are by no means as simple as that, however. Because of other factors, such as safety measures and ensuring that racing is both close, fast and fair to all involved, other sections (some more effective than others) are written into the regulations.

So when you read that a particular car has won the most recent 500 kilometre endurance race, how similar is that winning car to the car you can buy for the road?

Over the years I have driven many cars, both on the road and on the track, and currently I drive a Jaguar XJS 3.6 Coupé 5-speed on the road, and the TWR works Jaguar XJS 5.3 5-speed in the ETC when my Formula One Tyrrell commitments allow.

So what are the differences between the two "Pussycats"?

At a glance, my road car looks more racy than the racer because I have a TWR body pack incorporating Speedline wheels, front and rear spoilers, and skirts which are not homologated for the racer. I'm not sure what to think about the latest craze of hanging extra body panels on exotic cars (or even bread-and-butter models); some of them look like angular ice-cream wagons, but I think this TWR pack is very stylish. I can almost hear some of you groaning "Why spoil a nice car with plastic spoilers?" That's the wonderful thing about cars — we can express our individual tastes and I haven't seen much that I would swap my XJS for at the moment.

The race car is simply a standard bodyshell off the production line, which is built up using Jaguar parts. Under the regulations it is permissible to seam weld the bodyshell; that is reinforcing the manufacturer's welding of the chassis and body panels. This is necessary because of the extra grip and higher speeds of a racing car, and the constant driving on the limit. Take, for example, the Spa/Francorchamps 24 Hours touring car race. We will put far more stress on the XJS in 24 hours of circuit driving than you could in the lifetime of a roadgoing car, driving like a lunatic. You'll have to take my word for that because, as I'll explain later, the race cars have incredibly stiff suspension and the greatly improved grip of slick racing tyres, and until you have experienced such a phenomenon you cannot begin to comprehend the stress to which the whole car is subjected.

From the outside, the only other non-standard feature (apart from the stickers!) are the oil coolers situated under the front bumper. Extra cooling is allowed, but the bodywork must not be altered in any way to improve cooling. Controlling engine, gearbox and axle oil temperatures are a problem on a 175 mph 1½-ton projectile such as this, which is constantly revving and cornering on the limit for an extended period. The regulations allow extra coolers and pumps for this purpose.

Moving inside the car, here are the most noticeable differences. First my roadgoing XJS has that fabulous smell of quality leather. You sink into the seat and the walnut facia presents an array of gauges, stereo, heating, air-conditioning, trip computer and controls. You want to drive for hours!

When you jump into the race car for your stint halfway through a long-distance race the story is slightly different. The smell is now of hot machinery, and the Recaro race seat (rock hard, but incredibly comfortable) is hot and wet with sweat — your co-driver's sweat! Everything is matt black-painted and, as a fire precaution, there is no trim whatsoever.

In front of you is a metal facia full of essential temperature and pressure gauges. To your left is a bank of flick switches, growing out of the gearbox tunnel, with which you control the electrics and pumps. Also, you're surrounded by yards of roll-over protection bars. One common factor is simple; you want to drive it for hours . . .

So we've compared the road burner with the racer while stationary. How do they compare when rolling?

It is fair to say that the basic handling and feel characteristics of the two are similar — they both feel like an XJS — but there the similarity in driving terms ends.

The road car purrs almost silently, while the race car positively roars with its massive straight-through non-silenced exhaust system. The tuned camshafts make the engine positively rumble at low revs, but up to 7,000 rpm the smoothness of a 12-cylinder still comes through. My road car, being the new 3.6, is of course the new AJ6 six-cylinder unit which, in my example, is super-smooth, but not surprisingly it cannot quite match the Jaguar "twelve" that I have driven.

By far the largest difference is in the suspension or, more exactly, the shock absorber and spring stiffness. If you were to take the racer down your favourite country lane the chances are it would probably leap clean off the road and pounce on the nearest hedge. Strange, you may think: racing cars with slick tyres are supposed to handle well.

This is very much the case on a smoother, much wider, racing circuit when the "bald" racing slick tyres come up to temperature. The grip is phenomenal and the suspension needs to be much stiffer, otherwise the car would roll and wallow around uncontrollably. In contrast, if you made the road car as stiff as the racer it would bounce from bump to manhole cover, every slight undulation feeling like a "sleeping policeman", and your passengers would be less than impressed.

Another major shock comes from the steering. The race car, having had the power assistance disconnected from a normally well-balanced and light wheel, the steering is positively alive due to the very wide wheels and tyres and the stiff suspension exaggerating every bump on the track. To have power steering would simply leave the driver without enough "feel" of where the front wheels are pointing while sliding sideways, which happens at almost every corner during racing. Racing in the wet would be even more difficult. The effect of no power steering on the race car, however, produces the best bodybuilding course I've ever experienced.

This is especially so when you consider that the brakes on the race car have no servo assistance either. Remember, this car weighs one-and-a-half tons, will achieve 175 mph, and that braking system must cope with up to 24 hours of racing. The brake discs are massive and heavily ventilated, and the brake pads are of a very hard material which only works when hot. When you first go on to the track with cold brakes, the pedal is rock solid and it feels for all the world as if the brake pads are of solid oak and the brake discs are oiled. It takes a bit of confidence, but after a couple of laps the pads warm up and the brake pedal will begin to be of use to you. When the pad and disc is up to temperature the "Pussycat" will really dig its claws in. If, as does happen, you exceed the maximum temperatures, the brake pedal simply becomes something to brace yourself against, as it has little other effect. On the road you could not realistically emulate the enormous amount of heat generated, and the road car's brakes are less beefy but more than adequate.

To sum up, the two cars are totally different beasts at speed, so you could say that a race-winning car should not influence your next purchase. However, I think it would be wrong to say that. The regulations are sufficiently realistic to ensure that the strongest, fastest and most reliable cars will win. A good car will show its true strength, although it is fair to say that some models are not best suited to make the most of the technicalities.

A big "plus" is the enormous feedback of information to the manufacturers to help improve the road cars we can buy. I have experienced this at first hand, and it is very pleasing to see the manufacturers making full use of the opportunities created by saloon car racing.

"... currently I drive a Jaguar XJS 3.6 Coupé 5-speed on the road and the TWR works Jaguar XJS 5.3 5-speed in the ETC . . ." And Martin Brundle thinks his road car (above) looks more racy than the racer

COOL CAT
JAGUAR'S ROOFLESS RESTYLE

THE TURN-AROUND in Jaguar's fortunes has been nothing short of miraculous. From an ailing nationalised company struggling for survival, the big cat has good reason to purr once more.

Profits have been invested and a brand new all-alloy engine now powers the gentleman's high-speed conveyance — the XJS. Last year, just in time for Motor Fair, Jaguar revealed the gorgeous XJS Cabriolet — and what a show stopper it proved to be! It was the first convertible Jaguar since the demise of the E-type in 1974, and it was long overdue.

But it was that new 3.6-litre engine under the long power-bulging bonnet that let the knowledgeable see that a new era was dawning at Jaguar. Sawing off a roof in very limited numbers is one thing, but developing an engine is something else again. Indeed it can cost as much to develop an engine as it costs to develop the rest of a new car, which explains why so many manufacturers use last year's engine in next year's cars.

The new lightweight engine with pots of power will, hopefully, loom large in Jaguar's recovery. Clearly its potential for development has ▶

Jaguar XJS 3.6 CABRIOLET
£19,467 on the road

'Very smooth, almost silent, and quick enough for most people'

not yet been tapped, and we could not help wondering if a turbocharged, or better still supercharged, version could be just around the corner to end the reign of the V12 as well as the XK Six — all that Jaguar came up with in the years since the last war.

The XK Six was designed for saloons, although it made its first appearance in a sports car, and so the engine had to be quiet, smooth and behave well in traffic — as normal to drive as any mass-produced lump that rolls off the line in thousands.

The original engine also had to be easy to build so that it could be produced economically in high volume. Jaguar's founder, Sir William Lyons, added the proviso that the engine had to look good so that owners would be proud to lift their bonnets for the chap next door. In those days the envious neighbour could ogle what looked like a Grand Prix engine!

But we'll accept that many readers will be excited simply because the new car is a wind-in-the-hair machine.

How it goes

The new engine has been criticised by some magazines for its harshness, but one editor still seems more than happy to drive one around. We would be, too. On the road the XJSC may not have the smoothness that the best of BMW now offers, but it is very, very smooth, almost silent, and quick enough for most people.

At 120mph you can carp to your passenger about the wind rattling the material under the rear window without having to shout. And even at those speeds the air conditioning makes almost as much noise as the engine, and nothing really interferes with the superb Clarion stereo. Is it the only Japanese kit on this best of British car?

The AJ6 — as the new engine is known — has twin overhead camshafts operating four valves per cylinder in a head based on current Grand Prix practice, developing 225bhp at 5300rpm. The current six-cylinder develops 162bhp in its 3.4-litre version and 205bhp as a 4.2-litre. The new engine looks impressively sturdy, with a massively ribbed, deep skirted aluminium block that extends well below crankshaft level.

The traditional Jaguar practise of running the crankshaft in seven main bearings keeps things extra-smooth, and as these bearings are the same dimensions as those on the V12 there is lots of development potential.

Changes for the Cabriolet include the removal of the roof and rear quarter panels, and the fitment of new top rear wings, saddle (the panel forward of the boot lid), header rail, cant rails, B posts and centre bar. Associated changes involve new rear-quarter windows, a new luggage area, and the relocation of the fuel filler.

However, all that stiffening works a treat. Beneath the car, too, there is a new transmission tunnel stiffening panel and a rear cruciform (a cross-shaped member that provides two-dimensional stiffening).

Our track testing of the XJSC turned into a protracted affair. On our first visit the Cabriolet simply couldn't deliver the promised goodies, and the car went back with a request for a little tender loving care. Our best 0-60mph time of 8.3sec on that occasion was nothing special, and our top speed of 129½mph (and that with the air conditioning switched off)

COOL CAT

compared badly with Jaguar's own figures of 7.6sec and 142mph.

The weeks slipped by and we saw our Jaguar again at Millbrook with factory engineers looking for speed. 'Tell us when you reach 142mph and we'll put in a few quick laps ourselves,' we said. They later admitted to 131mph, and that with the mirrors folded back. Clearly the car had problems.

But everything comes to he who waits, and the Jaguar returned to the fold looking as lithesome as ever. Our testing, however, was to be done in 20mph gusts of wind — the forecast looked even bleaker.

However, with the Jaguar's ability to sit on the road like a fat leech, those maximum speed runs wouldn't call for nerves of steel. Out of interest we measured the speeds with the air conditioning on, then off, and — we couldn't resist it — with mirrors folded back. Those speeds for the flying lap were 129.3mph, 129.6mph and 133mph. On paper the car was no faster, but remember the wind. Note, too, that wing mirrors can waste more energy than air conditioning pumps! Our best ¼-mile with the air conditioning off and no other cheats worked out at 133.5mph. Our car was slow!

Our trip to the steering pad showed that an XJSC will understeer when driven in a clockwise direction and oversteer in an anticlockwise one. Our office expert muttered darkly about the gyroscopic action of the prop shaft, but we say no more. When the tail came out it proved very difficult to 'catch', with little information being transmitted through the power steering. If that sounds like rather nervous handling let's reassure you that this happens when you go beyond the car's tremendous adhesion limits — a feat that's nigh-on impossible on public roads.

High winds on the mile straights didn't help either, but there was no doubt the car was now faster. Even driving into the teeth of it we managed

It was a long time coming, but, after 10 years, Jaguar once again has a dashing drophead in its stable. With that powerful air conditioning and the targa top (opposite), it's easy to keep your cool in the XJSC no matter how high the mercury

0-60mph in 7.6sec to meet Jaguar's claim, and we say we did better than that. However, at the top end the Jaguar would struggle to meet the Supercar standard of 0-120mph in half a minute.

On the handling course, which is an exact copy of real corners on real roads, the Jaguar showed how safe it could be in a hurry. It's a realistic test and we were very impressed. It isn't in the Porsche 928S class when it comes to going quickly, but the things a Jaguar does well it does very well indeed.

How comfortable

The ride quality of an XJS is second to none, and a cross-country cruise at that level of luxury is certainly the lasting impression. If you can manage with a two seater and you are well heeled then you could do a lot worse.

Open the door and enjoy the distinctive smell of Connolly leather — a total of six hides are used in the interior — and the polished elm burr veneer of the facia/door inserts is finished to a level that only the British coachbuilders seem able to achieve. Whereas the XJS HE and 3.6 coupés are fully completed at the end of the Browns Lane track, the Cabriolet travels to Bedworth, Warwickshire, where Aston Martin Tickford fit the targa panels and the hood. The completed car then comes back to Browns Lane for final checks and valeting.

The Cabriolet is equipped with a targa roof which fits in front of the centre bar — the driver then has the choice of fitting either a 'half-hardtop' or folding hood to the rear of the centre bar. The half hard-top (which our car didn't have) is made from glass-reinforced plastic and incorporates a heated rear window, reducing wind noise and improving rigidity. The targa parts consists of two lightweight trimmed panels which weigh in at just 12 lb together. They proved easy to remove and less easy to replace. Clearly these things demand a knack.

On our test car it proved impossible to fit the tonneau cover over the folded rear hood. Of course, with the best air conditioning in the business you could always leave the roof alone.

What better system could you have than simply to dial a temperature and point another arrow to auto? No complaints there.

The front seats have yards of movement to accommodate the tallest driver, and a nice touch is that the steering wheel, too, can be moved fore and aft. However, moving the wheel forward a little to provide extra inches between wheelrim and thighs has the unfortunate effect of leaving the fingertip stalks quite a stretch away. Maybe a smaller steering wheel would improve matters …

From the driver's seat it can prove difficult to regard this Jaguar as a sports car, despite its turn of speed. At 15½ft it is just too big to be really nimble. That understeering behaviour we encountered on the test

SUPERTEST

The attractive Cabriolet treatment has smartened the rear end

No update here — facia is dated and seats are slippery

track could catch you out on the road if you are not the sort of driver to back off the loud pedal half way through a bend. However, if you do ease back the pace the nose tucks in safely and you are once more reassured. We were surprised at the way the meaty tyres scrubbed off speed through tight bends.

The all-disc brakes are confidence-inspiring and well up to the big job they are asked to do. We were also impressed by the lightness of the clutch — especially after climbing out of a Porsche 911.

The gearchange itself has been criticised for being heavy and notchy, but apart from an occasional reluctance to go into reverse without a good shove the Getrag box is arguably excellent. Fourth is a direct 1:1 ratio, giving a leisurely 3240rpm at 70mph, and fifth is very long-legged indeed with 28.4mph per 1000rpm. This means that at 70mph the engine turns at only 2464rpm — no wonder the silence is there.

We will criticise the visibility, however, and even with the roof off the folded hood can encroach into the rear view mirror's outlook. The slight roll on a fast left hander is enough to have you craning your neck to see round the corner.

Instrumentation looks a little dated now, despite those minor gauges reminding one of Dr McCoy's diagnostic machine on the *Starship Enterprise*. But everything — including the 160mph speedometer and the rev counter red-lined just before 6000rpm — is easy to see. The light switch calls for a little familiarity — hardly an owner's problem.

Push-push switches for the map-reading and interior lights can confuse, and it needs saying that the interior lamps on our car could simply drop out of their mountings whenever we closed the door. Two similar switches on the opposite side of the computer, control the hazard flashers and rear screen's heater (which, of course, this car doesn't have).

We approve of this computer because for once you don't need an MSc to operate it. The unit gives useful fuel consumption information as well as calculating average speed and distance covered. Its instant fuel consumption information is less use, changing every few seconds from 0.00mpg (when you stop), sailing into the 90+mpg region on a trailing throttle.

Useful to know that you will get 15mpg on a short trip to the office and 19mpg on a drive to the test track. We did see a best of 24mpg (gentle motorway pottering with the roof off).

Luggage space is good for a two-seater with a deep boot — despite the presence of an upright, hefty spare wheel — and plenty of stowage space inside the car. Instead of a rear seat there are two smaller luggage compartments topped off by a warning not to ride in the back for safety's sake.

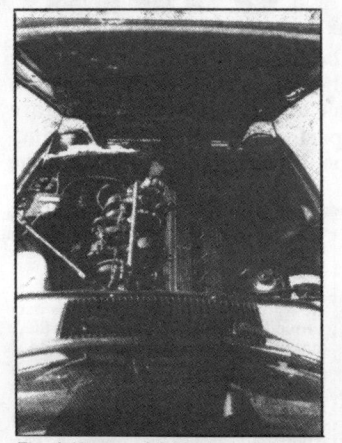

The AJ6 super 'six' makes all the right noises — at least to our ears

PERFORMANCE
Maximum speed 129.6mph (best ¼ mile 133.5mph), 127mph in 4th, 89mph in 3rd, 60mph in 2nd, 34mph in 1st

Acceleration from rest

mph	time (seconds)	speedo mph
30	2.6	31
40	4.3	42
50	5.4	52
60	7.3	62
70	9.9	74
80	12.7	85
90	15.9	96
100	20.0	108
110	24.7	118
120	29.8	129

Acceleration in gear

mph	4th (sec)	5th (sec)
20-40	7.8	11.8
30-50	7.3	11.5
40-60	7.5	11.3
50-70	7.5	11.8
60-80	7.4	12.8
70-90	7.3	13.7
80-100	7.7	14.6
90-110	8.8	15.0
100-120	11.7	

Standing ¼-mile 16.1sec
Terminal speed 92mph

ENGINE
Type and size front-mounted straight six, water cooled; 91mm bore x 92mm stroke = 3590cc; aluminium alloy head and block
Compression ratio 9.6:1
Valve gear twin overhead camshafts, duplex chain driven
Fuel system Lucas digital electronic fuel injection system with sensors for engine speed/temperature, manifold pressure, air temperature and throttle position. Auto cold start enrichment with cut-off on over-run
Max power (DIN) 225 bhp at 5300rpm
Max torque (DIN) 240 lb ft at 4000rpm

TRANSMISSION
Gearbox five-speed manual. Ratios: first 3.57:1, second 2.06:1, third 1.39:1, fourth 1.00:1 fifth 0.76:1, reverse 3.46:1, final drive 3.54:1 — Salisbury hypoid with 4-pin Powr-lok limited-slip differential
Mph/1000rpm 4th/5th gear 22.0mph/28.9mph

CHASSIS
Suspension front: independent, semi-trailing wishbones and coil springs; anti-dive geometry. Girling Monotube dampers, anti-roll bar. Rear: independent, lower transverse wishbones with drive shafts acting as upper links. Twin concentric coil spring and damper units for each wheel
Steering rack and pinion with power assistance. Column has fore and aft adjustment; 2.7 turns lock to lock, turning circle between kerbs 39.3ft
Wheels 6JK x 15 cast alloy, Dunlop D7 radial ply (215/70 VR 15) low-profile on test car
Brakes servo assisted four-wheel disc brakes, front ventilated. Safety split front and rear hydraulic circuits incorporating fluid loss sensor warning. Fly-off handbrake

VERDICT
Those expecting a gut-wrenching Supercar performance machine are going to be disappointed by the XJS Cabriolet. A gentleman's high-speed carriage is the description we heard most often from drivers who appreciate the finer things of life.

The engine that some testers criticised as harsh (not DRIVE testers) is very refined and makes just the right noises. More importantly, it is ripe for development: twin turbochargers perhaps?

Add to that a roof conversion which improves the looks of the XJS and you have a car which can't be matched for the money.

The Jaguar XJS in Cabriolet form with the new AJ6 engine remains a superb way to travel. It's the difference between a commuter special and the opulence of the Orient Express.

ROAD TEST
By MIKE McCARTHY

The V12 XJSC-HE has to be considered as the most desirable of Jaguar's XJS range, providing effortless, wind-in-the-hair motoring.

Coventry's future classic

The last open-topped Jaguar to be powered by a V12 engine, the Series III E-Type, departed the scene 10 years ago — and instantly became a classic. Once again, however, Jaguar are offering fresh-air fanatics V12 performance with the introduction of the XJSC V12. Will it, too, be considered a classic at the end of the next decade?

You could almost call the V12 Cabriolet 'the inevitable car': after all, the straight six AJ6 engine appeared in both Coupe and Cabrio, so it seemed only a matter of time before the V12 appeared in both as well. There was a time a few years back when the V12 seemed doomed to extinction: too big, too heavy and too thirsty for current conditions: one of the reasons why you could only get the Cabrio with the AJ6 engine. Now however, it's found its second breath, and if rumours from the Continent are true we should see engines with the same number of cylinders from BMW and perhaps even Mercedes-Benz.

In 1981 sales of the XJS range had dropped to 1199 units: last year it recorded a record 6082 sales worldwide. The expansion of the range via the V12 Cabrio can only help matters, and the Coventry company are looking for 8000 units to be sold in 1985.

Troubled development

The first couple of years of the six-cylinder Cabrio have been far from trouble-free. The engine itself proved somewhat troublesome, while the method of construction, which relied on an outside supplier (Aston Martin Tickford) to fit the targa top, meant that it was only available to special order, and there was a long waiting list. The decision was therefore taken to fit the roof entirely in-house.

Because of the Cabrio's relatively low volume production, it is constructed wrong-way round: a complete coupe body unit is built at Jaguar's Castle Bromwich body plant, then shipped to an outside contractor to have the roof removed. Painting and assembly take place at Jaguar's own factories.

Technically, the Cabrio V12 breaks no new ground. The V12 engine has been around for years, but it is still one of the, if not *the*, best power units in the world. It has been gradually modified over the years, most significantly in 1981 with the introduction of the Mays head which, along with the adoption of Lucas-Bosch Digital Electronic fuel injection, considerably improved consumption with little if any drop in performance. It is an all-alloy unit with a single overhead camshaft per bank, and delivers a healthy 295bhp (DIN) at 5500rpm, and a very lusty 320lb ft (DIN) torque at 3250rpm from its 5345cc. As we've seen from the tracks, though, it is capable of giving considerably more than this, so it must be regarded as lightly stressed.

The rest of the car is pure XJS Coupe, which means a three-speed automatic transmission, Power-Lok limited-slip diff, power-assisted rack and pinion steering, disc brakes all round (ventilated at the front), and big 215/70VR15 tyres on alloy wheels. Suspension is by wishbones and coils at the front and Jaguar's unique system of lower wishbones, with the drive-shafts acting as upper links and with twin coil springs per side, at the rear.

Because of the position of the roll-over bar, necessary to add strength when the top is removed, there are no rear seats in the Cabrio — it is strictly a two-seater.

The all-alloy V12 unit delivers its 295bhp (at 5500rpm) in a lusty, yet smooth manner. Below: The well equipped interior shouts 'trad Brit'. It is wonderful.

124

ROAD TEST

Naturally it comes comprehensively equipped, complete with air conditioning (which at first seems odd in an open car, but which soon proves its usefulness), acres of lovely leather upholstery and trim, and discreet burr elm on facia and doors. The top itself is in three sections, a two-part removable panel over the seats and a fold-down rear window section behind the roll-over bar.

At £27,000 the XJSC V12 is not cheap — but there are not many other cars which can offer 150mph and open-top motoring. There's the Aston Martin Volante at £63,000, the Bentley Continental convertible at a massive £82,000 (and its Rolls-Royce equivalent for another £1500), the charismatic Bristol Beaufighter at £50,000, the Mercedes 500SL which is close in image and price at £27,500, the Porsche Carrera Cabrio at £26,500, and the TVR 390SE at a relatively low £19,700. Anyone in the market for an expensive, fast convertible has quite a wide choice . . .

On-the-road performer

Where the Jaguar scores, though, is in its combination of price and performance. The factory claim a maximum speed of 150mph, and a 0-60mph time of 7.5s. The Porsche Carrera Cabrio will match it on top speed (152mph claimed) and accelerate much quicker to 60mph (6.2s) but it is a totally different sort of animal — refinement takes second place to sheer exhilaration. The Mercedes 500SL comes close to the Jaguar in refinement, matches it in accceleration (7.8 to 60mph) but is down on maximum at 137mph.

Paper figures are one thing though, actual on-the-road performance another. There is no doubt that the V12 power unit is superb in its effortlessness, and smooth but immensely powerful flow of energy. At any speed up to 120mph there are tremendous reserves of power, so a gentle caress of the throttle pedal has you wafting forward impressively — overtaking manoeuvres which, in most other cars would be a white-knuckle operation, are over and done with before you know where you are. It makes high speed look all so easy. Not that it is flawless however: in the test car it took quite a lot of churning to get it started, and there was a distinct shake when idling, neither of which we've come across before with this engine. But, it is economical: in spite of some very quick journeys, plus a fair amount of town work, we recorded a highly creditable 16.95mpg.

The sophisticated suspension set-up helps when recording fast cross-country journey times, with excellent road-holding, but as ever we have to criticise the steering. It is just too light, and too vague. A series of fast motorway curves taken at 130mph in a BMW M635 in the rain were very satisfying: in the Jaguar you were never quite sure what the front wheels were doing. This, in fact, is the biggest single criticism we can make of the car, and Jaguars in general, and where German machines with their ZF set-ups are so superior.

The loss of the roof, and the fitting of the three-part replacement, has also meant a loss of refinement. There were the occasional creaks and rattles from the structure when it was raised, while the opposition has caught up in on road noise suppression — we were surprised at how much bump-thump was transmitted to the interior. Wind noise, too, with the top up increased so that, at over 100mph, it was the dominating sound, drowning out the radio.

Converting from open to closed, and vise versa, is long-winded. The targa top panels are fiddly, though the rear section drops down quickly enough. When fully open, the Cabrio is surprisingly civilised — your hair doesn't get blown around for example — and with the air conditiononing working it is very enjoyable on a blazing hot day.

The interior trim is now looking distinctly old-fashioned, though the fit and finish are impressive. The row of square push buttons, the rotary light switch, and the vertical minor dials all have a distinctly seventies look. We must admit, though, that where the BMWs and Mercs may have better ergonomics and facia layout, we very much prefer the Jaguar's wood trim and leather upholstery — out of date maybe, but infinitely richer in feel.

Summing up, the XJSC V12 is a worthy addition to the Jaguar line up, and most of the criticisms that can be made of it lie in two areas: the (to us) unnecessary complication of the roof structure, and the fact that in many ways it is showing its age. Nevertheless, it is still a highly impressive motor car, and unmatched in the sheer ease and sensuousness of movement. Others, though, are catching up . . . ■

The roof comes (off) in three sections, giving one the choice of several degrees of open-top motoring.

Roof down or up, the XJSC looks far more graceful than the Coupe, giving the car a longer look.

JAGUAR XJSC-HE
£26,995

Specification

Cylinders/capacity	V12, 5345cc
Bore/stroke	90 × 70mm
Valve gear	Single ohc/bank
Fuel system	Lucas-Bosch Digital Electronic
Power/rpm	295bhp (DIN) at 5500rpm
Torque/rpm	320lb ft (DIN) at 3250rpm
Gear ratios	2.50, 1.50, 1.00:1
Final drive	2.88:1
Steering	Power-assisted discs, ventilated at front
Wheels	Light alloy, 6.5J × 15
Tyres	215/70VR15
Suspension (F)	Independent by double wishbones, coil springs, anti-roll bar
Suspension (R)	Independent by lower wishbones, driveshafts as upper arms, radius arms, twin coil springs/side

Dimensions

Length	187.6ins
Wheelbase	102.0ins
Track (F/R)	58.6/58ins
Width	70.6ins
Weight	35.4cwt

Performance
(Manufacturer's figures)

Maximum	150mph
0-60mph	7.5s
Fuel consumption (urban/56/75mph)	15.6/27.1/22.5mpg
Test consumption	16.95mpg

TWR JAGUAR SPORT

Subtle changes, dramatic improvements

BY THOS L. BRYANT
PHOTOS BY JOHN LAMM

To THOROUGHLY APPRECIATE a Jaguar XJ-S, you must drive it in England. Now, I'm not saying the big cat isn't fun here in America, for it is one of my personal favorites on any road, anywhere. But once you've had the opportunity to explore the back roads, narrow lanes and open motorways of the British Isles, you truly appreciate the excellent driving characteristics. The XJ-S may not rate as one of the most beautifully styled coupes in the world, with its almost Chevrolet Camaro-like angular aspect, but it is a design that has grown increasingly graceful with the passing of the years. Styling aside, few who have ever driven an XJ-S would complain of its performance and road-holding ability.

But the British enthusiast who found the XJ-S perfectly acceptable at home may have found himself challenged to keep up on trips to the Continent, particularly along the German *Autobahnen*. After all, that may be the world's ultimate automotive proving ground—miles and miles of high-speed highways without speed limits, filled with the fastest things on wheels. The XJ-S keeps up with little trouble, but it certainly doesn't walk off and leave the competition behind.

Enter Tom Walkinshaw. Born in Scotland in 1948, TW entered the racing world via Formula Ford in 1968. By 1976, Walkinshaw was a factory driver for BMW and was starting his own development engineering business. In just three years, Tom Walkinshaw Racing (TWR) was a known and proven name in European racing.

Working with Jaguar's full cooperation (after all, Walkinshaw has won a number of European racing championships for the company), TWR does just a few things to the marvelous 5.3-liter V-12 engine. Breathing is improved via a high-efficiency stainless steel exhaust system and increased-capacity intake manifold. The V-12's 262-bhp figure goes up by some 10 percent, but if that's not sufficient, TWR also offers a full engine modification program that results in a displacement increase to 6.0 liters and output in the 400-bhp range.

Perhaps more important than the engine tweaks is the TWR-developed close-ratio 5-speed gearbox. This ZF unit has shown its mettle in racing and gives the TWR Jaguar XJ-S electrifying performance. Walkinshaw claims 0–60 mph in 5.8 seconds and a top speed of 164 mph—numbers to give any *Autobahn* runner pause to think about challenging this long-legged Jaguar.

Walkinshaw's crew doesn't stop at the drivetrain; there is chassis work involved in making this superb sporting car even better. The TWR XJ-S gets beefed-up vented disc brakes front and rear, giving the car impressive stopping power. The suspension system also comes in for its share of attention, the Bilstein gas-pressurized shocks developed specially for TWR. The front springs are made a bit stiffer and the front-end geometry is adjusted to compensate for the changes.

It's not TWR's intent to turn the XJ-S into a race car, but rather a high-speed GT that will still offer comfort around town and on the road. The ride is quite good at low speeds; the Jaguar by TWR won't give you a case of teeth chatter over bumps. But when the speed builds and you start running along at 80, 90, even 100 mph or more, the TWR suspension modifications produce a hunkered-down sort of, well, cat-like grace.

To give the TWR Jaguar Sport that extra edge, Walkinshaw went to the wind tunnel to develop the body fairings, front and rear wraparound bumpers and a rear spoiler. There is a claimed 12.7-percent improvement in drag coefficient, which helps account for the 164-mph top speed.

Naturally, when a buyer is paying close to $50,000 for a special car, he may want some dolling up. TWR offers special paint schemes to the customer's order, along with black-out trim for window and headlamp surrounds, windshield wipers and door handles. Inside, Scottish tweed inserts are fitted to the centers of the front and rear seats, color coordinated with the exterior paint. Special carpeting and a leather-wrapped 4-spoke steering wheel are added.

Wheeling the TWR Jaguar XJ-S out onto a narrow road is as easy as driving the family sedan—it's that docile. The 5-speed is smooth and the shifts are easy, and the modified V-12 just purrs along. Hit the open stretch of clear road and nail the throttle—the Jag takes off like a scalded cat, covering distance with incredible speed, with barely a ripple to upset the occupants. Once into 5th gear, the engine settles to a quiet background sound and the TWR is a stable and secure GT car, whether cruising at 70 or 100 mph.

Off the line, there's plenty of power to spin the rear tires—Goodyear Eagle NCT 245/55VR-16s. You can confidently take on just about any other exotic car that crosses your path without fear of being left in the dust. The V-12 revs smoothly and quickly to its redline without hesitation, and the acceleration is enough to push you firmly against the backrest of your seat. If anyone asks, you can safely answer that, yes, racing certainly does improve the breed, particularly in the case of the TWR Jaguar XJ-S.

EUROPEAN SPECIFICATIONS

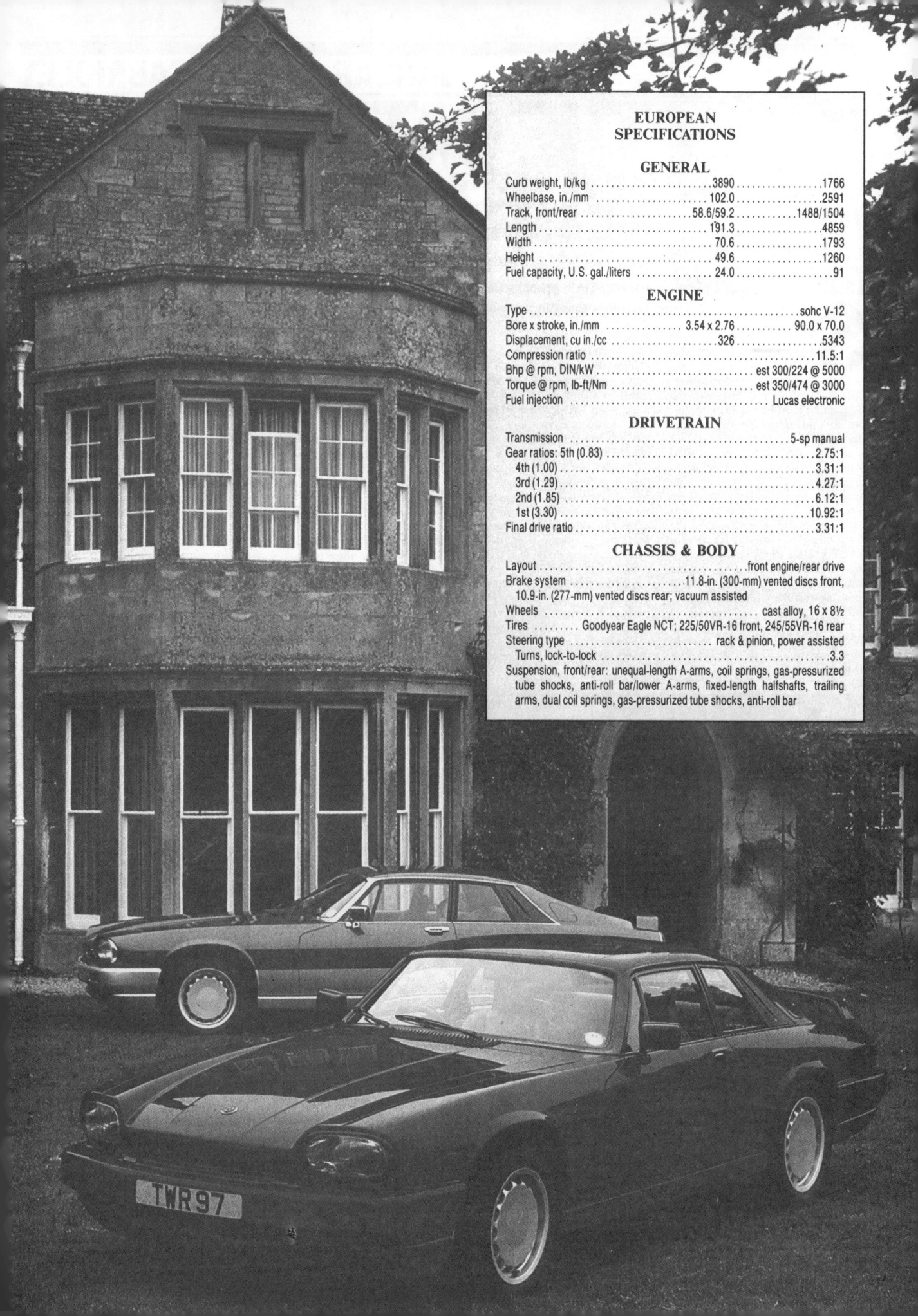

GENERAL

Curb weight, lb/kg	3890	1766
Wheelbase, in./mm	102.0	2591
Track, front/rear	58.6/59.2	1488/1504
Length	191.3	4859
Width	70.6	1793
Height	49.6	1260
Fuel capacity, U.S. gal./liters	24.0	91

ENGINE

Type		sohc V-12
Bore x stroke, in./mm	3.54 x 2.76	90.0 x 70.0
Displacement, cu in./cc	326	5343
Compression ratio		11.5:1
Bhp @ rpm, DIN/kW		est 300/224 @ 5000
Torque @ rpm, lb-ft/Nm		est 350/474 @ 3000
Fuel injection		Lucas electronic

DRIVETRAIN

Transmission		5-sp manual
Gear ratios: 5th (0.83)		2.75:1
4th (1.00)		3.31:1
3rd (1.29)		4.27:1
2nd (1.85)		6.12:1
1st (3.30)		10.92:1
Final drive ratio		3.31:1

CHASSIS & BODY

Layout	front engine/rear drive
Brake system	11.8-in. (300-mm) vented discs front, 10.9-in. (277-mm) vented discs rear; vacuum assisted
Wheels	cast alloy, 16 x 8½
Tires	Goodyear Eagle NCT; 225/50VR-16 front, 245/55VR-16 rear
Steering type	rack & pinion, power assisted
Turns, lock-to-lock	3.3
Suspension, front/rear:	unequal-length A-arms, coil springs, gas-pressurized tube shocks, anti-roll bar/lower A-arms, fixed-length halfshafts, trailing arms, dual coil springs, gas-pressurized tube shocks, anti-roll bar

JAGUAR XJ-S HE CABRIOLET

FOR:
REFINEMENT
PERFORMANCE

AGAINST:
FUEL CONSUMPTION

THE CAT'S WHISKERS

Now available with V12 power, the Jaguar XJ-S Cabriolet is perhaps the ultimate sports car. The Autocar test team has sampled one

Some people love the look of the Jaugar XJ-S, others hate it. There are very few split decisions. The flying buttresses of the coupé have been the source of much controversy, but Jaguar has held fast with the design first launched in 1975, preferring to update mechanically rather than change the sheet metal.

In 1982, the V12 engine received some fairly substantial changes to aid breathing efficiency through the adoption by Jaguar of the May 'Fireball' high turbulence, high compression lean burn principle on the now-classic 90×70mm 60 degree V12. It was this change that improved the somewhat thirsty 5.3-litre engine into the more fuel efficient — and more powerful version — we know today.

Jaguar unveiled a new engine in 1983 (advanced Jaguar 6) — the AJ6 — a 3.6-litre 24-valve six-cylinder in-line unit which was slotted into the XJ-S, but which is also destined to propel the new XJ40 scheduled to appear next year. At the same time, Jaguar — in conjunction with Aston Martin Tickford — also launched the Cabriolet version of the XJ-S, although customers have had to wait until 1985 before the mighty all-aluminium V12 power unit, boasting 295bhp at 5500rpm and 320lb ft torque at 3250rpm, has become available.

The new version, built in-house, is produced primarily for the German market in answer to strong demand, though is likely to find equally strong appeal in the US once federalised versions have been completed. Unlike the XJ-S saloon, the Cabriolet features a thin close-fitting hood; the thick rear three quarter pillars have gone, giving rise to a more elegant, perhaps less uncompromising appearance.

The body conversion is similar in design to the smaller-engined Cabriolet, featuring a folding rear cloth section with plastic window and twin cloth-covered targa style roof panels (stowed in a special boot pouch) which expose a large sunroof if desired.

From the outset, the car was designed as a two seater; unlike its coupé stablemate, there is no provision for rear seat passengers, Instead, the space is used to house two lockable storage boxes, each about the size of a briefcase. These are fully carpeted, as is the top section which is also equipped with a chrome retaining rail. The conversion from coupé to Cabriolet begins at Jaguar's Castle Bromwich plant. Here the body panels are assembled into a complete unit which is then trasferred to an outside contractor where the roof is removed and extra strengthening incorporated. The bodies are returned to Castle Bromwich to be painted, then transferred to Browns Lane for full assembly.

There was a time when Jaguar was seriously thinking of dropping the V12 power unit from production, but it appears that would have been a big mistake; other manufacturers, notably BMW, are currently thinking along the lines of a V12 for future development, putting Jaguar at a distinct advantage.

The XJ-S has always stood for effortless performance, and the Cabriolet 5.3 is further proof of this concept. On the one hand, the car is docile enough to be driven round town without showing any signs of discomfort; on the other, it can be used to provide silky smooth acceleration and supercar performance at will. The engine is what sets this car apart from any other on the road — it is superbly refined.

In comparative terms, acceleration is not as quick as the XJ-S HE (automatic) tested in April 1982, as we discovered on MIRA's horizontal straights, but it is the way the power comes in from, say, 60mph that really impresses. From rest and with the gear selector in D, the tyres can be coaxed to break traction on a full ▶

Cabrio's handling: *despite its size, the Jaguar XJ-SC HE has quite nimble cornering ability, although some understeer is apparent*

TEST UPDATE

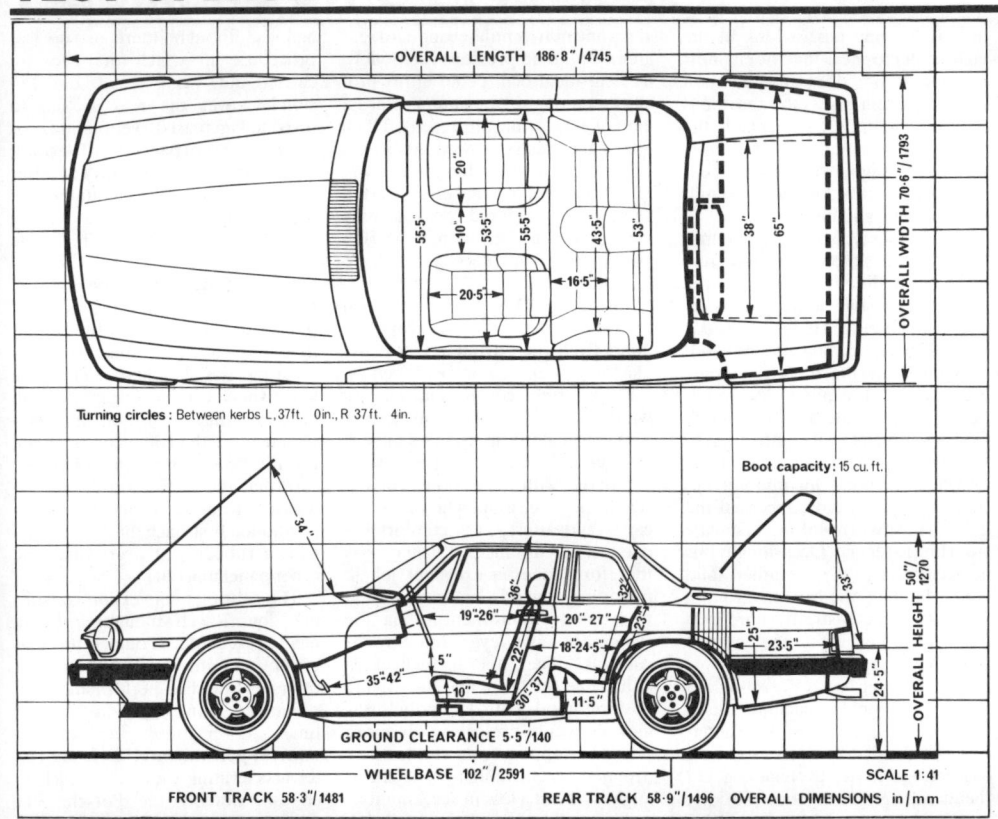

(Technical drawings of car — top view and side view with dimensions)

Overall length 186·8"/4745
Overall width 70·6"/1793
Turning circles: Between kerbs L, 37ft. 0in., R 37ft. 4in.
Boot capacity: 15 cu. ft.
Overall height 50"/1270
Ground clearance 5·5"/140
Wheelbase 102"/2591
Front track 58·3"/1481
Rear track 58·9"/1496
Scale 1:41
Overall dimensions in/mm

SPECIFICATION

ENGINE
Longways front, rear-wheel drive. Head/block al. alloy/al. alloy. 12 cylinders in 60 deg V, wet liners, 7 main bearings. Water cooled, viscous and electric fan.
Bore 90mm (3.54in), **stroke** 70.0mm (2.76in), **capacity** 5345cc (326 cu in).
Valve gear one ohc per bank, 2 valves per cylinder, chain camshaft drive. **Compression ratio** 12.5 to 1. Electronic breakerless ignition, Lucas-Bosch digital electronic fuel injection.
Max power 295bhp (PS-DIN) (21.7kW ISO) at 5500rpm. **Max torque** 320lb ft at 3250rpm.

TRANSMISSION
3-speed automatic.

Gear	Ratio	mph/1000rpm
Top	1.00-2.40	26.88
2nd	1.48-3.55	18.16
1st	2.48-5.95	10.84

Final drive: Hypoid bevel with Salisbury Powr-lok limited slip differential, ratio 2.88.

SUSPENSION
Front, independent, double wishbones, anti-dive geometry, coil springs, telescopic dampers, anti-roll bar.
Rear suspension, independent, lower transverse wishbones, driveshaft upper links, radius arms, coil springs, telescopic dampers.

STEERING
Rack and pinion, hydraulic power assistance. Steering wheel diameter 15.7in, 2.9 turns lock to lock.

BRAKES
Dual circuits, split front/rear. **Front** 11.18in (284mm) dia ventilated discs. **Rear** 10.38in (263.6mm) dia discs. Vacuum servo. Handbrake, side lever acting on rear discs.

WHEELS
Aluminium alloy 6½in rims. Radial ply, tubeless tyres (Dunlop SP Sport Super D7 on test car), size 215/70VR 15in, pressures F32 R30 psi (normal driving).

EQUIPMENT
Battery 12V, 68Ah. Alternator 67A. Headlamps 110/220W. Reversing lamp standard. 12 electric fuses. 2-speed, plus intermittent wipe screen wipers.

PERFORMANCE

MAXIMUM SPEEDS

Gear	mph	km/h	rpm
OD Top (Mean)	140	225	5200
(Best)	146	235	5450
2nd	118	190	6500
1st	70	113	6500

ACCELERATION FROM REST

True mph	Time (sec)	Speedo mph
30	3.2	31
40	4.7	41
50	6.2	52
60	7.7	62
70	9.7	73
80	12.4	84
90	15.4	93
100	18.7	105
110	23.9	113
120	32.2	125
130	—	138

Standing ¼-mile: 15.9sec, 91mph
Standing km: 28.5sec, 117mph

TEST CONDITIONS
Wind: 5.0mph
Temperature: 11deg C (51.8deg F)
Barometer: 29.6in Hg (1006mbar)
Humidity: 100 per cent
Surface: dry asphalt and concrete
Test distance: 1254miles
Figures taken at 5590 miles by our own staff at the Motor Industry Research Association proving ground at Nuneaton.
All *Autocar* test results are subject to world copyright and may not be reproduced in whole or in part without the Editor's written permission

IN EACH GEAR

mph	Top	2nd	1st
10-30	—	—	2.5
20-40	—	—	3.0
30-50	—	—	3.0
40-60	—	4.4	3.2
50-70	—	4.6	3.5
60-80	—	5.0	—
70-90	—	5.2	—
80-100	—	6.2	—
90-110	11.0	7.9	—
100-120	13.7	—	—
110-130	16.9	—	—

CONSUMPTION
FUEL: Overall mpg: 14.5 (19.4 litres/100km)
Autocar constant speed fuel consumption measuring equipment incompatible with fuel injection
Autocar formula: Hard 13.1mpg
Driving Average 16.0mpg
and conditions Gentle 18.9mpg
Grade of fuel: Premium, 4-star (97 RM)
Fuel tank: 20 Imp galls (91 litres)
Mileage recorder: 2.9 per cent long
Oil: (SAE 20W/50) 1000 miles/litre.

BRAKING
Fade (from 91mph in neutral)
Pedal load for 0.5g stops in lb

	start/end		start/end
1	40-25	6	40-38
2	45-30	7	45-40
3	45-30	8	45-40
4	40-35	9	45-45
5	40-40	10	45-40

Response (from 30mph in neutral)

Load	g	Distance
30lb	0.47	64ft
40lb	0.65	46ft
50lb	0.90	33ft
60lb	1.10	27ft
Handbrake	0.20	150ft

Max gradient: 1 in 3

WEIGHT
Kerb 35.1cwt/3933lb/1784kg
(Distribution F/R, 53/47)
Test 39.4cwt/4413lb/2001kg
Max payload 720lb/327kg

COSTS

Prices

Basic	£21668.23
Special Car Tax	£1805.69
VAT	£3521.08
Total (in GB)	**£26,995.00**
Licence	£100.00
Delivery charge (London)	£100.00
Number plates	£20.00
Total on the Road (excluding insurance)	**£27,215.00**
Insurance group	9

SERVICE & PARTS

Change	Interval 7500	15,000	30,000
Engine oil	Yes	Yes	Yes
Oil filter	Yes	Yes	Yes
Gearbox oil	No	No	Yes
Spark plugs	No	Yes	Yes
Air cleaner	No	Yes	Yes
Total cost	£120.31	£179.83	£204.12

(Assuming labour at £25.30 an hour inc VAT)

PARTS COST (inc VAT)

Brake pads (2 wheels) front	£34.13
Brake pads (2 wheels) rear	£18.72
Exhaust complete	£252.05
Tyre — each (typical)	£141.91
Windscreen	£136.85
Headlamp unit	£66.13
Front wing	£203.55
Rear bumper	£295.55

EQUIPMENT

Ammeter/Voltmeter	●
Automatic	●
Cruise control	●
Economy gauge	N/A
Electronic ignition	●
Five speed	N/A
Limited slip differential	●
Power steering	●
Rev counter	●
Self-levelling suspension	N/A
Steering wheel rake adjustment	N/A
Steering wheel reach adjustment	●
Trip computer	●
Headrests front	●
Heated seats	N/A
Height adjustment	N/A
Lumbar adjustment	N/A
Rear seat belts	N/A
Seat back recline	●
Seat cushion tilt	N/A
Seat tilt	N/A
Split rear seats	N/A
Electric window	●
Heated rear window	●
Interior adjustable headlamps	N/A
Sunroof	●
Tinted glass	●
Headlamp wash/wipe	●
Tailgate wash/wipe	N/A
Alloy wheels	●
Low profile tyres	N/A
Tyre size	215/70 VR15
Central locking	●
Child proof locks	N/A
Cigar lighter	●
Clock	●
Fog lamps	N/A
Internal boot release	N/A
Locking fuel cap	●
Luggage cover	N/A
Metallic paint	●
Radio/cassette	●
Speakers	●
Aerial	●

● Standard ○ Optional at extra cost
N/A Not applicable † Part of option package DO Dealer option

WARRANTY
12 months/unlimited mileage, ?-year anti-corrosion

PRODUCED AND SOLD BY:
Jaguar Cars, Browns Lane, Allesley, Coventry.

JAGUAR XJ-S HE CABRIOLET

blooded open throttle start with 30mph appearing in 3.2secs. By the time 60mph is reached, in 7.7secs, the engine, changing from first to second gear at 64mph and 5900rpm, is just beginning to get into its stride. From here on the V12 really begins to make headway so that the 0-to-100mph figure is just 18.7secs. The change from second to third occurs at precisely 100mph and, if the throttle is held open, the maximum speed of 140mph (146mph best) occurs soon after.

Changing gear manually, to make full use of engine revs, did not improve the performance at all, but simply served to highlight the maximum speeds in gears, which are 70 and 118mph respectively. The General Motors three-speed automatic gearbox clearly soaks up a certain amount of available power but performance is not affected to such an extent that the Jaguar ever feels as though it is suffering unduly. Instead, the automatic feels well-suited to the power unit and offers more than acceptable mid-range acceleration for overtaking. For instance, the 50 to 70mph increment in second gear takes a mere 4.6secs, and 70 to 90mph in the same gear 5.2secs.

In comparison with the XJ-S coupé, with its 153mph top speed and 0-60mph acceleration figure of 6.5secs, the Cabriolet looks slow; though, in fairness, the soft top version is appreciably heavier with a kerb weight of 3933lb compared with 3824lb for the coupé; it also has slightly inferior aerodynamics.

A driver looking for an ecomical sports car might shy away from the XJ-SC 5.3. Our test vehicle netted an overall fuel consumption figure of 14.5mpg over the 1254-mile test distance, slightly down on the 16mpg figure for the coupé. However, as the old observation goes, a driver who can afford an XJ-SC 5.3 is unlikely to be overly worried about fuel consumption. Clearly one has to pay for the performance but, in actual fact, economy is no worse than many other supercars.

If fuel consumption is the Achilles' heel, then refinement is the *pièce de résistance*. Without doubt, the XJ-SC 5.3 is a superbly refined motor car. On tickover, engine noise seems not to exist — one has to look at the tachometer to check that the engine is running — nor does any vibration enter the cabin unless the throttle is blipped violently. At speed the situation is very much the same. Only when using higher revs in each gear does a distinctive, yet well muted, howl become audible.

Road noise is also low, but amplified by the absence of any real mechanical din; it is, in fact, the biggest intrusion until speeds in excess of 80mph are reached. At this point, if one is heading into the wind, buffeting round the soft top markedly increases; if travelling downwind, however, the same does not apply. Consquently, interior comfort, when travelling at speed, depends largely on wind direction. One minor irritation we experienced concerned the gear change when shifting manually from first to second. Occasionally this is accompanied by a rather loud clunk, lurching the car forward in a most uncharacteristically unrefined manner rather than the normal smooth take-up of power.

For such a big beast, the XJ-SC 5.3 is quite a nimble machine round country lanes, thanks to the responsive — even slightly over-responsive — steering and viceless handling. On the handling circuit the Cabriolet exhibited considerable understeer (much more than the XJ-S coupé) but even driven under this constraint it is still quite a satisfying machine. It is almost impossible to dislodge the rear into oversteer unless negotiating very tight corners or driving in wet conditions. There is very little body roll, even *in extremis*. Ride befits a car of this class, being possessed of a firmish feel, so that one feels confident when cornering quickly. At the same time, it is very supple when it comes to soaking up road imperfections. Under hard acceleration the nose pitches up detectably and gives one a slight floating-on-air feeling.

Ground clearance, although not a real problem with the XJ-S, appears slightly worse in Cabriolet guise, the result of extra bracing round the back axle deemed necessary to ensure adequate stiffness for the body.

There was no cause for complaint with this Jaguar's brakes, which are essentially the same ventilated front discs and non-ventilated inboard rear items found on previous XJ models. A near-ideal 60lb pedal effort produced creditable maximum retardation of 1.1g, while our repeated high speed fade tests showed the brakes easily capable of handling emergency situations. Initial pedal effort started out at 40-25lb and peaked only fractionally higher with a 45-50lb pedal effort while decelerating from 91mph at 0.5g.

Inside, the XJ-SC 5.3 is identical in trim and fittings to the XJ-S coupé — apart from the obvious sheet metal changes — and comes with central locking, cruise control and electric windows. Though the seats do not look particularly special and appear to lack adjustment control when compared with many other current seats, they are, as all who drove the car would attest, very comfortable over long distances. Jaguar has therefore done an excellent job in providing a good compromise seat whose only adjustment is via the customary rake lever and fore-aft controls. The steering wheel is adjustable for reach and the driving position good. Our only criticism is that we would prefer slightly more sideways support from the leather-trimmed seats, though perhaps this might come at a loss in accessibility.

Air conditioning is fortunately, is a standard feature of the XJ-SC 5.3, and in warm weather with the top up is especially useful in preventing the seats from becoming sticky. Hood removal is achieved in two separate actions, though there is no set procedure in which the two are carried out. As a result, one can simply lower the rear section to provide increased ventilation, or remove the roof panels too if desired. The rear section is lowered after disengaging the centre-mounted lock mechanism and pulling loose the velcro fastenings by the rear side windows.

Erecting the rear section is a little fiddly until one gets the hang of pulling the hood against the central mounting hoop so that the two securing rods click home before the lock wheel can be engaged. With practice, this need not take two people more than 30 seconds. When fitting the roof panels, one must make sure that the section on the left — looking from frent to rear — is positioned first since this is equipped with a rubber seal over which the other panel must fit.

The cabriolet conversion is, without doubt, well thought out and neatly executed. Coupled to that creamy-smooth V12 power unit, it offers outstanding performance and refinement along with — for the first time — versatility. In terms of outright performance, there are faster accelerating cars from which to choose, notably the Porsche 911 Carrera Cabriolet and Aston Martin Volante Convertible, but where the Jaguar makes its mark is with its unique, and quite exceptional, blend of powertrain plus the inherent benefits offered by a cabriolet. ■

Lusty engine: *5.3-litre V12 produces maximum power of 295bhp*

Diminutive boot: *with the targa panels stowed away it is small indeed*

Luxury interior: *trimmed in leather, the Cabrio is an opulent beast*

ROAD TEST

Jaguar XJ-SC Cabriolet

The return of the pop-top cat.

• Eleven years have passed since Jaguar's last E-type two-seater ragtops were imported into this country. England's famous maker of luxurious sports cars and sumptuous sedans has since made do with steel roofs, and the grand-touring world has been the poorer.

One way or another, ragtop Jags never shorted you on emotion. Even in the midst of mechanical and electrical travails, they left you longing for short nights and sunny days. Now comes the XJ-SC Cabriolet, a worthy addition to the long if sporadic line of open-air cats from Coventry.

Stop at a light in the XJ-SC and people grin, whoop, and holler, "Hey, that's a *beautiful* car!" When the roof is up, we agree. Surprisingly, though, onlookers often blurt compliments even when the top's panels and flaps are stowed away. Swivel two levers and the two canvas-covered, pop-off roof panels will unlatch and slip out, to be neatly pouched in the deep trunk. To lower the rear section of the roof, unlatch its round lock from the inside, peel the Velcro-fastened edges of the fabric away from its framework, and fold the fabric and the plastic window down, snapping them beneath a snug boot. Unfortunately, you are now left with an array of roof hoops rising starkly naked above the swoop of the lower body. (Pausing to change lenses, Aaron Kiley gazed at the framework and mused, "It looks like it's under construction.") The staff consensus is that the XJ-SC looks best with its toupee in place, yet none of us feels truly repelled by its aesthetics when everything is stripped down to let the outside in. And it does come in.

Head for the country for grand touring after everyday life has tethered you for a while and you'll feel the beauty of this striking device as clearly as others see it. Grand touring means different things to different people, but it always means something wonderful. In an XJ-SC, the pleasures of touring grandly are smooth and satisfying. Jaguar always skips the sharp aggression of such traditional rivals as the six-cylinder BMW coupe and the V-8 coupes from Mercedes and Porsche, not to mention the rowdy soft- and hard-topped Corvettes. The XJ-SC offers not only open-air motoring but a very different accent on sporting transportation.

Underneath, the usual Jaguar charms and drawbacks abide. The XJ-SC's long, suggestive hoodline houses a huge V-12 engine whose horsepower proves a suitable match for the motor's broad-shouldered appearance. Just as important, though it is blessed with the bloodlines of its forebears, today's XJ may be exhibiting new signs of civility and respectability. According to the factory, quality control is vastly improved and great strides have been made in reliability. Maintenance requirements are lower than ever. The XJ's fuel economy has never been much to write home about—we're dealing with two tons of high performance here—but the Michael May lean-burn combustion chambers, introduced in 1982, do help advance the cause of efficiency. EPA city economy

JAGUAR XJ-SC

holds the fort at 13 mpg, and our own mixed use returned 14 mpg.

Luckily for Jaguar's reputation as a producer of road burners, Michael May's sculpturing of the V-12's heads returns much more than improved fuel economy. With a whopping 11.5:1 compression ratio burning beneath its single-overhead-cam heart, the 5.3-liter twelve whirs out an effortless 262 horsepower. Operating on super-unleaded fuel, the engine adroitly belies the mass of machinery within, never whimpering on the way to its impressive 6500-rpm redline. This V-12 remains the basis of Jaguar's mid-engined IMSA GTP batmobile, and every time you step into the throttle hard enough to force an automatic kickdown, you are reminded why the V-12 represented the ultimate roadgoing engine for so many years.

When production of the old ragtop ceased, however, some of the wonder went out of Jaguars. We need not have worried, it turns out: as long as there is an England, there will be ragtops on the horizon. Ragtops are Great Britain's nose-thumbing gesture at the sodden skies that often sulk there. Logical people laboring under such murk would never fall in love with convertibles, but of course logic and the British are often at odds. Basically, Brits are tickled pink anytime their beloved ragtops are close at hand.

Over the past few years, Americans have grown happier to have Jags at hand. As is often said, much of the credit goes to John Egan, Jaguar's chairman and chief executive officer, who signed on in 1980 and led the company's resurgence. Sales in America alone soared from 3029 cars in 1980 to 18,000 in 1984. Publicly owned for the past two years, Jaguar enjoys continued sales-and-service success in the States. We snapped up more than half the 1985 production run of 38,000 cars, and the XJ-S coupe has broken sales records here for three years running.

Mechanically, the Cabriolet is identical to the XJ-S coupe. Structurally, however, the convertible has been injected with new beef. The roof framework houses two substantial steel rods that make a minor contribution to stiffness, but the major beefing comes from a boxed section that takes the place of the coupe's rear seats. Intended primarily to offset the torsional stiffness lost when the full roof was eliminated, the box wears handsome carpeting and houses two lockable storage bins. Trimmed with luggage slats and a restraint bar, the cat box (ahem) stands ready to support any traveling cases, haversacks, miniature steamer trunks, or other traveling caboodle carried in the cabin, but definitely not any plus-two passengers. In terms of their structural contribution, the reinforcements don't quite make up for the skylights in the roof. The lengthy Jaguar exhibits a few shakes and shudders not found in, say, the stubby Porsche 911 Cabriolet, though the Jag is certainly far from the shakiest pop-top around.

We only wish it were as easy to open up as the others. In our test car, the knob intended to release the framed rear roof section almost didn't: it proved nearly impossible to turn. And the trim, tight boot stubbornly resisted the snapping-down operation.

Inside, the new roofline nips off a bit of the coupe's headroom, but comfort seems to suffer little, if at all. The seats' lumbar padding has been improved, and the subtle shaping of their bolsters delivers better support than their flatness suggests. Delicious-smelling leather upholstery, although often a slippery hindrance to staying behind the wheel, in this case has sufficient texture to assist the bolsters in keeping you in place. The steering wheel is also covered in reasonably grippy leather, but the rim is too skinny to provide a genuinely secure grip.

Driver confidence also suffers because of the distraction created by the XJ's scattered control layout. The main on/off/resume cruise-control button sits obtusely on the console just behind the shifter, while the "set" button is tucked in the tip of the turn-signal lever, far to the left. In another lapse of logic, the climate system's controls are split by the radio. Worse, our

SC's ventilation fan quit, the oil-pressure gauge threw fits of erratic readings, and the roof panels sprang a leak when we tested the weather sealing by running the car through an automatic wash—not good omens for Jaguar's recently improved reputation. (Jaguar responds that our test car was built as a pilot model—even though the Cabriolet has been on the U.K. market for two years—and that the top panels should fit better in U.S. production models.)

Having grumbled our gripes, a number of niceties stand out, too. A handy trip computer provides pertinent poop on time, distance, average speed, and fuel use. And beneath the computer sits the head unit of an Alpine sound system, whose sweet musicality, to our ears, ranks high among today's factory-supplied audio systems. It is said that the British don't like music, but they like the noise it makes. You could have fooled us: this system easily and smoothly delivers music, not noise. The radio's layout is good, the AM performance is better than most, and the FM section sounds fine. The cassette player's performance, with tapes recorded on a good deck from records or compact discs, absolutely caresses the ears. Given the XJ-SC's lowish wind noise (for a convertible), Jaguar should add a CD player, because plenty of musical detail comes through without strain even at high speeds.

We're talking 135-mph potential. Too bad the three-speed automatic and the tall gearing take the sting out of the V-12 off the line. In this price category, a four- or five-speed gearbox should be aboard to liven things up. Moving off isn't slow, but as our 8.4-second 0-to-60 time suggests, the beast doesn't bust loose until the smooth, long-legged automatic finally lets the revs get up to business.

The big four-wheel discs, ventilated in front, turn out to be strong points, smoothly snubbing this 4016-pound pilgrim's progress without fade, though stops from 70 mph require a longish 213 feet. Squeezing down into tight corners from high speeds under hard braking cre-

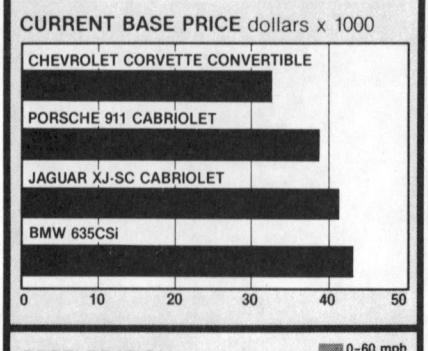

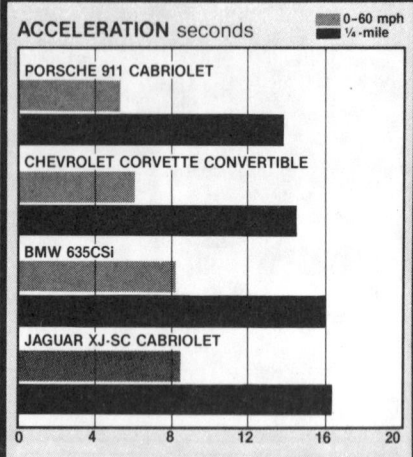

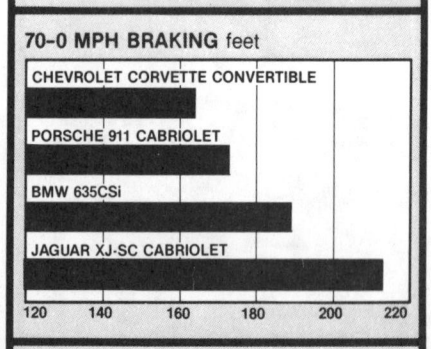

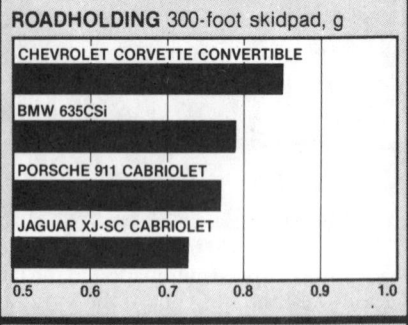

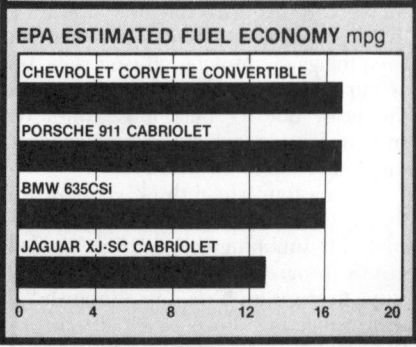

JAGUAR XJ-SC

CURRENT BASE PRICE dollars x 1000
- CHEVROLET CORVETTE CONVERTIBLE
- PORSCHE 911 CABRIOLET
- JAGUAR XJ-SC CABRIOLET
- BMW 635CSi

ACCELERATION seconds (0–60 mph, ¼-mile)
- PORSCHE 911 CABRIOLET
- CHEVROLET CORVETTE CONVERTIBLE
- BMW 635CSi
- JAGUAR XJ-SC CABRIOLET

70-0 MPH BRAKING feet
- CHEVROLET CORVETTE CONVERTIBLE
- PORSCHE 911 CABRIOLET
- BMW 635CSi
- JAGUAR XJ-SC CABRIOLET

ROADHOLDING 300-foot skidpad, g
- CHEVROLET CORVETTE CONVERTIBLE
- BMW 635CSi
- PORSCHE 911 CABRIOLET
- JAGUAR XJ-SC CABRIOLET

EPA ESTIMATED FUEL ECONOMY mpg
- CHEVROLET CORVETTE CONVERTIBLE
- PORSCHE 911 CABRIOLET
- BMW 635CSi
- JAGUAR XJ-SC CABRIOLET

Vehicle type: front-engine, rear-wheel-drive, 2-passenger, 2-door cabriolet

Price as tested: $43,500

Options on test car: base Jaguar XJ-SC Cabriolet, $41,500; gas-guzzler tax, $1500; freight, $500

Standard accessories: power steering, windows, and locks, A/C, cruise control

Sound system: Alpine AM/FM-stereo radio/cassette, 4 speakers

ENGINE
Type .... V-12, aluminum block and heads
Bore x stroke .... 3.54 x 2.76 in, 90.0 x 70.0mm
Displacement .... 326 cu in, 5344cc
Compression ratio 11.5:1
Fuel system .... Lucas electronic fuel injection
Emissions controls ...... four 3-way catalytic converters, feedback fuel-air-ratio control, auxiliary air pump
Valve gear .... chain-driven single overhead cams, hydraulic lifters
Power (SAE net) .... 262 bhp @ 5000 rpm
Torque (SAE net) .... 290 lb-ft @ 3000 rpm
Redline .... 6500 rpm

DRIVETRAIN
Transmission .... 3-speed automatic
Final-drive ratio .... 2.88:1, limited slip

Gear	Ratio	Mph/1000 rpm	Max. test speed
I	2.48	10.8	70 mph (6500 rpm)
II	1.48	18.1	118 mph (6500 rpm)
III	1.00	26.8	135 mph (5050 rpm)

DIMENSIONS AND CAPACITIES
Wheelbase .... 102.0 in
Track, F/R .... 58.6/59.2 in
Length .... 191.7 in
Width .... 70.6 in
Height .... 47.8 in
Ground clearance .... 4.5 in
Curb weight .... 4016 lb
Weight distribution, F/R 55.1/44.9%
Fuel capacity .... 24.0 gal
Oil capacity .... 11.4 qt
Water capacity .... 21.0 qt

CHASSIS/BODY
Type .... unit construction with 2 rubber-isolated crossmembers
Body material .... welded steel stampings

INTERIOR
SAE volume, front seat .. 51 cu ft
  luggage space .... 16 cu ft
Front seats .... bucket
Seat adjustments .... fore and aft, seatback angle
General comfort .... poor fair **good** excellent
Fore-and-aft support .... poor fair **good** excellent
Lateral support .... poor **fair** good excellent

SUSPENSION
F: .... ind, unequal-length control arms, coil springs, anti-roll bar
R: .... ind, fixed-length half-shaft, 1 control arm, 1 trailing link, and 2 coil-shock units per side

STEERING
Type .... rack-and-pinion, power-assisted
Turns lock-to-lock 2.8
Turning circle curb-to-curb 39.4 ft

BRAKES
F: .... 11.1 x 1.0-in vented disc
R: .... 10.3 x 0.5-in disc
Power assist .... vacuum

WHEELS AND TIRES
Wheel size .... 6.0 x 15 in
Wheel type .... cast aluminum
Tires .... Pirelli P5, 215/70VR-15
Test inflation pressures, F/R 32/30 psi

CAR AND DRIVER TEST RESULTS

ACCELERATION   Seconds
Zero to 30 mph 3.2
  40 mph 4.8
  50 mph 6.5
  60 mph 8.4
  70 mph 10.6
  80 mph 13.8
  90 mph 17.3
  100 mph 22.0
  110 mph 28.9
  120 mph 39.8
Top-gear passing time, 30–50 mph 3.8
  50–70 mph 6.1
Standing ¼-mile .... 16.3 sec @ 88 mph
Top speed .... 135 mph

BRAKING
70-0 mph @ impending lockup .... 213 ft
Modulation .... poor **fair** good excellent
Fade .... **none** moderate heavy
Front-rear balance .... poor **fair** good

HANDLING
Roadholding, 300-ft-dia skidpad .... 0.73 g
Understeer .... minimal **moderate** excessive

COAST-DOWN MEASUREMENTS
Road horsepower @ 30 mph .... 7.0 hp
  50 mph .... 18.0 hp
  70 mph .... 39.0 hp

FUEL ECONOMY
EPA city driving .... 13 mpg
EPA highway driving .... 17 mpg
C/D observed fuel economy .... 14 mpg

INTERIOR SOUND LEVEL
Idle .... 43 dBA
Full-throttle acceleration .... 75 dBA
70-mph cruising .... 70 dBA
70-mph coasting .... 69 dBA

ates no fuss whatever. As you feather off the pedal, unloading the nose, the steering takes up the transition as calmly as warm milk puts you to sleep.

In the harsh light of skidpad testing, the XJ's all-independent, coil-sprung suspension and comparatively slim wheels and tires limit roadholding to 0.73 g. But the car's behavior in the real world is satisfying enough to appeal to the buyers Jaguar has in mind. The shift lever looks delicate and feels clunky, and the throttle stiffly resists forays into downshift territory, but somehow the lasting impression is that of a basically blissful drivetrain. Some testers, however, feel it is *too* blissful and yearn for a more enthusiastic response from within.

The XJ-SC's handling follows British practice in providing light, benign rack-and-pinion power steering, plus tire grip that feels more sporting than that of the typical American luxury sportster—though not as sporting as the fine bite and firm bump-buffering provided by Germany's hard-running machines. On the other hand, the Jaguar provides levels of ride comfort that German engineers express little interest in supplying, and a degree of suspension control that is rare in American-made machinery. Released by the throttle to do your bidding, the Jag hurls at your senses an orderly world, all veddy British, the V-12's thunder insistent but distant, pleasantly muted even as you scissor across sweeping corners in a rush. The sensational engine gathers the car's haunches to spew you onto the following straights in great, lubricious spendings of energy. If you have the patience to listen for the distant thunder, Jaguar rumbles the language of power, and you can hear it in your bones.

What your gleeful senses total up may equal the perfect answer for the high-dollar grand-touring market. The XJ-SC Cabriolet is just basic enough to provide deep thrills, but more than subtle enough to cuddle with. Its behavioral envelope will probably suit many of those concerned with finding proper conveyances for their golden years of grand touring. As long as their fingers are still strong enough to pop the top, that is.

As a halfway step toward true open-air motoring, the XJ-SC Cabriolet proves two things: First, major amounts of sunshine can be admitted to the interior of the XJ without sacrificing any of the car's basic goodness. And second, although the XJ-SC is a noble and carefully engineered alternative to a steel-roofed coupe, it's no substitute at all for a true convertible.
—*Larry Griffin*

COUNTERPOINT

• I'm afraid there's nothing but disappointment in the new XJ-SC for a thrill seeker like myself. Aside from Jaguar's unconventional means of letting the great outdoors in, time has stood still for this lowrider.

I've long hoped that Jaguar would firm the XJ-S up into a real sports machine, but, alas, that was not to be. The XJ-SC still uses the pillowy suspension settings that make the XJ6 such a joyous rider—but it's way too loose-jointed for a spirited two-seater. When I call on a hard corner, the body makes like a yo-yo. No change here.

Then there's the mighty V-12, which feels all bottled up at the low end. It's needed a good four-speed automatic to set it free for ages.

It's been reported that Jaguar has improved its quality control. Unfortunately, our test car suffered from a balky folding top, a climate control that went belly-up, and a bad case of the stalls.

When I slide into a low-slung, broad-shouldered V-12 sports car, I expect a kick in the pants. The XJ-SC may be fine for wealthy men twenty years my senior, but it's far too much of a gentleman for me. —*Rich Ceppos*

It's hard to rationalize a car like the Jaguar XJ-SC. For far less money, there are any number of cars that perform as well as or better than this machine. Few, however, can match the Jag's individuality and panache.

In a time of look-alike car designs, the XJ-SC is a welcome break from the norm. It is leaner and swoopier than its photos suggest; at times its look is almost exotic. Of course, as with almost any exotic car, one pays a price for good looks and distinction. The Jag's seats are limited in their range of adjustment and offer little support, many of its ergonomic features leave much to be desired, and our test car suffered a number of nagging failures.

In compensation, however, the XJ-SC offers a smooth and strong engine, effortless cruising at any speed, and a ride that is almost unsurpassed in suppleness. Add to these virtues the car's handsome shape, and this Jaguar becomes an inviting alternative for someone who can afford to indulge in its unique blend of individuality.
—*Arthur St. Antoine*

If your idea of automotive luxury is centered on a profligate use of the world's resources, the Jaguar XJ-SC is the car for you. A mid-sixties, full-size American sedan is a model of efficiency compared with the Coventry cat. I can't think of another two-seater with such a large and massive body. Neither can I recall a car that extracts so little performance from such a large and lavishly conceived powerplant.

To be sure, the XJ-SC goes about its business quietly and smoothly, leading one to conclude that passenger comfort was the goal of its extravagant design. However, the absence of such creature comforts as power-adjustable seats and a power-operated roof belies that conclusion. That leaves sheer conspicuous consumption as the main thrust of this car, and that's just too anachronistic a concept, even for a luxury cruiser, in 1986. —*Csaba Csere*

JAGUAR v JAGUAR

CAT FIGHT

It is a strange experience of contrasting qualities. At first you're bemused by the inferiority of the miserable little beer-barrel seat and awkward driving position. Was it really this bad? How could Jaguar get it so wrong? Next, you're marvelling at the car's amazing ability and superb fluidity. Despite the hefty powertrain, the unassisted steering is light, precise, communicative; the handling taut, tidy, sharp, the ride smooth, unruffled. There's no need to extend the engine. Squeeze the throttle at idling revs and the car surges ahead with a gruff growl and terrific verve. Modern high-speed, short stroke engines, aided and abetted by the exhaust-driven turbocharger, will erase from our memory what slugging tractability is really about if we're not careful. Without the occasional refresher, we are in danger of forgetting just how good the Jaguar E-type was.

It's an opportune moment to look back and take stock of the enduring E-type, marking its 25th anniversary this year. To look forward as well, through the decade-old XJS coupé, spiritual (if overweight) successor to the E-type in 3.6 cabriolet guise, to the upcoming XJ40 saloon. For the past three years, Jaguar have been honing their AJ6 engine in the XJS coupé. They have not been wasted, those 36 months.

Whatever your viewpoint, this year will be a landmark in Jaguar's history, a year when practically the last tangible link with their old classics will be severed by the arrival of the XJ40. Beyond that, there is no place in the Browns Lane camp for the XK twincam that's powered most of Jaguar's six-cylinder cars, E-type and racers included, since 1948.

Ergonomically, the E-type feels every bit its age and more. Stepping from the luxurious, air-conditioned XJS into Ken Bell's stark 1962 3.8 roadster – an unrestored 36,000 miler that's used sparingly only in the summer months – it certainly felt like a 25-year time warp, even though the coupé's much roomier cabin is far from avant garde: it was hardly that in 1975, and it has changed little since.

Taking care not to gouge your eyes on the lethal window pillar, you clamber through the E-type's vestigial door, over the wide sill that helps give it such good torsional rigidity, into a cramped, utilitarian cockpit. There's no need to reach for the seat adjuster. The shapeless little bucket will almost certainly be hard against its rearmost stop anyway. The only way to untangle your knees from the thin, wood-rimmed steering wheel is to loosen its ring lock and telescope it out, close to your chest.

Within a year or so of the E-type's launch, Jaguar relieved cockpit cramping by extending the footwells and relocating the high-up hanging pedals. Hunched or not, with the lid off you get a magnificent view out over the outrageously long bonnet, bulged in the middle to accommodate the engine's superstructure, louvred at the back to help cool it.

The XJS hums into life at the twist of a key. In the E-type, you push a button to fire the engine, which is much more satisfying. The one purrs sveltely, the other growls malevolently. Even allowing for age and mileage, the ancient XK engine of this E-type was never a match for the new AJ6 when it comes for smoothness. Nor should it be. Not now. The low-rev snatch and top-end vibro that marred the original 3.6 debutantes in '83 have been smoothed away, which augurs well for the XJ40. Whether you could balance a pencil on the cam covers at 5000 rpm, I doubt, but the AJ6 makes the E-type's XK feel comparatively coarse and rough, especially when extended or shut down on the over-run. It

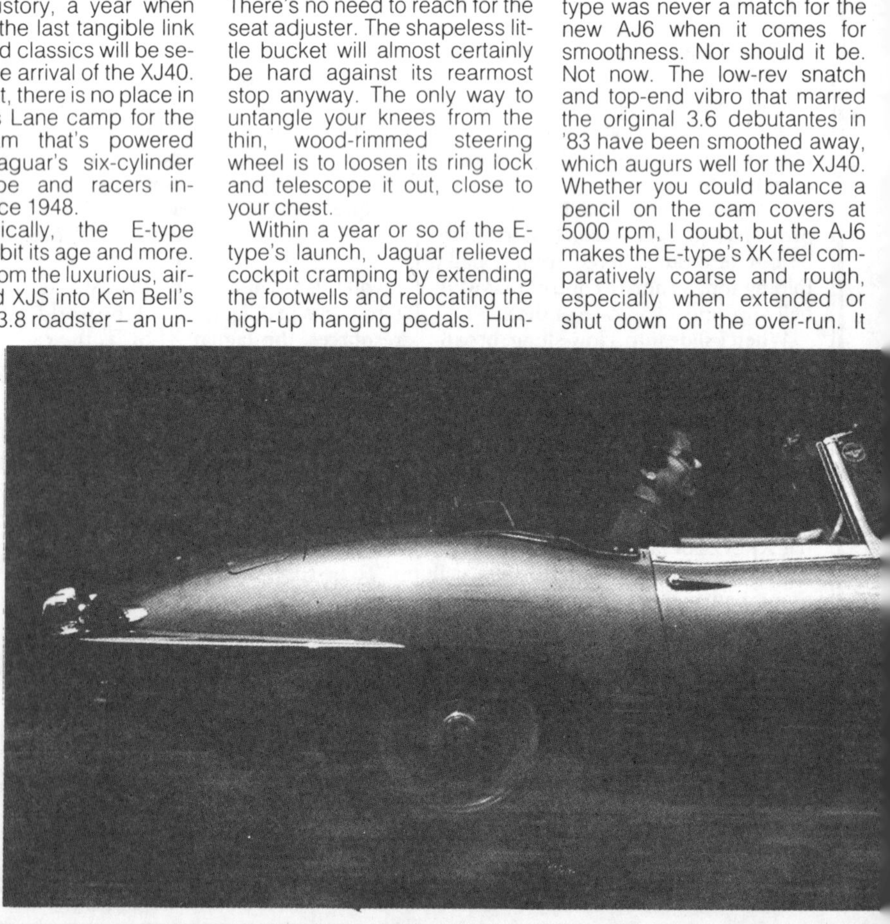

JAGUAR v JAGUAR

Is the XJS Cabrio the Eighties' equivalent of the classic drop-head E-type? Roger Bell drives the big, beautiful Jaguar "sixes" and assesses their strengths

XJS looks better in open form, but it's not the most elegant of Jaguars. Homely instrument panel is barely changed from 1975, apart from gaining burr walnut

hardly belittles its output, though.

With a top speed of 140 mph, the XJS is no sluggard. Its acceleration is strong and positive especially above 80 mph. It reels in the horizon swiftly and peacefully but not entirely without effort. Real vigour comes only by extending the engine to high revs in the lower ratios. Not so in the 3.8 E-type, which can hit 100 mph from rest in 16 seconds, against the XJS 3.6's 20. More to the point, it will lunge from 50 to 70 mph in top in less than half the time taken by the XJS: 5.3 seconds against 11.5.

Motor's Test No 10/61, of March, 1961, still shows one of the most remarkable sets of top-gear acceleration figures we've recorded: every 20 mph increment in the 3.8 E, from 10-30 mph to 80-100 mph, was on the right side of six seconds. It is this huge spread of collosal pulling power that gives the E-type its whopping performance. Flex your right foot and it's away, instantly opening up ground on an XJS caught napping in the wrong gear, never mind anything else with a faltering turbo, off boost with its gauge in the doldrums. That's what throttle response is all about.

Not that all the E-type's data should be regarded as hallowed material. Forget the 265 bhp claimed by Jaguar for the 12-valve SU-fed 3.8E (and the subsequent 4.2 for that matter). That was a gross figure, not comparable with the 24-valve injected 3.6's honest 225 bhp DIN. No, the E-type's decisive performance edge is down to traction-engine torque and a huge 356 kg (7 cwt) weight advantage, rather than to innately superior output. Lower gearing plays its part, too. Not that every production E-type could have matched the 150 mph top speed of Jaguar's two original blueprinted press demonstrators, which were always regarded as suspiciously fast. None could have done so safely without Dunlop R5 racing tyres, initially offered as optional equipment.

We certainly didn't attempt any high-speed runs in Ken Bell's original-condition 3.8, which started life in British Racing Green but for over two decades now has sported Golden Sands paintwork. In the early Sixties, E-types were in such demand that many well-heeled customers grabbed what car they could, then had it resprayed in their first-choice colour.

Punching through the Surrey lanes, it was enough just to feel the old car's potential, to revel in its thumping pickup. Mr Bell, technical luminary of the young and burgeoning Jaguar Enthusiasts Club, says that the front torsion bar springs have settled a bit, that we'd have noticed a little undulating float at high speeds, evidence of wilting dampers. Maybe. At our modest pace, I was aware only of supple springs and firm control. Modern low-profile tyres and the stiff suspension needed to keep them flat on the road have had a pernicious effect on the ride quality of many subsequent performance cars. It has crept up on us, this unfortunate decline, and it takes an E-type, free from jarring thumps and jiggle on its soft springs and tall Dunlops, to bring it into sharp focus.

Expecting the worst of the notorious four-speed Moss gearbox, the E-type's shift was something of a welcome surprise. Make no mistake, it's not good: snatch the stiff lever through its narrow gate with the engine singing and grating dogs will signal their objection. At modest revs, though, you can pamper the lever silently past weak synchro. Downshifts always call for nifty double-declutching, which is immensely satisfying when you get it right, a strain when you don't. A drivers' gearbox, make no mistake.

The '64 4.2 E-type brought

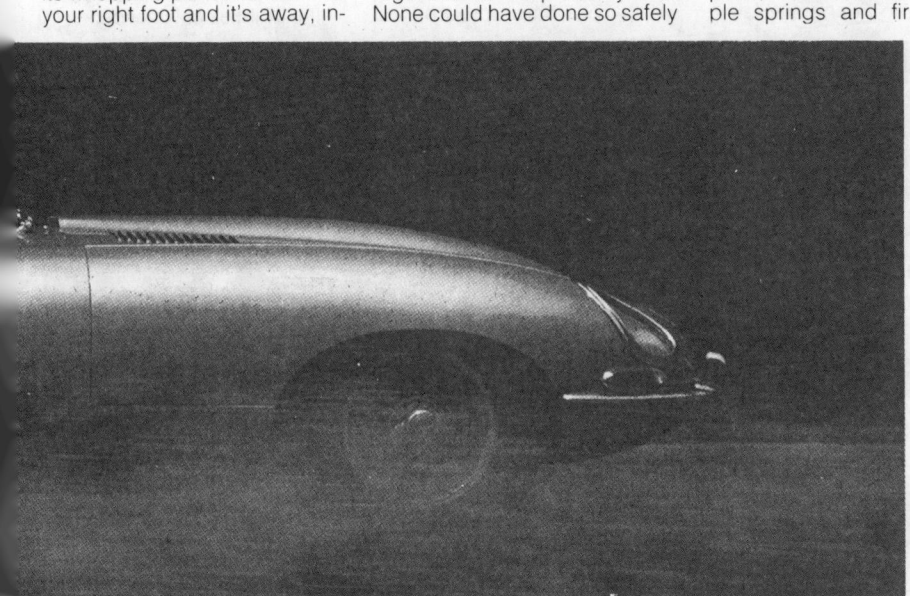

Some thought the E-type's shape suggestive, but there's no denying its sleek, striking aggression. Badge sits in the small, low air intake of a pure racer; wood-rim wheel fronts uncompromisingly functional facia

JAGUAR v JAGUAR

XJS's leather-clad seats are short on support, but otherwise luxury is the cockpit's keynote. AJ6 engine, refined over earlier examples, is a worthy successor to the venerable XK unit; looks good, too

with it a vastly superior all synchro Jaguar box that was far easier to use but perversely less rewarding than the sirened-first agricultural Moss affair of the original 3.8. The Getrag five-speeder of the XJS is cast in the same mould as the 4.2's. By modern standards, it's still a rather meaty shift that needs a very firm hand if you really must rush it. Although the tiresome low-speed shunting of the early 3.6s has been cured (by removing the Lucas fuel injection's over-run cutoff), delicate throttle control is still needed when feeding in the vague clutch to avoid kangaroo lurch. The rather juddery E-type's clutch has seen better days.

Like its predecessor, the XJS rides superbly, with none of the harsh thudding of a modern Merc or Porsche. However, being a heavier, bulkier car, as befits its grand touring status, it dips and tilts on its soft springs rather more than the skateboard E-type. Even so, the XJS is marvellously stable and secure. What it lacks in agility and sharpness, which isn't much, it makes up in strong adhesion, permitting greater liberties than the rather soft, boulevard handling suggests is prudent.

Steering precision and feedback have been sacrificed for fingertip control through a servo mechanism that the E-type manages without. Considering the great cast-iron lump supported by the front wheels, one can only marvel at the ease and directness of the E's steering, so full of feel on broken surfaces you can sense the rack nudging its pinion to and fro without actually kicking the wheel. Response is certainly crisp, but maybe not so sharp as the 2.5 turns from lock to lock would have you believe: the turning circle is a cumbrous 11.3 m (37 ft).

Later E-types, which grew in girth and weight to something approaching XJS corpulence, didn't point so cleanly and faithfully as the early 3.8. That's one reason why aficionados still rate the original E-type as the best driving machine of the series, and certainly the quickest. Its strong, smooth, relatively free-revving engine – a purer power-plant than the subsequent 4.2 – was another. Despite praise from our testers in 1961, its brakes were not really up to the performance, though. Those of the 4.2 were very much better, and the XJS's – lighter, less prone to fade with ventilated discs – better still.

When it comes to refinement and comfort, the old 3.8, stark and cramped and bereft of creature comforts, is no match for the XJS grand tourer. Nor would you expect it to be. The 3.6 XJS cabrio deserves more supportive seats, just as the early E-type did. Its dash and switchgear and quaintly old-fashioned decor are not of this decade, either. Even so, given a long trans-contintental journey, I would be hard pressed to choose a more agreeable means of all-weather transport than this.

Ken Bell praises his E-type roadster's simple up-and-over headgear just as our testers did in '61. That of the XJS is a lot more elaborate, if not so quick to operate. Still, there are three distinct levels of al fresco motoring here (with two roof panels and a folding back hood). Just as important, wind noise with the top up is remarkably low.

While the E-type values steadily climb (Bell's '62 3.8 roadster, without any mechanical restoration, has an insurance value of £15,000), those of the £22,395 XJS 3.6 cabrio will continue to plummet through normal depreciation for years to come. It's hard to see another production Jaguar even approaching the E-type's stature or appeal, let alone matching it. The quantum leap forward it made in the early Sixties is too big a stride for Jaguar to repeat in the Eighties or Nineties. M

Leather in the E-type, too, but seats are skimpier and the ambience utilitarian. This early E-type has the 3781 cc version of the XK engine with triple SU carburetters. E-type's monocoque finishes at the bulkhead; spaceframe front end can just be seen

Jaguar XJ-SC Cabriolet
The top cat goes topless
by Bob Nagy
PHOTOGRAPHY BY WILLIAM CLAXTON

Jaguar can rightfully lay claim to a long and distinguished history of producing upscale open-air 2-seaters. However, the tradition that had previously yielded the likes of the SS100, XK-120, and XK-140 appeared in grave danger of ending during the 1970s. Proposed new auto regulations regarding rollover safety seriously threatened to eliminate the convertible as a vehicle type. Faced with the prospect of incurring sizeable tooling costs that could ultimately go for naught, Jaguar elected to forego the drop-top variant when it replaced its XK-E model with the XJ-S in 1975. The anticipated legislation never materialized. In mid-1984, the company made what many considered an inevitable move and introduced the XJ-SC Cabriolet. Unfortunately, the striking wind-in-the-hair variant was only available in Eurospec. Now, some 12 long years after the last E-Type roadster rolled off the production line and across the Atlantic, Jaguar has decided to introduce its XJ-SC to the lucrative American market.

The latest addition to Jaguar's U.S. lineup is being built on a limited-production basis. Fewer than 600 units will be offered for sale here in 1986. The most luxurious, best appointed model ever to wear a Jaguar badge, the XJ-SC carries a $42,700 pricetag, some $3500 above the XJ-S coupe. The Cabriolet does arrive the way any self-respecting luxo-tourer should, fully loaded. Simply put, it offers no options save for a modest selection of interior and exterior colors. Numbered among the XJ-SC's comprehensive list of standards are power assists on virtually every functional system, an electronic climate control package, and a six-function trip computer. For 1986, a new 40-watt Alpine AM/FM/cassette stereo has also been factored into the package.

Jaguar developed the XJ-SC as a 2-passenger car from the start. Construction initially begins at the firm's Castle Bromwich body plant near Birmingham. The basic shell is then shipped to the nearby Park Sheetmetal Company, where the roof modifications are made. The finished body goes back to Castle Bromwich for paint before being sent on to Jaguar's Browns Lane plant in Coventry for mating with the drivetrain and final assembly. The latter process requires each car to successfully undergo a thorough test-drive regimen before receiving its final sign-off.

Although sharing the same basic platform as the XJ-S Coupe, the Cabriolet incorporates a number of significant structural modifications to help it compensate for the lack of a full steel roof. Besides supplemental gussetting in the floorpan, the XJ-SC carries an extra rear crossmember and additional bulkhead bracing just ahead of the trunk. The roof hoop contains a pair of ¾-in. steel rods to further bolster torsional rigidity, while the A-pillar and top of the windscreen are also beefed up. Last links in the support system are the longitudinal rails on either side of the greenhouse, which extend down and wrap around the back of the fixed rear quarter glass. Even though the structure detracts somewhat from the sensual lines of a pure convertible, this additional bracing adds materially to the solidity and strength of the XJ-SC. While not nearly as tight as the coupe, the Cabriolet displays considerably less of the dreaded cowl shake than most of its open-air contemporaries.

The Cabriolet arrives the way any self-respecting luxo-tourer should, fully loaded

The XJ-SC's removable top is a combination of solid fiberglass panels and a traditional canvas folding rear section. The front panels pop in and out easily enough with a quarter turn of the latch handles. They are stored in the two-pocket pouch found in the Jag's 10.7-cu-ft trunk, which also contains a full-size spare tire. Lowering the rear portion of the top also requires only a quarter turn of a knob. However, due to its location in the center of the roof hoop, operating the rear latch mechanism entails a somewhat greater challenge. Properly fitting the canvas boot is another exercise that demands some degree of effort. It's a case where being elegant mandates being patient as well. Even so, an individual can easily complete the entire process in well under five minutes. Though it passed our car-wash leak test with flying colors, the XJ-SC's top assembly did strike us as about a half cut below the rest of the vehicle in overall fit and finish. There is already some speculation that the folding rear portion may be replaced by a fixed metal section within a couple years.

Residing beneath the XJ-SC's long, flowing hood is Jaguar's 5.3-liter H.E. V-12. The all-aluminum SOHC engine uses the "Fireball" 2-valve cylinder head developed by Swiss engineer Michael May. Its patented "High Efficiency" lean-burn combustion chamber design allows Jaguar to bump the compression ratio to 11.5:1 and reduce emissions while enhancing fuel economy. The silky-smooth powerplant makes 262 hp at 5000 rpm and 290 lb-ft of torque at 3000 rpm. It revs eagerly to its posted 6500-rpm redline and beyond, sufficient to give the car a factory-advertised top end of 140 mph. Like all Jaguars, it's quite content to motor along

Inside the XJ-SC, one enters a world of fine Connolly leather and burled walnut

at triple-digit touring speeds for extended periods of time and space.

The potent V-12 ships power to the rear wheels through a THM400 3-speed automatic supplied by General Motors. The non-lockup, non-O/D transmission drives through a limited-slip differential with a 2.88:1 final ratio and delivers shifts that are positive without being unduly harsh. The matchup gives the 4040-lb Cabriolet catlike acceleration, allowing it to

A potent pussycat, Jaguar's new Cabriolet carries the same silky-smooth 262-hp V-12 engine that powers the XJ-S coupe. The XJ-SC offers fresh-air fanatics a choice of several different drop-top configurations.

slip into motion smoothly and gather speed and momentum quickly. During our formal testing, the XJ-SC clawed its way from 0-60 mph in 8.66 sec and covered the quarter mile in 16.70 sec at 89.1 mph. In daily use, the Jaguar proved as adept at dealing with the normal stop-and-go regimen as it was when unleashed on the open road. The XJ-SC's 13/17-mpg city/highway ratings hardly make it a favorite with the gang at EPA. But during our flogging, it returned a 16.4-mpg average, a figure we consider downright respectable given the nature of the vehicle.

The Cabriolet's fully independent suspension has been carried over intact from the Coupe. Up front, it employs unequal length upper and lower A-arms with coil springs and hydraulic dampers. The rear incorporates Jaguar's classic twin coil-over shocks working in conjunction with angled trailing arms, a single lower control arm, and fixed-length half-shafts that act as an upper lateral link. The racecar-like design is trimmed out with anti-roll bars fore and aft. The result is a chassis tuning geared to provide a supple, well-controlled ride, albeit one heavily skewed toward the civilized end of the scale.

The XJ-SC is shod with 215/70VR15 Pirelli P5 steel radials. Chosen more for compliance value than for their absolute cornering power, they still do a creditable job of keeping two tons of snarling Big Cat hooked up to the pavement, as confirmed by the Cabriolet's 0.77 g skid-pad reading. The power-assisted rack-and-pinion steering provides a very liveable tradeoff between boost and overall road feel. The 102.0-in. wheelbase contributes to the XJ-SC's superb tracking ability, and its 47/53% weight distribution endows it with basically neutral balance.

Like the Coupe, the Cabriolet is definitely happiest when taking on high-speed sweepers. Negotiating tight corners elicits a fair degree of understeer as the laws of physics begin to take precedence over the best efforts of the suspension and tires. Premium level stopping power is ensured by the huge 4-wheel power disc brakes, 11.1 in. vented up front and 10.3 in. solid in the rear. Good modulation with minimal fade characterize their performance, and stability is enhanced through the use of anti-dive geometry in the front end. A 134-ft 60-0-mph stopping distance provides the best index of their absolute capabilities.

Slipping through the XJ-SC's low-slung doors, one enters a world of fine Connolly leathers and burled walnut veneer, all matched with meticulous care. Its well-contoured bucket seats lack several of the more exotic adjustments possessed by their Japanese counterparts. But they do afford reasonably good support and a high degree of long-haul comfort. The rear portion of the passenger compartment has been completely redone as part of the Cabriolet makeover. The area behind the seats now contains a fully carpeted bay set off by a pair of lockable storage bins.

In keeping with longstanding tradition, Jaguar has fitted its newest S-Type with a full complement of classic white-on-black analog instruments. A handsome leather-wrapped steering wheel nicely frames these highly legible gauges, however its relatively large diameter and thin rim won't readily win the hearts and hands of most enthusiasts. A related item that also drew some criticism was the column itself, which telescopes but still lacks any sort of provision for tilt adjustment. Despite these and a few other ergonomic quirks—including the need for no fewer than four different keys to operate the various locking mechanisms—the XJ-SC Cabriolet is one car we could easily learn to live with on a daily basis.

Those who value absolute security more than the ability to pick up a tan on the fly may still find the XJ-S Coupe a more desirable alternative. And part-time sunshine fans can elect to spring for its $1300 power sunroof option. But for fanatic sun worshippers, there's really no contest. Ditto for those who thrive on high-profiling, as the XJ-SC also demonstrated an uncanny ability to elicit favorable comments from numerous unsolicited admirers. Thanks to intensified corporate efforts that brought about huge improvements in the areas of quality and reliabilty, Jaguar has found itself on an especially strong roll in the U.S. for the past several years. We think the arrival of the new XJ-SC Cabriolet will only help continue that saga. **MT**

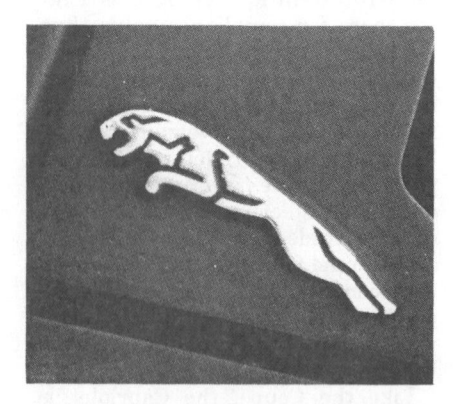

TECH DATA

Jaguar XJ-SC Cabriolet

GENERAL
Vehicle mfr. ....Jaguar Cars, Ltd., Coventry, England
Vehicle importer............Jaguar Cars, Inc., Leonia, N.J.
Body type ....2-door cabriolet
Drive system ....Front engine, rear drive
Base price......................$42,700
Major options on test car......................................None
Price as tested...............$42,700

ENGINE
Type ....V-12, water cooled, cast alloy block and heads
Displacement.................5344 cc (326.0 cu in.)
Bore & stroke ....90.0 x 70.0 mm (3.54 x 2.76 in.)
Compression ratio 11.5:1
Induction system...........Electronic fuel injection
Valvetrain......................SOHC, 2 valves/cylinder
Crankshaft ....Forged, 7 main bearings
Max. engine speed.........6500 rpm
Max. power (SAE net).....262 hp @ 5000 rpm
Max. torque (SAE net)....290 lb-ft @ 3000 rpm
Emission control ....3-way catalyst
Recommended fuel ....Regular unleaded

DRIVETRAIN
Transmission.................3-sp. auto.
Transmission ratios
  (1st)...........2.50:1
  (2nd)..........1.50:1
  (3rd) 1.00:1
Axle ratio 2.88:1
Final drive ratio 2.88:1

CAPACITIES
Crankcase ....10.6 L (2.9 gal)
Cooling system..............19.5 L (5.2 gal)
Fuel tank ....84.0 L (22.2 gal)
Luggage (trunk).............303.0 L (10.7 cu ft)

SUSPENSION
Front..............................Independent, A-arms, coil springs, hydraulic shocks, anti-roll bar
Rear...............................Independent, trailing arms with single lower A-arm, coil springs, hydraulic shocks, anti-roll bar

STEERING
Type ....Rack and pinion, power assist
Ratio 3.53:1
Turns, lock to lock 2.7

BRAKES
Front..............................11.1-in. vented disc, power assist
Rear...............................10.3-in. disc

WHEELS AND TIRES
Wheel size.....................15.0 x 6.5 in.
Wheel type....................Cast alloy
Tire size.........................215/70VR15
Tire mfr. & model..........Pirelli P5
Tire construction...........Steel-belted radial

DIMENSIONS
Curb weight ....1832.5 kg (4040 lb)
Weight distribution, f/r (%)..........................47/53
Wheelbase ....2591 mm (102.0 in.)
Overall length................4869 mm (191.7 in.)
Overall width ....1793 mm (70.6 in.)
Overall height................1214 mm (47.8 in.)
Track, f/r.......................1488/1504 mm (58.6/59.2 in.)
Min. ground clearance ..114.0 mm (4.5 in.)

CALCULATED DATA
Power-to-weight ratio ....15.4 lb/hp
Top speed......................140 mph (est.)
Drag coefficient.............0.39

SKIDPAD
Lateral acceleration.......0.77 g

FUEL ECONOMY (mpg)
EPA rating, city/hwy.......13/17
Test average ....16.4 mpg

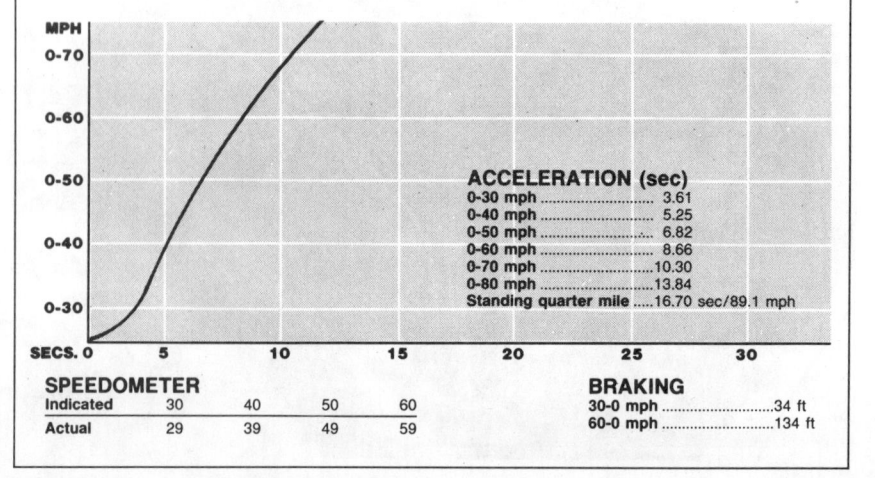

ACCELERATION (sec)
0-30 mph 3.61
0-40 mph 5.25
0-50 mph 6.82
0-60 mph 8.66
0-70 mph 10.30
0-80 mph 13.84
Standing quarter mile ....16.70 sec/89.1 mph

SPEEDOMETER
Indicated	30	40	50	60
Actual	29	39	49	59

BRAKING
30-0 mph ....34 ft
60-0 mph ....134 ft

BUYING SECONDHAND

JAGUAR XJ-S

A CLASSIC OPPORTUNITY

The Jaguar XJ-S has always been a big, heavy, fast and mechanically complex coupé and it is now available at derisory prices on the secondhand market. But that doesn't necessarily mean it's a bargain. We point out the problems a prospective buyer might encounter

Some cars go on selling for a long time, yet somehow fade from the limelight. Only last month, for instance, Jaguar announced further changes to the XJ-S range, and it was a shock to realise that this successful model has already been on sale for more than 11 years. Not only that, but Jaguar's chairman, Sir John Egan, recently said: "the XJ-S still has a lot of life in it — there is an awful lot we want to do yet. There are no plans to stop production."

Even though the XJ-S has always been a big, heavy, fast *and* mechanically complex coupé, it is now available at derisory prices on the secondhand market. But that doesn't necessarily mean that it is a bargain — especially when you take service and parts prices, and fuel economy, into account. The sad fact, too, is that many old XJ-Ss are now with their fourth or fifth owners, and are by no means in peak condition.

A secondhand Jaguar XJ-S, in other words, should never be bought without a great deal of thought — but the car does offer a great deal of performance for a remarkably low purchase (if not running) cost.

BACKGROUND

A new range of Jaguars — the original XJ6 family — was launched in the autumn of 1968, and was expanded progressively in the next few years. Not only were Jaguar *and* Daimler versions offered, but the XJ12/Double Six cars, with the now-famous light-alloy V12 engine, arrived in 1972, and a long-wheelbase version was put on sale in 1973.

The XJ-S Coupé was revealed in the autumn of 1975, and was a 2+2 seater based on a shortened version of the XJ12's floorpan, with the same front and rear suspensions, brakes and steering gear.

Until 1983, all XJ-Ss had V12 engines, this then being joined by the all-new 24-valve twin-cam Type AJ6 six-cylinder unit.

Until 1983, every XJ-S had fixed-head coupé bodywork, with those controversial sail panels at the rear quarters, but a Cabriolet version became available, and both types are still built to this day. In 11 years there have been only detail styling changes, though many mechanical improvements have been made in the same period.

ENGINES

The vast majority of all XJ-S cars have the famous 5.3-litre fuel-injected single-cam V12 engine. In its original guise this was rated at 285bhp. From the summer of 1981, the HE (High Efficiency) version of the engine was standardised, at which point the car itself was renamed XJ-S HE.

This car was not only more fuel efficient than before, but was also more powerful, rated at 299bhp. The engine was subsequently re-rated at 295bhp and is still in use today.

The XJ-S 3.6, announced in the autumn of 1983, was the first Jaguar to use the new-generation AJ6 twin-cam engine, with 24 valves, and 3590cc. This was rated at 225bhp. It was not until the autumn of 1986 that this engine found another home in the new Jaguar/Daimler saloons.

TRANSMISSIONS

This is quite a complex area for such a simple range. When the V12-engined car was announced it was available with a choice of transmissions — the old Jaguar all-synchromesh four-speed manual gearbox (with no overdrive, even as an option), or with a Borg-Warner Type 12 three-speed automatic.

The manual gearbox was always right at the limit of its stress capabilities, and Jaguar was always quite ▶

Buyers of early examples *should beware of corrosion in the usual places and leaking door seals which cause increased wind noise*

143

BUYING SECONDHAND

Six-cylinder *cars are still rare on the secondhand market*

◀ reluctant to build this version of the car. After being in very short supply for some time, it was finally discontinued in the spring of 1979. By then the Borg-Warner Type 12 automatic had also been dropped in favour of the more modern General Motors THM 400 three-speed transmission, which it has to this day.

Overall gearing was raised in the summer of 1981, by the fitment of a 2.88:1 instead of 3.08:1 final drive ratio, to coincide with the adoption of the HE engine.

Before 1987, the AJ6-engined six-cylinder cars had never been offered with automatic transmission of any type, but the four-speed ZF automatic is now optional. All cars built before the end of 1986 were fitted with the 'overdrive' type five-speed transmission from Getrag, this is the 'box now being offered on the new range of XJ6 (XJ40) saloons. It is not, however, offered on the V12-engined cars.

BODY STYLES

Each and every XJ-S has been built on the same 8ft 6ins floor pan, and with the same basic styling shape, but there are several variations on that theme.

The original car was a close-coupled four-seater (or shall we say a generous 2+2) coupé, with a small, near-vertical, rear window, flanked by two long sloping sail panels. It had all the usual Jaguar creature comforts, including leather-trimmed seats, electric windows, and full air-conditioning. UK-market cars always had two large headlamps, whereas those sold in the USA had four circular headlamps.

Except in detail, that style has not changed in 11 years. From the summer of 1981 (to coincide with the arrival of the HE engine), there was a wooden-panelled facia instead of the original leather one, different wheel styles, and other cosmetic changes.

The coupé body style was available either with the V12 engine (from the start), or with the new six-cylinder engine (from the autumn of 1983). You recognise a six-cylinder XJ-S by the more prominent 'power bulge' in the bonnet panel.

From the autumn of 1983, the Cabriolet was offered, at first available only with the six-cylinder engine, and then with the V12 from the summer of 1985.

Unlike the Coupé, this was a two-seater model, with fixed side-rails above and behind the side windows, and a permanent 'targa bar' above and behind the (front) seats. The car could be used with permanent rigid panels either in front of or behind the Targa bar, or with a cabriolet top.

TWR sells a '+2' seating conversion to provide vestigial rear accommodation for this body style. Incidentally, there are persistent rumours that Jaguar plans to produce a fully convertible XJ-S, and that this will be launched in 1988.

ROAD BEHAVIOUR

No-one should ever have to complain about a lack of performance in these cars. The 'small-engined' 3.6-litre cars have top speeds of more than 140mph, while that of the V12 engined cars is around 150mph. In either case, there is rapid acceleration to match.

Six-cylinder cars, which are still rare on the secondhand market, record around 18-20mpg in everyday use, whereas V12-engined cars are considerably more thirsty. Pre-HE models with automatic transmission, might be good for only 14-16mpg, whereas later models are two or even three mpg more fuel efficient.

Although the XJ-S is a large and heavy car (the V12-engined cars, with only the driver on board, tip the scales at more than 4000lb), it is surprisingly nimble. The combination of well-developed all-independent suspension and very light (some say too light) steering, makes it easy to urge the cars along very quickly indeed where conditions allow. If there is a parking problem with the XJ-S, it is caused by poor rearward visibility in the coupés, and by sheer bulk, not by the effort required to slot in and out of spaces.

Both the XJ-Ss are refined GT, rather than rorty sports, cars, the sort that can be driven with little effort, in air-conditioned comfort. The only pain, in the case of the 12-cylinder examples, is that on a long, fast, journey you need to stop every 250 miles (three hours or so), and spend another £30 on petrol...

AVAILABLITY

Even though a high proportion of XJ-S cars have always been sold overseas, there is still a good supply on the secondhand car market. The most numerous of the various types is undoubtedly the automatic transmission V12 coupé, while the rarest are probably the manual transmission, or cabriolet, versions of the same car.

Because the six-cylinder-engined models came on to the market only at the end of 1983, and because deliveries were at first very limited, secondhand versions of these types are still rare.

World-wide deliveries of XJ-S models up to mid-March 1987 were 47,067 V12s and 3332 six-cylinder models.

Not only because of the after-effects of the huge rise in petrol prices in 1979 and 1980, but because Jaguar was going through a well-publicised bad patch in product quality at the time, sales of XJ-S cars slumped considerably in the two or three years *before* the launch of the HE, so few 1979-1981 models seem to be around.

Production has been steadily rising ever since, and quite a number of 1983-1985 models are to be found.

The XJ-S is now selling better than ever before, and is still in full production at Brown's Lane, so there should be no problem in finding parts and service expertise to keep older cars running. However, don't forget that *current* XJ-S new-car prices vary from £21,500 to £29,500 — and that parts prices are pitched accordingly; anyone about to close the deal on a good-looking secondhand XJ-S costing less than £5000 should bear this in mind. There are 130 Jaguar dealers in the whole of the UK.

WHAT TO LOOK FOR

The best overall advice we can offer to prospective XJ-S buyers is to find the car with the lowest possible mileage for its year, and one which is showing no signs of corrosion. Cost of ownership — spares and service prices, in particular — should loom large in any XJ-S owner's budgeting.

We ought to state, right away, that the XJ-S is not a car for the home mechanic, even the dedicated DIY motorist. Take one look under the bonnet of the V12-engined car, and you will see what we mean.

More than in most other surveys of this type, therefore, we recommend that you be sure of the car's previous history before you take the decision, and sign the cheque, to buy it. An XJ-S which has been regularly and properly serviced is a far better bet than one whose antecedents are clouded.

MECHANICAL

We once wrote about the V12 engines as used in Jaguar saloons: 'The bad news is the engines are costly to repair if they go wrong, but the good news is that they have a fine reputation'. The same advice applies to the unit in the XJ-S.

Because they have mainly alloy castings, it is most important that ▶

Cabriolet *was introduced in 1983 and seats two passengers only*

144

JAGUAR XJ-S

The 3.6 engine is just as expensive to maintain as the V12

Wood-faced dashboard was introduced in 1981. Manual 'box is rare

Prospective buyers of V12 cars should look for oil or water leaks

Access to rear seats is awkward and space very cramped

MODELS AVAILABLE

September 1975: Original XJ-S Coupé announced, on 102ins wheelbase, with coupé body style, and 285bhp injected 5.3-litre V12 engine. Choice of four-speed manual or three-speed automatic transmission.
Spring 1979: Rare manual-transmission option discontinued.
July 1981: Original version dropped, and replaced by new 'HE' (High Efficiency) version. Same engine size, but 299bhp with improved fuel efficiency. Style changes included higher overall gearing, revised gear selection, wider-rim wheels of new pattern, and wood (was leather) faced facia, plus detail style changes.
October 1983: Widening of range, with launch of new 3.6-litre twin-cam AJ6 family of engines, allied to five-speed (Rover) manual transmission, no automatic version at first. Optional Cabriolet body style also launched, available with 3.6-litre engine only. Range for 1984: XJ-S 3.6, manual gearbox 6-cyl 225bhp; XJ-SC 3.6, manual gearbox 6-cyl 225bhp; XJ-S HE, automatic gearbox V12-cyl 299bhp.
July 1985: Cabriolet version of XJ-S HE launched, mechanically as XJ-S Coupé.
March 1987: Range continues: XJ-S 3.6 Coupé 6-cyl 225bhp; XJ-SC 3.6 Cabriolet 6-cyl 225bhp; XJ-S HE Coupé V12-cyl 299bhp; XJ-S HE Cabriolet V12-cyl 299bhp — four-speed ZF automatic transmission now available on 3.6-litre models.

SPECIFICATION AND PERFORMANCE

Specification	Manual Coupé	Auto Coupé	Auto Coupé	Auto HE Coupé	Auto HE Cabrio	6-cyl Manual Cabrio
Engine size (cc)	5343	5343	5343	5343	5343	3590
Engine layout	V12ohc	V12ohc	V12ohc	V12ohc	V12ohc	ohc 6
Engine power	285	285	296	299	295	225
Car length	— 15ft 6.8ins —					
width	— 5ft 10.6ins —					
height	— 4ft 2ins —					
Boot (cu ft)	— 15cu ft —					
Turning circle (kerbs)	— 37ft 0ins approx —					
Unladen weight (lb)	3902	3866	3890	3824	3933	3577
Max payload (lb)	720	720	720	720	720	771
Performance:						
Tested in *Autocar*:	7 Feb 1976	28 May 1977	24 Jan 1981	24 Apr 1982	25 Sep 1985	21 Apr 1984
Top speed (mph)	153	142	151	153	140	141
0-60mph	6.9	7.5	6.6	6.5	7.7	7.4
Overall mpg	15.4	14.0	14.3	16.0	14.5	17.6

APPROXIMATE VALUES

Price range	3.6 Coupé	3.6 Cabrio	V12 Coupé
£2300-£2500			1976
£2800-£3000			1977
£3600-£3900			1978
£4300-£4600			1979
£5000-£5300			1980
£6000-£6300			1981 (pre-HE)
£8400-£8800			1981 (HE)
£10,400-£10,800			1982
£13,000-£13,700			1983
£13,700-£14,200	1984		
£15,200-£15,700		1984	
£16,200-£16,700			1984
£16,800-£17,300	1985		
£18,200-£18,700	1986		
£19,000-£20,000		1985	1985
£21,000-£22,000		1986	
£21,500-£22,500			1986

Note: V12 Cabriolets are still very rare on the secondhand scene — expect to pay at least £3000 more than the equivalent Coupé. Automatic transmission was not available on 3.6-litre six-cylinder versions built before 1987. It is now an option.

PARTS PRICES

All prices include VAT at 15 per cent	6-cyl 3.6	V12-cyl 5.3
Engine assembly — bare (exchange)	£1704.30	£2129.80
Gearbox assembly (new)	£1064.90	N/A
Clutch, complete (new)	£151.11	N/A
Automatic transmission with convertor (exchange)	N/A	£730.25
Brake pads — front set (new)	£36.80	£36.80
Suspension dampers — front (each)	£28.75	£28.75
Water radiator assembly (new)	£225.40	£232.30
Tyre price, typical	£142.03	£142.03
Alternator (exchange)	£62.10	£62.10
Starter motor (exchange)	£138.00	£138.00
Headlamp unit — each	£76.48	£76.48
Front wing panel	£227.70	£227.70
Bumper, front, complete	£429.64	£429.64
Windscreen, laminated	£157.55	£157.55
Exhaust system, complete	£331.78	£305.90

BUYING SECONDHAND
JAGUAR XJ-S

Boot houses *tools, spare wheel and bag which takes targa panels*

On older cars *check for rust around headlamp cut outs and on arches*

Locking compartments *give useful stowage space in cabriolet*

Cabrio *is stylish and versatile, but still rare on the used market*

the engines should not be seen to overheat, or cylinder head warping could speedily result. Keep a close eye on cooling water and lubricating oil condition and levels. Treat any signs of fluid loss with grave suspicion, and look for leaks from one of the myriad hoses you will find in this complex and well-filled engine bay.

Be sure that the correct coolant inhibitor is still being used in the water (coolant freezing, or internal corrosion, can both be horrifyingly expensive), and that the engine is not using more than about one pint of oil for every 500 miles of motoring. A very low oil level can result in big end bearings suffering starvation, and damage. Look for oil leaks from the rear main area — but these are not as prevalent as they used to be with Jaguar XK engines.

Engine chain tensioners sometimes give trouble after a life of 50,000 miles, and some engine water pump problems have been known — this may have been due to over-tightening of drive belts overloading the bearings. Changing the various drive belts is not too difficult but — as with all things connected with this engine — it seems to take longer than expected.

Thank goodness the valve clearance don't often need adjusting — even to get at the cams and the shims, means that the entire induction system, manifolding, and fuel rails have to be removed first.

Very little is yet known about long-life prospects for the AJ6 twin-cam six-cylinder engines. Cars with this engine are still very rare indeed on the secondhand market.

There are very few V12-engined XJ-Ss with manual transmission, although the 'boxes seem to be well up to their job. They should not lose synchromesh or become noisy. Some cars seem to suffer from clutch drag, although this can usually be adjusted. Six-cylinder-engined models have Getrag transmissions, none of which should yet be in need of major attention.

Of the automatic transmissions found in these cars, the GM THM400 type should be very smooth and, if properly maintained and kept well supplied with fluid, extremely reliable.

The earlier Borg Warner gearbox was near the limit of its torque capabilities, and may be a bit jerky and less refined than the later type; a conversion to the GM 'box is not possible as a different engine crankshaft is required. In both cases, be sure that there is no leakage from pipes to and from the front-mounted oil coolers.

Propeller shaft centre bearings sometimes wear and rumble, but the chassis-mounted differentials are very strong. They might leak oil through bearings on to the inboard disc brakes, but unless they have been run with too little oil they should be quiet, and neither should the limited-slip differentials be noisy.

In the front suspension, look for worn bottom ball joints, and wear in the various suspension bushes. On hard-driven cars, look for scored brake discs at 50,000-mile intervals (and, in any case, expect to have to replace pads more frequently than you might expect). Steering rack bushes tend to tear out of their positions, especially if the XJ-S in question has had a lot of town use; look also for power steering oil leaks into the rack boots.

Rear suspension bushes tend to wear their radius arm bushes by the time a 50,000/60,000-mile interval is reached, while sub-frame mountings and axle mountings may have been worn, or partially destroyed, after five years or so.

Note that there are 16 grease points on the chassis — this is by no means a high-tech layout — which must not be neglected. Signs of neglect are highlighted by the rear lower wishbone bushes, which might have dried out, be noisy, and also be worn.

It pays, too, to check that all the car's electrical systems are working properly. Not for nothing was Lucas nicknamed 'the Prince of Darkness' by some frustrated American customers. In particular, check the electric window operation, the air-conditioning, and Similar auxiliaries.

Incidentally, a measure of the car's complication is that a regular 30,000-mile service takes eight hours.

BODY

In the late 1970s, and early in the 1980s, Jaguar had two basic problems — one was a severe lack of morale among the workforce (the company seemed to have lost its independence in the massive BL combine), and this was manifested in a general decline of vehicle build quality.

The result was that for a time cars like the XJ-S got a bad name. Even though this was, in part, overdone, there is no doubt that the general quality was not always acceptable. As the 1980s continued, it progressively improved. The low point was the 1979-1981 period in particular; along with the arrival of the second 'oil price shock' at the same time, sales began to sag alarmingly.

Paint quality suffered when the new Castle Bromwich body plant was still finding its feet, and this can lead to rusting in what might be called the 'conventional' places on the body shell. Look for rust on cars of more than four years old inside the front wheel arches, on the front wings above the wheels, and around the headlamp cut outs. There might be rusting along the bolted line of outer to inner front wings, on the sills, on the various sharp edges of the body panels, and around door locks, handles, and badges. There are body seams, too, which harbour water, and which might begin to corrode away. Nevertheless, the XJ-S is a world away from being as corrosion prone as the preceding E-type.

Inside the car, check for damaged leather seat facings, wear on the carpets, and any damage to the decorative wood on the facia and instrument panels. There is some tendency towards wind noise and leaking around the doors, but this can usually be sorted out by rehanging the doors or replacing the damaged rubber seals.

Take heart, though — the XJ-S has an immensely strong monocoque, and no normal amount of corrosion should weaken it significantly. But, make no mistake — converting a poor XJ-S into a good one is going to cost a lots of money so it makes sense to buy a good one. ■

TEST UPDATE

THE CAT STRIKES BACK

Jaguar's improved XJ-S is making its debut, complete with beefed-up suspension, new seats and steering wheel. We found it transformed

At speeds just the right side of recklessness the car behaved like only a very few can. Through sweeping tree-lined bends it cornered with precision, precious little roll and tenacious grip — like a front-engined Porsche perhaps, or an M-series BMW.

This was an XJS. Not the softly-sprung *autobahn* express that the car has been since its September '75 launch, but the latest 3.6-litre coupe — with a package of suspension mods aimed by Jaguar at the 'younger, more sporting driver'. Those who presumably fall into the older, less sporting category can still have the old suspension, the 5.3-litre V12 and a choice of coupé or cabrio bodies, but no more soft-tops will be produced with the 3.6-litre six.

The XJS's metamorphosis to a true sports car is down to the skill of Jaguar's chassis engineers. They are quite capable of weaving magic with subtle changes to springs and dampers, but the XJS needed a more radical fix. Front spring rates are up a whopping 43 per cent to 99.6lb ft/in and rears 34 per cent stiffer at 180lb ft/in, with new Boge dampers tweaked to suit. The front anti-roll bar goes up in diameter to 22mm, making it 42 per cent stiffer, while a smaller rear anti-roll bar is fitted for the first time.

Changes in spring rate of that magnitude hint at much-reduced wheel travel. That in turn means less camber change as the suspension moves and the possibility of using larger tyres — Jaguar opted for massive 235/60 VR15 Pirelli P600s on the new 3.6, wider than a Porsche 911 or a Ferrari 328. The P600 rubber needed wider rims, and new 6.5ins 'spoked' alloy wheels grace the XJS 3.6.

Jags have been criticised for over-light steering in the past, and with good reason — an untimely sneeze resulted in an unplanned lane change. The extra rubber on the road alone would have increased steering effort on the new XJS, but Jaguar has gone further than that. Not only has the level of hydraulic assistance been reduced but new, stiffer, rack mountings are intended to provide much more feel. The old thin-rimmed steering wheel has been consigned to the scrap bin, replaced by a meatier leather-covered item with sculpted hand grips — it's still a full 16 inches across, though.

The born-again XJS still weighs well over a ton and a half, yet turns into corners like a GTi half its weight. The weighting of the steering at speed is firm and reassuring, and at 2.6 turns lock to lock the gearing is ideal. Put the big cat on line and it stays there, the P600s doing all you could ask in the dry and the chassis refusing to be deflected from its chosen course by bumps or camber changes. It's the damping performance that's most remarkable, the Boge shocks allowing the springs to absorb the bumps then rebound enough to keep the

Steering wheel *is now leather-trimmed, but still too large*

Classic *black and white dials flank minor control 'drums'*

All-encompassing *trip computer is optional at £305*

24-valve AJ6 *engine is now more refined than ever*

Pirelli P600s *do all they can in the dry; chassis refuses to be deflected by bumps*

Damping performance *is remarkable, keeping wheels in contact with ground at all times*

wheel in contact with the ground under the severest provocation.

This same property endows the Jag with exceptional stability, tracking arrow true at a ton over crests and dips alike. Yet the ride remains relatively smooth and compliant, with hardly any tyre or bump noise transmitted to the cabin — the XJS is much better than comparable German cars in this respect.

The steering wheel isn't the only change on the inside of the XJS — the leather-faced cloth seats are new too. The original seats were like the old car's suspension — comfortable and perfectly adequate but lacking the support needed for serious cornering. The Jag's seats are now among the very best — perfectly shaped, firm and with body-hugging side cushions that keep you in place during the most enthusiastic bend-swinging. Our car had the optional electric lumbar adjustment and heating package, the lumbar support moving over a wider range than I've ever come across before. You could climb out of seats like this unaided and ache-free after a non-stop dash to Milan.

The rest of the interior is typical Jaguar — plenty of polished wood, leather trim and chrome switches scattered around in a fairly random manner. Two steering column stalks look after the usual indicator, dip-switch and wiper functions. A large ribbed collar on the column allows the wheel to telescope a couple of inches — that brings it out far enough for drivers with the seats fully back, but also takes the stalks out of fingertip reach.

Speedo and rev counter are classic white on black, flanking four vertical gauges that resemble the drums of a fruit machine — these actually show coolant temperature, oil pressure, fuel contents and battery voltage. Fully automatic temperature control by the combined heating and air conditioning system is standard, as is a stereo cassette player — the test car had the optional Clarion digital unit (£140 extra) and a trip computer (another £305). You do get electric remote door mirrors (although you can't see much out of the nearside one thanks to its flat glass) and electric windows included in the XJS's basic spec.

The XJS is quite roomy for a 2+2, and four adults would not have to be too friendly for short distances at least. The boot is large, as you have a right to expect in a car well over 15 feet long, although it isn't very deep and the central locking doesn't extend to the boot catch.

No driveline changes were made to the new XJS, but earlier this year the 24-valve AJ6 engine gained the engine management system from the XJ6 saloon. At the same time the ▶

147

TEST UPDATE

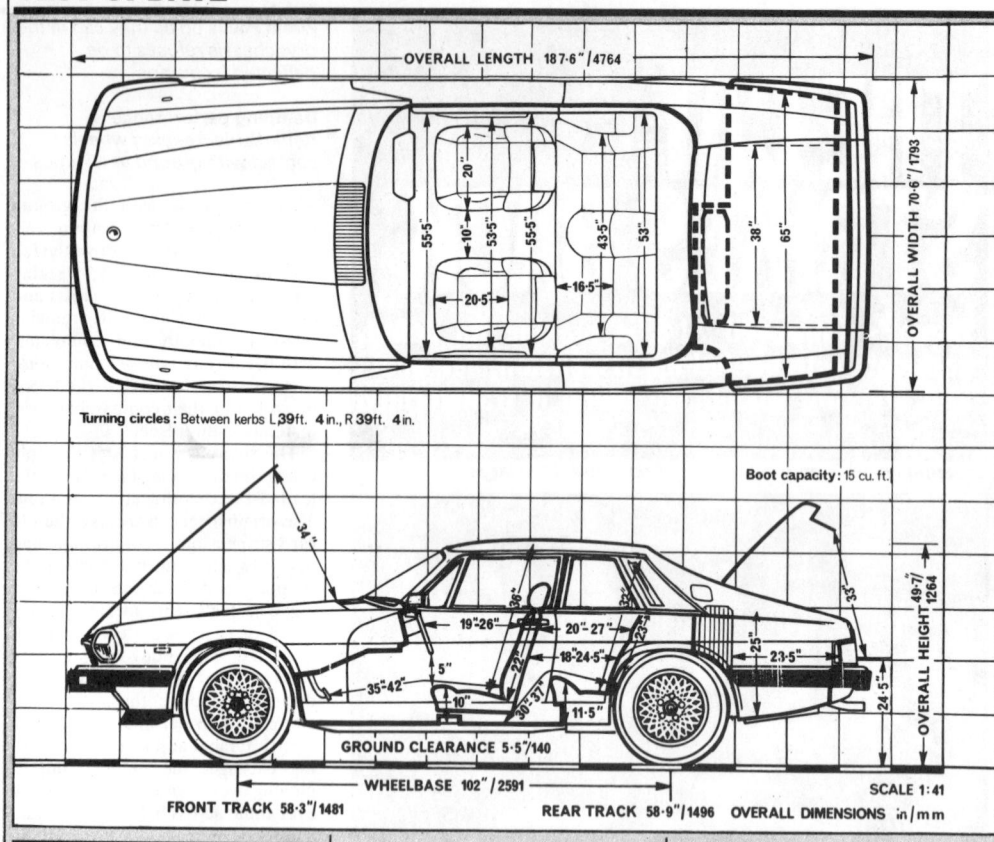

Overall length 187·6"/4764
Overall width 70·6"/1793
Overall height 49·7"/1264
Wheelbase 102"/2591
Front track 58·3"/1481
Rear track 58·9"/1496
Ground clearance 5·5"/140
Boot capacity 15 cu. ft.
Turning circles: Between kerbs L 39ft. 4in., R 39ft. 4in.
SCALE 1:41
OVERALL DIMENSIONS in/mm

MODEL

JAGUAR XJS AUTOMATIC 3.6
PRODUCED BY:
Jaguar plc,
Browns Lane,
Allesley,
Coventry

SPECIFICATION

ENGINE
Longways, front, rear-wheel drive. Head/block al alloy. al. alloy. 6 cylinders in line, dry liners, 7 main bearings. Water cooled, viscous fan. **Bore** 91.0mm (3.58in), **stroke** 92.0mm (3.62in), **capacity** 3590cc (219 cu in). **Valve gear** 2 ohc, 4 valves per cylinder, chain camshaft drive. **Compression ratio** 9.6 to 1. Lucas fully-mapped electronic ignition and fuel injection management system. **Max power** 221bhp (PS-DIN) (165kW ISO) at 5250rpm. **Max torque** 248lb ft at 4000rpm.

TRANSMISSION
4-speed automatic, torque converter.

Gear	Ratio	mph/1000rpm
Top	0.730	29.1
3rd	1.000	21.3
2nd	1.480	14.4
1st	2.480	8.6

Final drive: hypoid bevel, ratio 3.54:1.

SUSPENSION
Front, independent, dual wishbone, coil springs, telescopic dampers, anti-roll bar.
Rear, independent, lower wishbones, driveshaft upper links, radius arms, coil springs, telescopic dampers, anti-roll bar.

STEERING
Rack and pinion, hydraulic power assistance. Steering wheel diameter 16in, 2.6 turns lock to lock.

BRAKES
Dual circuits, split front/rear. **Front** 11.18in (284mm) dia ventilated discs. **Rear** 10.38in (264mm) dia discs. Vacuum servo. Handbrake, side lever acting on rear discs.

WHEELS
Al alloy, 6.5in rims. Radial ply tyres (Pirelli P600 on test car), size 235/60VR15, pressures F32 R32 psi (normal driving).

EQUIPMENT
Battery 12V, 68Ah. Alternator 75A. Headlamps 110/120W. 21 electric fuses. 2-speed plus intermittent screen wipers. Electric screen washer. Air blending interior heater; air conditioning standard. Cloth/leather seats, cloth headlining. Carpet with heel mat, floor covering. Scissor jack; 2 jacking points each side.

PERFORMANCE

MAXIMUM SPEEDS

Gear	mph (Mean)	km/h	rpm
OD Top (Mean)	134	216	4600
(Best)	134	216	4600
3rd	121	195	5700
2nd	82	132	5700
1st	49	79	5700

ACCELERATION FROM REST

True mph	Time (sec)	Speedo mph
30	3.1	32
40	4.4	43
50	5.8	54
60	7.8	65
70	10.0	75
80	12.6	85
90	16.3	96
100	20.4	107
110	26.1	118
120	37.2	129

Standing ¼-mile: 16.0sec, 89mph
Standing km: 29.4sec, 112mph

IN EACH GEAR

mph	Top	3rd	2nd
10-30	—	—	—
20-40	—	—	—
30-50	—	—	—
40-60	—	—	4.1
50-70	—	—	4.2
60-80	—	—	4.9
70-90	—	6.8	—
80-100	—	7.8	—
90-120	—	9.7	—

FUEL CONSUMPTION
Overall mpg: 18.0 (15.7 litres/100km) 4.0mpl
Autocar formula: Hard 16.2mpg
Driving Average 19.8mpg
and conditions Gentle 23.4mpg
Grade of fuel: Premium, 4-star (97 RM)
Fuel tank: 20 Imp galls (91 litres)
Mileage recorder: 0.2 per cent long
Oil: (SAE API/SE) negligible

BRAKING
Fade (from 89mph in neutral)
Pedal load for 0.5g stops in lb

start/end		start/end
1 30-35	6	30-38
2 30-35	7	35-40
3 30-35	8	35-40
4 30-35	9	30-35
5 30-35	10	30-35

Response (from 30mph in neutral)

Load	g	Distance
10lb	0.10	301ft
20lb	0.35	86ft
30lb	0.58	52ft
40lb	0.90	33ft
50lb	1.10	27ft
Handbrake	0.35	86ft

Max gradient: 1 in 3

WEIGHT
Kerb 31.6cwt/3542lb/1610kg
(Distribution F/R, 55.0/45.0)
Test 35.1cwt/3927lb/1785kg
Max payload 771lb/350kg

COSTS

Prices
Basic	£18,862.87
Special Car Tax	£1571.91
VAT	£3065.22
Total (in GB)	**£23,500**
Licence	£100.00
Delivery charge (London)	£201.25
Number plates	£20.00
Total on the Road	**£23,821.25**
(excluding insurance)	
Insurance group	0A

EXTRAS (fitted to test car)
Trip computer	£305
Fog lamps	£140
Electric audio	£140
Heated seats/lumbar supports	£400
Automatic gearbox	£740
Total as tested on the road	**£25,546.25**

SERVICE & PARTS

Change	Interval 7500	15,000
Engine oil	Yes	Yes
Oil filter	Yes	Yes
Gearbox oil	No	No
Spark plugs	No	Yes
Air cleaner	No	Yes
Total cost	£102.97	£156.10

(Assuming labour at £26.50 an hour inc VAT)

PARTS COST (inc VAT)
Brake pads (2 wheels) front	£39.10
Brake shoes (2 wheels) rear	£20.70
Exhaust complete	£339.14
Tyre — each (typical)	£183.95
Windscreen	£152.55
Headlamp unit	£79.93
Front wing	£227.70
Rear bumper	£366.85

WARRANTY
12 months/unlimited mileage

EQUIPMENT
Ammeter/Voltmeter	●
Cruise control	£425
Economy gauge	N/A
Limited slip differential	●
Power steering	●
Rev counter	●
Self-levelling suspension	N/A
Steering rake adjustment	N/A
Steering reach adjustment	●
Headrests front	●
Height adjustment	N/A
Seat back recline	●
Seat cushion tilt	N/A
Seat tilt	N/A
Split rear seats	N/A
Door mirror remote control	●
Electric windows	●
Heated rear window	●
Interior adjustable headlamps	N/A
Sunroof	£1115
Tinted glass	●
Headlamp wash/wipe	£305
Central locking	●
Clock	●
Fog lamps	£140
Internal boot release	N/A
Locking fuel cap	●
Metallic paint	●
Radio/cassette	●
Aerial	●
Speakers	●

● Standard ONC Optional at no extra cost N/A Not applicable

TEST CONDITIONS
Wind:	0-5mph
Temperature:	15deg C (59deg F)
Barometer:	29.9in Hg (1015mbar)
Humidity:	70 per cent
Surface:	dry asphalt and concrete
Test distance:	780 miles

Figures taken at 3280 miles at the General Motors proving ground at Millbrook.
All Autocar test results are subject to world copyright and may not be reproduced in whole or part without the Editor's written permission.

JAGUAR XJ-S 3.6 AUTO

Transformation: *the XJ-S is still a big, heavy car, but now it can handle the backroads like a 928S*

◀ZF four-speed automatic used in the XJ6 became an option to the Getrag five-speed manual box — the same HP22 transmission with its lock-up torque converter is also used by BMW, Range Rover and Lotus.

The all-alloy 3.6 litre engine has undergone a lot of careful development over the years, and the twin overhead camshaft four-valves-per-cylinder unit is now as smooth and refined as the best — certainly there is no trace of the on/off idle roughness of early examples and the AJ6 unit only begins to sound busy as its 5700rpm redline is approached. The microprocessor-controlled ignition/fuel injection system is a Lucas/Bosch unit integrating the two functions according to the obligatory 'maps' in its memory. It seems to have been responsible for a slight loss in peak power, down from 225bhp at 5300rpm to 221bhp at 5250rpm, but peak torque is up 8lb ft to 248lb ft.

For a car that scales 1610kg, the 3.6 XJS is rapid. Not supercar quick, mind you, but 0-60 in 7.8 seconds isn't bad for an automatic. The ZF box shifts right on the 5700rpm red line under full throttle acceleration, passing the ¼-mile mark in 16 seconds at 89mph and reaches the kilo at 29.4 seconds/112mph. When we last tested a manual 3.6 we recorded 7.4 seconds to 60, a 15.9 second quarter and the kilometre in 28.6 seconds. Top speed with the ZF auto was a little down on Jaguar's claimed 140mph — we managed 134 with the Millbrook bowl's attendant tyre scrub, compared to 141 for the manual version.

Front seats: *now among the very best for support*

Rear seats: *only for the very small...*

Even though a manual 'box would be more fun, you lose little apart from its £740 cost by opting for the auto. The ZF 'box will kick down rapidly when the throttle is floored, changing down two gears at 60 to give plenty of overtaking punch. It can be held in third or second using the shift lever — this has a strong detent to prevent upshifts overshooting into neutral, but the shift handle itself is a curious little toggle that looks like a refugee from an RN-issue duffle coat.

You needn't worry about stopping the XJS. Its all-disc braking system (ventilated at the front) proved virtually free of fade during our (severe) test.

Jaguar's package of chassis mods have transformed the XJS. It's still a big, heavy car, make no mistake, but it can now handle the backroads like a 928S. On the motorway it is much quieter and more refined than the Porsche, although the Jag can't match the shattering performance provided by 5 litres of 32-valve V8.

With the ZF automatic the 3.6-litre engine performs well enough, but the five-speed Getrag 'box would be more in keeping with the car as well as delivering a tad more oomph. The XJS is not only very definitely a sports car now, at £23,500 the 3.6 coupe has to be considered something of a bargain — it's now the car the XJS always should have been. ■

LISTER·JAGUAR XJ·S NAS

Here, kitty—nice, kitty...

BY TED WEST

L ISTER-JAGUAR. To anyone who was seriously interested in sports car racing in the Fifties, the name conjures magic. Lister-Jaguars were great roaring machines with enormous, sensuously flowing fenders and fully faired headrests—brutally fast race cars devised by Brian Lister, who shamelessly held that he could build faster Jaguars than Coventry. And the list of greats who chose to drive his cars—Clark, Moss, McLaren, Hansgen and such—argued his point eloquently. Lister was among the pioneers of the modern "Modified" (later "Sports-Racer"), a breeding line that led, via Bruce McLaren, directly to that singular apparition, the late-Sixties big-bore Can-Am car.

But it's 1987 now and Brian Lister still isn't finished. He continues to believe he can go one-up on Jaguar—and with the current Lister-Jaguar XJ-S NAS, just maybe he can.

One shiny black example of this car was delivered to us by Gary W. Bartlett, president of Lister of North America, a concern located in that madcap maelstrom of import car fever,

Muncie, Indiana. Speaking in a sunny, ingratiatingly broad Hoosier twang that gives the word "might" two distinct syllables—*may-eet*—Gary explained that ever since he's been old enough not to know better, he has been a Jaguar mechanic, parts supplier and incurable Coventryphile. And now, thanks to Mr Lister's XJ-S, Gary's future prospects in the Jaguar business look to be just *faw-een*.

It should come as no great surprise, perhaps, that some wizard aftermarket tuner like Lister has finally gone after the XJ-S. It has always been a lovely car... but maybe just a tad *too* lovely. Refined and mannerly it undeniably is, yet sometimes it seems the XJ-S could benefit from a strong cup of black coffee and some stern talk, just to get its attention.

Well, that's exactly what happened to the Lister XJ-S NAS (for North American Specification, since you asked). Though the economics of a small operation like Lister of North America rule out anything so ambitious as EPA's certifying a heavily modified V-12 engine (in Blighty, Lister offers a breathed-on 400-bhp version), all other systems fundamental to the XJ-S's performance have received very careful attention indeed.

First and most significant, with the Lister XJ-S we at long last have a Jaguar V-12 with a 5-speed! It's not just *any* 5-speed, either, but the same superb Getrag 5-speed found in the BMW 7-Series. Even though, as noted, the car's engine is nearly stock (it *does* have larger tuned air intakes and a large-diameter stainless steel dual-exhaust system for better flow, though no power increase is claimed), the inclusion of a 5-speed utterly transforms the performance of this car. The XJ-S engine has always had good flexibility, but with the Getrag you can locate the sweetest spots in its generous powerband almost at will.

And to make these sweet spots still more fun to exploit, the Lister-Jag is also fitted with a lower, quicker 3.54:1 final drive. (The stock automatic trans version uses a 2.88:1.) Yet amidst all this parts-bin tomfoolery, the car never loses sight of its XJ-S heritage. When cruising along in top gear, its drivetrain yields all the gentle repose you'd expect from such an elegant critter out on the prowl.

We had but two complaints with the drivetrain combination. First, as installed in our example, the Getrag had an insufficient reverse detent, and we had to be very discriminating to avoid reverse whilst going for first. Second, for a car of such

The XJ-S is already a massive, aggressive-looking machine; front air dam, rear spoiler and 16 x 7-in. wheels announce Lister's V-12, 5-speed package in no uncertain terms.

generally civilized demeanor, we found the engine and drivetrain noise just a bit loud at cruising speed.

The Lister-Jag's chassis and handling have received very serious attention; if there is one quality in the standard XJ-S that disappoints more than all others, it is the rather floaty, flaccid nature of its ride control when driven con brio. Therefore the Lister gets the works—adjustable Koni gas-pressurized shocks, a rear anti-roll bar, steel mounting bushings for the rack and pinion, 16- by 7-in. wheels shod with superlow profile, supersticky P225/50VR-16S Goodyear Eagle Gatorbacks and Repco Metal Master brake pads.

And the impact of all this new hardware is no less heady than the inclusion of the 5-speed. The car still retains all of Jaguar's nigh-wondrous suppleness in unruly terrain, yet in more aggressive driving its body roll and general ride control are vastly improved. Compared with the standard XJ-S's often disconcerting body motions, this Lister version tells you it's in deadly earnest the instant you turn the wheel. Turn the wheel more, squeeze on more power, and the steering feel, much aided by the steel mounting bushings, tells you exactly where you are and what you can do about it.

And we repeat, all of this comes with virtually no degradation in the refinement of the ride. Utterly first class, as befits a car of the XJ-S's dignified athleticism.

In the English leather department, Lister fits an excellent padded-leather 14-in. steering wheel. The wheel feels great, though unfortunately it blocks one's view of the turn signal lights and high-beam indicator. The horn button's location under the leather in the center of the wheel is also vague and didn't actuate the horn as cleanly as it should—a minor assembly problem, we were assured, not a design problem.

Externally, Lister adds the full complement of front air dams, side skirts, rear aprons and polyurethane rear deck winglets. These are claimed to reduce lift and drag, and no doubt they do slightly. Realistically speaking, though, at anywhere near American touring speeds their primary function is cosmetic. However, given the fact that the XJ-S shape is now going on 13, like any new teen-ager, it will require some special attention. Accordingly, these various aerodynamic gubbins lend the Lister an updated contemporary look that invariably snaps necks . . . while at this late date a standard XJ-S is nigh invisible.

Finally, in a fond gesture to British civility, the Lister-Jag can be ordered with a full 6-piece set of custom-designed Swaine-Adeney-Brigg luggage (makers to the Queen, of course), each piece lined with moiré silk and built to fit the XJ-S trunk perfectly. We don't know what the Queen pays, but you'll pay a lordly $3190.

The Lister-Jaguar XJ-S NAS is an honest improvement upon an honestly winsome automobile. For XJ-S buyers who want something just a taste more aggressive than the standard item and for those who just want something *different*, Lister of North America may have the answer. A 5-speed transmission and more alert behavior make the Lister-Jaguar the sporty GT coupe many XJ-S admirers have wanted all along. It's a truly slick car—for which you'll have to pay a truly slick bill. With all the equipment noted, you'll need just under $55,500—$15,000 over showroom stock. But if visual impact has a high dollar value for you—and you want robust roadworthiness to back it up—the Lister-Jag may be no bad buy at all.

CONTINUED FROM PAGE 93

driver's seat of the XJ-S was a warm and wonderful place to be.

And that, of course, is why the XJ-S is a magnificent motorcar. The power from the 5.3-liter engine is enormous by 1982 standards, coming on at about 2000 rpm, peaking at 5000, but still available in huge quantities at the V-12's redline of 6500 rpm. Driving through the recalibrated THM 400 automatic, the engine can pull the 3900-lb. XJ-S from stop to 60 mph in less than 7.5 secs., even with anemic 2.88:1 gearing. But beyond quickness and seemingly endless power, there is an almost supernatural smoothness and quietness about the Jaguar V-12 that is not to be found in any other engine anywhere. It enables driver and passenger to carry on a perfectly normal conversation while the car is gathering up motorway at a rate well in excess of two miles per minute. At more reasonable speeds the only sound the driver hears is the occasional cycling of the Jaguar's thermostatically controlled heat/air conditioning system, which is excellent. In the two cars we drove there wasn't the tiniest hint of wind noise or poor sealing.

With 50 years of Jaguar ride and handling development behind it, and 50 years of tire development by Jaguar suppliers, the XJ-S is nearly perfect. The power rack-and-pinion steering is quick (three turns lock-to-lock), very precise, and designed to give the driver excellent feel without roughness. The double-wishbone, coil-spring front suspension and fully independent rear suspension with inboard disc brakes is a Jaguar hallmark system, and it is simply excellent. The XJ-S goes exactly where it is pointed save for a tiny trace of understeer, and pointing it at straights, dips, curves, rocks, or potholes doesn't affect the car's balance at all: It soldiers on without looking back. There is enough shock absorption and wheel travel available at each corner, coupled with constant camber geometry, to keep the car flat and properly aimed over any reasonable road surface. And, although the XJ-S weighs nearly two tons, the distribution is close to 56/44 in a car that's actually slightly smaller than one of the new GM front-drive intermediates and has a 3-in. shorter wheelbase. With these specifications, the XJ-S is either the nimblest heavy car or the heaviest nimble car on the market. In tight situations this nimbleness is enhanced by the automatic transmission, which allows the driver to keep both eyes on the road and both hands on the wheel, and to downshift with the right foot to bring on as much or as little power as he needs.

With the added equipment and the 10% increase in fuel efficiency, the XJ-S will go into the mid-'80s as the complete luxury GT that critics have been demanding of Jaguar for years, and fill that spot until the all-new XJ-40 sports car debuts. Although the body shell design is seven years old, there is still nothing else on the road that looks even remotely like an XJ-S. And even with the record sales year that Jaguar Rover Triumph is expecting, there will be few enough XJ-S cars sold (at a starting price of $32,100) to guarantee some measure of exclusivity for the seeker of a magnificent motorcar.

UNTAMED

For a marque with such an illustrious sporting history, Jaguar have made little of the potential of their XJ-S, which is more long distance cruiser than sports car. The same cannot be said of the 460bhp Lister tested here. Is this the car Jaguar should have built? Peter Dron thinks so . . .

SUPER TEST

CAN A car that costs £55,000 be considered to offer good value? Before answering, suspend memories of your last bank statement and pretend that all the figures on it had five extra noughts on the end, followed by the letters "CR" rather than "DR".

That is Proposition One concerning the Lister Jaguar XJ-S from WP Automotive in Leatherhead, in the heart of the industrial South of England. Proposition Two is less complicated:

Anyone who loves the silken smoothness, near-silent performance and boulevard ride of the standard XJ-S is likely to hate this car.

By the same token, anyone who lists the Porsche 928S4 and Aston Martin Vantage among his favourite cars will be surprised and impressed by the six-litre Lister. That is where the 'good value' claim holds up. The 928S4 is less expensive at present, costing £50,306, while the Aston will sting you for £70,000. That the near doubling in price over the standard XJ-S can, as we believe, be justified, is a tribute to excellent engineering. In any case, as we shall explain, there are ways of reducing the initial cost of a Lister converted car.

There is nothing that can match a well-tuned V12 for its combination of smoothness and excitement. Enlarged from its standard capacity of 5,343cc to 5,995cc by lengthening the stroke from 70 to 78.5mm (the bore is unchanged at 90mm), the Lister's engine is formidable.

One of the characteristics of the standard product is its deceptive performance. There is *nothing* deceptive about this car: it has the engine and running gear of a muscle car, and it pretends to be nothing else. In Aston style, a rocking motion is transmitted to the body at tickover, though otherwise the engine is every bit as smooth as standard. To describe it as substantially louder understates the case by several decibels, but the sound will be music to most ears – a Vantage orchestra with an extra brass section.

In this form the power output is 482bhp at 6,000 rpm (up from 299/5,000) with peak torque of 490lb ft at 4,250rpm (from 391/3,250). It is no wonder, then, that there is a vigorous improvement in performance throughout the range.

Of course, the Lister has the added advantage over the XJ-S V12 HE of a five-speed manual gearbox instead of an automatic, but the top speed increase, from about 145 to something like 170mph (WP's claim is credible, but not verifiable even at Millbrook, Britain's best test track) indicates that transmission differences are no more than a minor element in the transformation from fast luxury tourer to all-out high-performance car.

We recorded a 0-60mph time of 5.4sec, passing the 100mph mark in 12.5sec. Equivalent figures for the 928S4 (top speed 167mph) are approximately 5.5 and 14sec (the forthcoming SE version should be a fair bit quicker), and for the Aston Vantage 5.4 and 12.2sec. The Aston's maximum is around 164mph.

With its massive torque, the Lister will pull without complaint, if you *really* want it to, from 1,000rpm in top gear, and the acceleration figures in top and fourth gears, unsurprisingly, are sufficient to blow most of the opposition away.

Since we drove the car on several non-consecutive days, we were not able to construct an accurate fuel consumption check. However, the brief check that we did make gave a figure of just under 12mpg. Before you gasp in horror, we should point out that we were making use of the extra performance, that a standard XJ-S won't, when driven hard, get you to the good side of 13mpg, and anyway, most ways of having fun cost money. Given a small measure of restraint, the 20.5-gallon tank should take you and the Jaguar more than 250 miles between fill-ups.

As far as transmission is concerned, buyers of V12 Jaguars are given no option by the factory: the three-speed GM 400 is specified and if, like us, you are eccentric and/or old-fashioned enough to prefer a manual gearbox on a high-powered car, you are out of luck.

The five-speed in the Lister is from Getrag, with the 'normal' gate, that is with fifth on a dog-leg. Change quality of this 'box, if a little heavy, as are virtually all gearboxes attached to engines of the stump-pulling variety, is positive and rapid. The clutch gives excellent service once on the move, but can be rather tiring in traffic, not only because of its weight but also because it is rather sharp on initial take-up from rest.

The ratios are ideally spaced, and the car allows its driver the option of using the engine's torque by changing up relatively early. It's more fun, though, to rev to 6,000rpm in every gear. Top gear gives 26.6mph/1,000rpm, and maximum speeds in the intermediates are 36, 64, 94 and 131mph.

Ever since the XJ-S was launched, more than a dozen years ago, journalists have written of it with a mixture of admiration and frustration: it is undeniably a very fine car, smooth and pleasantly deceiving, usually travelling faster than it seems to be. Only when driven in convoy with rival high-performance coupés, most of which come from

SUPER TEST

Germany, do its full abilities become apparent. Yet, if only the suspension could be biased more towards handling than ride quality, if only the steering could be given some real 'feel', if only the XJ-S were a bit more the sports car and a bit less the long-distance cruiser . . .

Jaguar have at last pandered but a little to this kind of opinion with the latest, stiffer version of the 3.6-litre XJ-S. At the top of the range, though, the V12 automatic is left to soldier on alone as the softly-suspended cruiser, frustrating those who wish to extend what many *still* consider to be the world's smoothest road car engine.

All the standard suspension mountings are used, but the Koni dampers are uprated by 30 per cent on rebound, with standard bounce, while the springs are 100 per cent stiffer at the front and 50 per cent stiffer at the rear. Various different makes of tyre have been tested but none has been found to outperform the Pirelli P700s on this car. Sizes are 245/45VR16 on 8.5J three-piece aluminium rims, built to WP's specifications by Compomotive.

For sheer roadholding, even the best of the mid-engined road cars would be given a run for their money by this Jaguar. But it is not all grip and then goodbye: the chassis characteristics ensure that the driver is aware of a steady build-up of side forces, and an equally gradual loss of adhesion as the limit is exceeded, so that there is time to make adjustments when necessary.

The essentially sound geometry of the XJ-S is at the core of the Lister's cornering behaviour, but the two cars could hardly be more different in character. On a bumpy road, the standard model's damping control is satisfactory up to a point, but beyond that point is flaccid rather than sporting. Its body rolls, it begins to lose its poise; the sporting driver will inevitably be saying "If only . . ." to himself.

The Lister hardly rolls at all, understeers gently under power except when the tail is brutally kicked out of line through a tight bend, and all the while the steering gives the driver messages about what is happening, without excessive kickback (though, as in many fat-tyred cars, there is a tendency to follow white lines).

When we drove the car, its suspension settings were near the end of many months of development, but WP Automotive were still not entirely satisfied. It was easy to identify the problem: when the car dropped into a dip while cornering, its suspension was a little slow to right itself. The stiffer-rebound Konis now installed have apparently solved this problem.

Lifting off in mid-bend induces noticeable but progressive tuck-in to the inside of the curve. This is a car with very high limits thanks to plenty of grip and remarkable traction even in the wet, but it has forgiving behaviour when those limits are approached, a practical vindication of the view (which we share) that the best configuration for a high-performance road car (as opposed to racing requirements) is to have the engine at the front and the rear wheels driven.

As may be expected, the WP Jaguar loses much of the factory car's remarkable ride qualities. The severity of bumps and pot holes driven over at low speed is more noticeable to the occupants, but not to the extent that any real discomfort is felt. A tendency to tremble over ridges and ruts around town is in any case a small price to pay for the greatly improved control at higher speeds.

The braking system of the XJ-S V12 is more than able to handle that car's performance, but would be over-stressed in the Lister. To cope with the extra power, discs which are cross-drilled as well as internally ventilated are used. Developed for racing versions of the Lister, they are 13.2in × 1.5in front and 11.5in × 1in rear, with four-pot aluminium calipers and aluminium bells. The servo is standard.

Servo assistance is only enough to take the edge off the required pedal pressure in traffic; quite a firm push is necessary but the resulting retardation is reassuring. We detected no tendency to fade. What was particularly impressive was the ability to brake deep into a corner, 'feather off' and then squeeze on the power gradually, the limited pitch and squat of the chassis, the action of the brakes and the smooth throttle response all conspiring to flatter the driver's ability and technique.

Other than the extra pedal, an appropriately smaller, fatter-rimmed steering wheel, and different seats, the interior is as standard. The driving position is excellent, with room to rest the clutch foot, and a relationship between brake and throttle which makes heel-and-toe changes easy. The standard XJ-S seats have recently, at long last, been provided with better shaping and padding, but the Lister's (leather-trimmed Recaros in this car) are even better, figure-hugging and giving good support to all areas. When you have a car with this performance it is important to be wedged firmly into the seat during fast cornering.

The XJ-S's instrument display, though a bit dated (and never actually a standard-setter) does the job adequately, and certainly the two main dials are large and clear. The air conditioning is also adequate, excellent in hot weather (when it works, as they do seem to have some reliability problems), very efficent at demisting, good at warming the cabin from a cold start, but not fully able to deal with the mixed requirements which British weather conditions impose.

As we mentioned earlier, this is not a quiet car. Dominated by throttle opening is the best way to describe cabin noise: ease off and in the intermediate gears there is a fair level of whine. Put your foot down hard and there is none of the remote purring of a contented pussycat: this is a snarling, ferocious, wild animal, and you'll either love the sound it makes or consider it excessive and anti-social. We loved it.

The price quoted at the beginning of this article, £55,000, was not exactly plucked out of the air, as it is the price of the car as tested, based on a brand-new XJ-S V12. However, as with many converted cars, the price you actually pay depends very much on what you specify. The simplest and most obvious way of cutting the cost is to start with a secondhand car, and these days you can pick up a good secondhand 'S' for well under £20,000.

Our test car came loaded with just about every option. The important thing to remember is that you can buy the whole kit or simply select parts of it.

The body styling kit, for example, is available in several forms. The version tested, with its extended arches, front air dam, side skirts, rear valance and bootlid spoiler costs £7,705 including a full repaint (all prices quoted here include VAT), but a simpler version is available for under £2,200. The twin headlamp conversion costs £684.25, and the modified door mirrors £454.25.

The engine is the area in which there is most scope for varied specifications, right up to a seven-litre version that'll hit you for more than £15,000. The "Stage 3" six-litre with its steel conrods and forged pistons from Cosworth, gas-flowed and polished cylinder heads, with a compression ratio of 11.7:1, special valve springs, lightweight steel followers, re-profiled camshafts, modified induction system with four throttle bodies, reprogrammed injection and adjusted ignition,

ENGINE
Cylinders	V12
Capacity, cc	5,995
Bore/stroke, mm	90/78.5
Camshaft	Chain-driven sohc per bank, two valves per cylinder
Compression ratio	11.7:1
Fuel system	Lucas digital fuel injection with Bosch injector nozzles
Maximum power, bhp/rpm	482/6,000
Maximum torque, lb ft/rpm	490/4,500

TRANSMISSION
Type	Five-speed synchromesh driving the rear wheels through Salisbury limited-slip differential
Internal ratios and mph/1,000rpm	
Fifth	0.760:1
Fourth	1.000:1
Third	1.391:1
Second	2.056:1
First	3.573:1
Final drive	3.54:1

SUSPENSION – STEERING – BRAKES
Front	Independent by double wishbones, uprated coil springs and Koni dampers, anti-roll bar
Rear	Independent by lower wishbones and fixed-length driveshafts, uprated twin concentric coil spring/damper units each side, reaction bar
Steering	Power-assisted rack and pinion with adjustable pressure valve
Brakes, front/rear	Vented front discs, 13.2in × 1.5in, vented rear discs 11.5in × 0.85in

WHEELS – TYRES
Wheels	Three-piece aluminium alloy, 9J × 16in
Tyres	245/45 VR16 Pirelli P700

DIMENSIONS
Length, in	Width, in	Height, in	Wheelbase, in	Front/rear track, in	Fuel tank, gall	Kerb weight, cwt
187.6	70.6	49.6	102.0	58.3/58.9	20.0	35 (approx)

PERFORMANCE
Maximum speed, mph (estimated) 165

Acceleration through gears, sec

0-30mph	0-40mph	0-50mph	0-60mph	0-70mph	0-80mph	0-90mph	0-100mph	0-110mph	0-120mph
2.5	3.5	4.4	5.4	7.0	8.5	10.1	12.5	15.0	17.9

Acceleration in fourth, sec

30-50mph	40-60mph	50-70mph	60-80mph	70-90mph	80-100mph	90-110mph	100-120mph
5.0	5.0	5.2	5.2	4.9	4.7	4.8	5.3

Acceleration in fifth, sec

30-50mph	40-60mph	50-70mph	60-80mph	70-90mph	80-100mph	90-110mph	100-120mph
7.6	6.8	7.1	7.8	8.7	8.9	9.0	—

FUEL CONSUMPTION
Overall test figures, mpg — See text

Makers — Pearce-Lister, WP Automotive Ltd, 10 Mole Business Park, Leatherhead, Surrey KT22 7AG. Tel: Leatherhead (0372) 377474. **Price £55,000** (see text)

costs £10,148.75 plus £534.75 fitting.

To go with that engine, you'll need the tubular exhaust manifolds specially fabricated for efficient breathing and the large-bore stainless-steel exhaust system (with or without central silencers); together these cost £1,943.50 plus £276 fitting.

A good idea at this stage is to add the high-flow water radiator, claimed to improve cooling effect by 40 per cent, and priced at £282.33.

WP do not supply a standard suspension package, as they feel it would not suit all tastes. Instead, they specially manufacture springs to individual requirements (different spring rates do not add to the cost). The basic spring/damper kit, fitted, costs £914.25, the reaction bar (to improve rear axle location and prevent tramp) adds £382.38, and the rear anti-roll bar £306.25.

An adjustable power steering valve can be supplied and fitted for £342.70, so you can have as much or as little resistance as you choose. To improve steering stability, you can have special rack bushes, fitted for £103.50.

A range of wheels and tyres is offered, but the best package is probably the one fitted to the test car: the P700s cost £274.85 each, and the wheels £204.70 each. The front brake system costs £2,127.50 fitted, and the rear brakes £632.50.

An essential part of the package, in our opinion, is the manual gearbox, which is £4,542.50 fitted. The 3.54:1 final drive, with uprated Salisbury Power-lock limited-slip differential, costs £759 plus £276 fitting.

If you still have some spare change left over after all this, a pair of Recaro seats trimmed in leather to your personal requirements costs £2,280.50 plus £414 fitting, and a complete re-trim, of the door panels, rear section etc, £3,220. Lastly, the 'sports' steering wheel adds £149.50. Oh, and you don't *have* to have left-hand drive.

You could keep your XJ-S looking absolutely standard, and really worry owners of Ferraris and Porsches.

However, although we generally make rather sniffy, disapproving noises about body kits, in the case of the XJ-S we are only too happy to see the dull side elevation altered by wheel arch extensions. They significantly change the visual balance of the car, we think for the better. We'd prefer the arches to be more curved, as they were in Lister's first conversions, but apparently the current shape is more popular with buyers. The finish, even of the well-used development car that we drove, is well in keeping with the price.

The Lister is much more of a sports car than is the 928S4, though it will be interesting to see how the Special Equipment version of the Porsche shapes up.

Does it rival the 'factory acknowledged' TWR range (*Fast Lane*, August 1987)? Surprisingly perhaps, the Lister is at least as well executed. It's unashamedly sportier and takes the whole conversion one stage further. TWR pay more attention to items like super quiet exhaust systems and their treatment, at the moment, is perhaps more coherent as a complete package. Lister, however, offer a more individualistic approach to suit each customer.

Maybe Sir William Lyons would not have approved of this modified Jaguar. Although Jim Randle, Jaguar's present chassis engineer, would almost certainly enjoy driving it, he might well say, "But it's not a Jaguar". However, it is exactly the type of car that many present owners of Ferraris, Porsches and Aston Martins, who find the standard XJ-S lacking in appeal, might consider.

As far as we are concerned, this is the type of XJ-S that Jaguar should have built, not to replace the standard car, but to widen its appeal and enhance the company's prestige.

NEWSWEEK

Jaguar's first full convertible since the E-Type roadster will go on sale in the UK within weeks. But more than half of the convertibles built at Coventry are US-bound, where the Americans are clamouring to buy

XJ-S Convertible a snip at £40,000

ONE OF the major attractions — if not *the* major attraction — at this week's Geneva Motor Show will be the new Jaguar XJ-S V-12 Convertible which will go on sale in May with a £35,000-plus price sticker. Jaguar are already hailing their forthcoming two-seater offering as one of the world's fastest and most refined fully open-top cars, with only three or four genuine rivals. Where Jaguar will have the upper hand on these rivals will be in price: the new XJ-S Convertible will certainly not be more than £40,000 and will therefore be comparatively inexpensive.

A power-operated fabric tailored hood, fully lined and insulated, can be raised or lowered at the touch of a button in 12 seconds flat. A novel feature is the tinted glass and heated rear window (instead of the more usual flexible rear window featured on so many lesser convertibles).

As revealed in *Motor* (w/e Feb 27), the Teves anti-lock braking system is now fitted as standard throughout the XJ-S range, and the 155 mph Convertible version will, of course, receive the Teves treatment.

Also featured as standard equipment when the car appears in UK showrooms for the first time in May will be full air conditioning, top quality hide upholstery, and matched walnut veneers. Although the car is still a couple of months away from making its UK showroom debut, orders are flooding in, according to a Jaguar spokesman. The fact that the new offering is Jaguar's first full convertible since the E-Type roadster — which ceased production in 1974 — is just one of many factors excited potential buyers have firmly in mind.

Around 4000 to 5000 Convertibles will be produced by Jaguar this year, and well over half of the cars built at Coventry will be shipped to the USA.

Only limited official details are being released on the car at this stage, Geneva being used merely as an occasion to whet appetites. Full details and the exact pricing will be announced just prior to the official launch in May.

What seems clear, though, is that Karmann, the German coachbuilder, were responsible for the engineering work and designing the hood mechanism on the new Convertible. The hood design is said to be remarkably simple: two catches secure it to the windscreen and after being released a switch on the dashboard causes the electric top to fold back and fit snugly into a recess behind the car's two seats. The Porsche 911 is the only other European sports car to have a power top.

Remember where you saw it first: Motor's first-ever shot of Jaguar's new XJ-S Convertible (disguised by fake B-pillars) early last year

What a difference a year makes. Twelve months ago in the wake of the Big Bang, life was rich for those so-called young urban professionals who earned fortunes in the city, drank champagne at lunchtime and raced around the M25 during the night.

A company BMW, or better still, a Porsche was an essential part of the brief, along with those other lifestyle trappings – a Filofax, mobile Cellphone and swish London apartment. While the really high achievers could quite often be seen behind the wheel of a 928 or 911, those further down the ladder (but who were still doing well, mind) would opt instead for one of the four cylinder cars – a 944 perhaps, or failing that, a 924S.

For without doubt, in terms of status, prestige and style, the Porsche badge said it all. If you wanted to show you'd arrived, but weren't yet ready for that move up to the S-Class Mercedes, you just had to have one.

But now, during the first months of 1988, the picture doesn't look so rosy. We've had Black Monday and the resultant stock market crash. The tabloids are predicting financial gloom and despondency and even going so far as to forecast the imminent demise of the Yuppie, the very phenomenon they actively helped to create.

For Porsche too, the signs are currently not so promising. A well publicised down-turn in sales of its staple four cylinder 924/944 models, coupled with the shock departure of its chief executive Peter Schutz has stirred the normally calm, prosperous waters of the Stuttgart sports car maker. As even the cheapest Porsche now costs well over £20,000, some buyers are giving up on the prospect of Porsche ownership and opting instead for one of the less exalted but still highly rated Japanese sports coupes, such as the Toyota Supra, or even the smaller Celica as well as cars like the Honda Prelude.

But no matter how competent the products from the fast-rising east or the prevailing state of the worldwide financial markets, there will always be a demand for exotics in the Porsche mould – for high quality cars with the 'right' aura and class, designed for customers for whom price isn't necessarily an overriding consideration.

At this quality end of the market, you're buying, above all, a prestige name. In the sports bracket, the front runners are indisputably (and alphabetically) BMW, Ferrari, Jaguar and Porsche. However a look at their product list reveals a curious fact – it's not often they compete directly with one another, model for model. You won't find a £25,000 Ferrari for instance challenging a Porsche coupe, nor a V12 Cabriolet BMW offered for sale (not yet anyway). They might sound good but such cars don't exist.

Focusing on that £25,000 figure, though, we do find two coupes of comparable class and character – Porsche's 944 and the Jaguar XJS 3.6. On the surface, a perhaps unlikely pairing but in reality, two cars which find a surprising amount in common.

Both for example derive from mid-seventies designs (the 944 evolved from the lesser 924) yet both have managed to ride the years with panache. True, the styling is different as are the basic dimensions yet both stand as comfortable, well equipped upmarket high performance coupes, aimed at the same kind of enthusiastic, image-conscious driver who, as we said, has money to spend on something a little extravagant, on something that boasts style *and* performance.

Other similarities? In each case, the mechanical layout is one of front engine-rear drive, all independent suspension, all-disc braking and power steering. Both will power onto and past 135mph with ease (according to the publicity material) and sprint to 60mph in around 7-8 seconds so they're fast, no question.

And yet, it wasn't until quite recently that we would have considered pairing 944 with XJS despite all of the similarities mentioned above. That's because the Jaguar has always been too softly sprung to be rated a true sporting car – too much of a boulevardier for the type of driver who would normally go for a Porsche.

However last autumn, Jaguar made some important changes to the steering and suspension of the 3.6-litre XJS, significantly raising its spring and roll bar rates and specifying new Pirelli P600 low profile tyres too. The power

> Power, performance and prestige are the elements that count, as Jaguar's XJS 3.6, now with uprated steering and suspension, challenges Porsche's 944 in the luxury coupe class. Which is our favourite?

PRESTIGE

PERFORMANCE

rack and pinion steering gained extra weighting for improved 'feel'.

The factory also made some worthwhile changes to the Jaguar's opulent interior, bringing in sports front seating and a thicker, leather-rimmed wheel. All this was in keeping with the XJS's new role, as that of the car for the 'younger, more sporting driver.'

The 3.6-litre XJS, at £23,821, comes in at the foot of Jaguar's coupe range which extends upwards to the quicker V12 versions that are available in either open or closed forms. Running with the six cylinder, 24 valve DOHC engine that also appears in the XJ6 saloon, the coupe-only Jaguar XJS 3.6 produces a hefty 221bhp at 5100rpm coupled with an excellent torque figure of 249lb ft, achieved at 4000rpm.

The power and size of the Jaguar engine (not to mention the car itself) certainly contrasts with that of the 944 which, in its latest guise, sports a 160bhp version of the four cylinder 2.5-litre engine that also features in the £21,030 baseline Porsche, the 924S.

Above the £24,237 944, there's a 16 valve option in the form of the 944S (190bhp, £26,994), then the sensational Turbo, costing £36,080 and producing 220bhp. In the pipeline is a 944 Cabriolet but that is still some time away from British showrooms.

Like the Jaguar unit, the 944 engine has its electronic fuel injection and ignition circuits controlled by a highly complex under bonnet 'brain'. Unusually though, the Porsche's engine block carries twin balancer shafts to 'iron out' the unwanted dynamic forces inherent in such a big capacity four cylinder engine.

The Porsche also sites its standard five-speed manual gearbox at the back for good theoretical weight distribution. The XJS, by contrast, has its transmission mounted conventionally at the front with the engine. A four-speed ZF auto is a £780 option on the Jaguar while Porsche offer a three-speed automatic facility on the 944 for £980.

TWIN TEST

Though the Porsche may lose out to the XJS under the power and performance headings, it scores impressively elsewhere. Not only is it lighter, more aerodynamic and compact than the heavyweight 2-dr XJS, it wins on government economy and warranty cover too, offering a two year policy to the Jaguar's one. On luxury equipment, both cars fare well but as regards ABS brakes, Jaguar doesn't offer anti lock circuitry at all on the XJS while on the Porsche, ABS is a very expensive £2022 option.

PERFORMANCE
JAGUAR	●●●●●
PORSCHE	●●●●○

Two very quick cars but it's the XJS that's the faster of the pair, both in outright speed and initial acceleration. Behind the wheel, the Jaguar *feels* fast too, as it should do with over 220bhp on tap beneath that massive bonnet.

That's not to decry the Porsche, though. Its 2479cc engine, newly engineered so that it produces the same power whether or not it's fitted with an exhaust catalysor (for emission-conscious markets such as the US and Germany) is a gem, spinning smoothly and sweetly throughout its peak power band. As a result of the modifications, it's lost 3bhp, dropping it to 160bhp but gained an extra 4lb ft of torque for improved pulling power. However as that maximum torque reading now occurs 1500rpm further up the rev scale (at 4500rpm), the benefits of the higher torque seem to be all but cancelled out.

Indeed, the revised engine makes little or no difference to the Porsche's performance. In fact, if anything, the 944 is actually a little slower than it used to be, which may be attributable to its slightly increased weight.

The best we could squeeze out of our test car was 130mph – still wildly in excess of the national limit of course but not quite the 135mph figure that Porsche quotes in its brochure. No problems with standing start acceleration: there the 944 really does live up to expectations, sprinting from 0-60mph in a forceful 8.0secs and touching 100mph in under 24secs.

It pulls vigorously in third, fourth and fifth too and cruises easily at motorway speeds. But not as serenely as the Jaguar whose twin cam engine is ticking over at a gentle 2500rpm at the legal limit. The XJS is the quieter of the two in terms of noise suppression and also the one with the greater speed potential thanks to its bigger capacity, more powerful engine.

But remember the Jaguar's power advantage is compromised by a much heavier kerb weight and poorer drag. Study the figures and it emerges that as far as the respective power/weight ratios are concerned, the Porsche and Jaguar are in fact pretty evenly matched, at around 130bhp/ton.

Still, there's no holding the XJS on acceleration where its 61bhp advantage comes into its own along the test straight. Our Pieseler speed equipment clocked the XJS from 0-60mph in just 7.1secs which beats Jaguar's own figure by 0.3sec. That's good going but around the high speed banking, the XJS's timed maximum of 136mph wasn't quite what we were expecting. The factory claims 142mph for the manual 3.6.

Figures aside, the Jaguar impresses with its raciness and refinement. Its AJ6 engine, originally disappointing in terms of top end behaviour is now a reformed character. Gone is the former coarseness at highish revs, to be replaced by a degree of smoothness that parallels that of its majestic V12 cousin. It revs very freely – indeed, it pays to keep a sharp eye on the tacho to prevent over-revving past the unusually low 5700rpm red line – and like the sophisticated Porsche engine, the AJ6 only begins to sound strained when it's working hard.

Third is a useful overtaking ratio with the XJS, particularly from 50mph onwards at which point its pick-up is little short of sensational. There, its huge reserves of torque coupled with the free-breathing benefits of its four valve per cylinder head really come to the fore but lower down, it's just as flexible. Running right alongside however is the Porsche whose gear-for-gear overtaking times are, remarkably, an almost exact match for the Jaguar's – but not quite as sharp as those of the first 944 we tested back in 1983.

The enthusiast driver will by tradition opt for a manual gearchange wherever possible. Best of these two is the Porsche's whose short, stout lever moves in a firm but positive manner through the tight five-speed gate. It's a well-engineered shift but in no way as light and easy as the Japanese can produce. The notchy XJS transmission is even harder work and it's this, coupled with a heavier clutch action that might – whisper it – even prompt us to go for the auto route instead.

HANDLING AND RIDE
PORSCHE	●●●●○
JAGUAR	●●●○○

There are many hair raising anecdotes about the 911, the traditionalist's Porsche, but very few about the 944. That's because the newer, front-engined car is easier to drive and has always had more predictable handling. Overcook it in a 911 and you could be in trouble. In a 944, you actively have to *try* hard to get in the same situation.

Typically, the Porsche, with MacPherson strut geometry at the front and semi trailing arms behind, drives in a neat, competent fashion with plenty of tyre grip promoting high cornering power. It feels light and manoeuvrable, thanks to quite excellent power steering that at 3.3 turns lock to lock, is ideally geared and which provides good weighting and feel too.

On dry roads, the Porsche corners neutrally but push hard and the front wheels will eventually start to lose grip through understeer. Obviously in the wet, this phenomenon happens at a much lower level but in each case,

JAGUAR XJS 3·6

New style Jaguar XJS gains stiffened suspension plus new Pirelli tyres for sportier handling. Cornering is sharp, grip strong but suspect steering spoils dynamic package. Potent six cylinder twin cam engine produces 221bhp, giving excellent performance. Traditional wood and leather is attractive but headroom limited

simply backing off on the throttle sees the car step back into line in a gentle, progressive manner.

What happens if you lift off in a hurry? Not much. To drive the Porsche's heavy tail end out shape calls for some degree of provocation. Then, catching those wayward rear wheels and preventing a spin can ultimately be tricky. Normally though, this side of its character remains safely hidden and the 944 can be cornered hard with plenty of confidence.

As we've said in the past, the four cylinder Porsches (with the exception of the 944 Turbo) could do with firmer damping to eliminate some of the body jinking that crops up under hard braking and acceleration. A set of sports dampers, available for an extra £260 would probably do the trick. That said, we'd be wary of firming the suspension up too much for fear of making the ride even harder than is the case already.

The 944 isn't actually uncomfortable but on most road surfaces, its low profile tyres patter noisily and the suspension seems to make unnecessarily hard work of minor bumps and ruts, let alone big ones. There's a fair amount of bump thump to endure and this presumably would be even more evident should the 944 be wearing the optional ultra-low profile tyres that cost the best part of £1000.

With its tougher springs and gumball Pirelli tyres, the new XJS is a very different animal from its soft-riding predecessor. For a start, it's much more fun to drive, offering much sharper cornering coupled with quite astonishing levels of roadholding. Gone is the characteristic floatiness through fast bends and lightness of the power steering system, both hallmarks of the old car. Instead, the XJS has become far more responsive, virtually to the point of becoming nervous, which, it has to be said, doesn't engender an immediate feeling of driver security, of the kind you get straightaway behind the wheel of the four cylinder Porsche.

The steering, for example, despite its improved weighting, is still too light and doesn't possess the precision of the 944 system, which allows the car to be placed on the road with almost millimetre accuracy.

The high geared XJS steering is free of the wheel kickbacks that can unsettle the Porsche on bad roads, but still, for all that, it doesn't feel right: over-sensitive one moment on initial turn-in to a sharp bend yet slack in the central position and sluggish in response to small correctional inputs. A wide turning circle doesn't help matters either when manoeuvring.

However the Jaguar's basic handling is very good and few will get near to finding out how it will behave *in extremis* – not with sane driving on public roads anyway, for its cornering limits are extremely high. The tyres hold on well, the car feels well balanced and through fast, sweeping bends the Jaguar remains poised, staying on line in convincing fashion.

Be careful with that throttle pedal

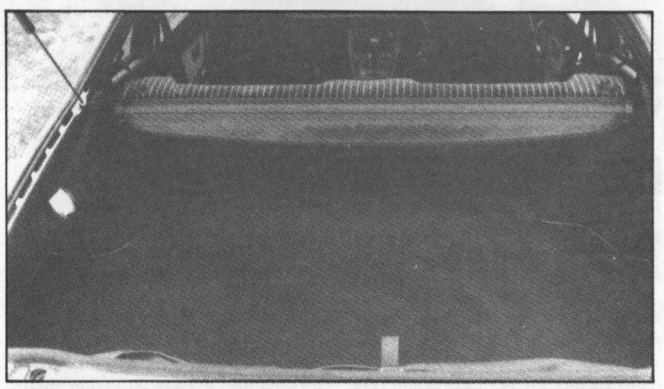

Though smaller than the XJS and running with just four cylinders to the Jaguar's six, the £25,000 944 still costs more than the British car. However, for style, sophistication and impeccable build quality, the Porsche remains as convincing as ever. Interior is compact, particularly in the rear. Revised engine provides 160bhp

though, for the Jaguar, big powerful car that it is, can be a real handful with careless driving. It corners predictably with progressive understeer as its normal trait but suffice to say, when it does finally let go, say, if too much power is applied before the exit of a tight, wet surfaced bend, then it takes skill and quick reactions to stop the resultant tail end breakaway. The Porsche is much harder to unstick.

Those special Pirelli tyres must also play a part in the Jaguar's ride quality which is comfortable at high speeds (more so than the Porsche) but pretty bumpy around town. Granted, it's knobbly rather than crashy but still not as compliant as one might expect from a luxury car like the Jaguar.

Braking on both cars was firm and progressive but the Jaguar, we found, had a habit of braking to the left.

ACCOMMODATION

JAGUAR	●●●●
PORSCHE	●●●

The XJS is much the bigger of the two cars, standing almost *2ft* longer than the Porsche. It's also 2in wider. But for such an extravagantly long car, it's short on interior space to an almost ridiculous degree.

The cabin is relatively small but at least there is room for four to travel in comfort. Not so with the Porsche which qualifies as a 2+2, nothing more, its token rear seats being suitable for very small children only.

There are further differences. The Jaguar, its narrow side windows promoting a marked feeling of claustrophobia, can initially feel daunting. Driver and passenger sit either side of a wide transmission tunnel: ahead, there's the impressive view out over the Jaguar's huge curving bonnet – behind, restricted visibility through the Jaguar's narrow rear glass. The Porsche, by contrast, feels more manageable, more intimate.

Inside the XJS, there are front sport seats as standard, trimmed in cloth and leather. Full leather upholstery is an option, together with, on the driver's side, electric lumbar support and internal heating. The seats are quite narrow yet support well, look good and prove comfortable over long journeys.

Restricted headroom is a problem for the tall XJS driver (especially if the car comes with the optional £1115 sunroof). If the seat base had a tilt facility, it would help but it's not offered. Still, there's plenty of front seat travel on offer plus the familiar Jaguar telescopic adjustment for the new style, chunkier steering wheel.

In the back, the XJS is on the cramped side, kneeroom in particular being restricted along with interior width. Passengers sit close together on thickly bolstered seats – but what a contrast to the Porsche whose rear seating seems to be for show only. There, the steeply raked rear screen virtually eliminates headroom altogether while available kneespace is also at a premium.

It's much better news in the front of

■ TWIN TEST   Jaguar XJS 3.6, Porsche 944 ■

the Porsche where there are a pair of rather gaudy, high backed sports seats; the driver's getting electric tilt adjustment for the cushion as standard. For the passenger's to be so equipped, costs £618 extra.

Some testers found the Porsche's extrovert seating to be initially uncomfortable due to the arching of the backrest across the shoulder blades. But before long, you do get used to it. Certainly the softly cushioned 944 seats provide excellent support for the thighs and sides of the back, the result being that during fast cornering, you're not thrown about.

The driving position is a good one and most drivers should be able to sit comfortably in the Porsche even though the wheel doesn't adjust in any way. Some would prefer the wheel slightly higher than it is. You sit close to the floor and leg and headroom are both generous. For ease of entry/exit though, it's the Jaguar that's the more practical car.

Boot space? The honours must go the way of the XJS although frankly, there's not much in it. The 944 offers a flat, shallow load bay beneath its hatchback tail. By folding forward the top section of the rear backrests, the quoted capacity is virtually doubled to close on 8cu ft and the load floor is usefully extended. But even so, the 944 can't carry that much so holiday bag and baggage will have to be limited.

A useful touch in the 944 is the fold across screen to keep what luggage you are carrying out of public view. Another is the availability of a split rear backrest for an additional £350, thus adding to the versatility of the Porsche's luggage area.

The XJS boot is short but deep and served by a conveniently low load sill. At 13.3cu ft, its capacity is that much bigger than the Porsche's but considering the sheer size of the Jaguar's graceful tail, you would have thought the boot be even more accommodating than it really is. Again, a sad example of wasteful packaging.

LIVING WITH THE CARS	
PORSCHE	●●●●
JAGUAR	●●●●

Contrasting styles here, with the Jaguar offering the traditional British wood and leather luxury treatment while the Porsche eschews such olde worlde trappings for an altogether more clinical yet still luxurious, close fitting interior decor.

You're greeted by that familiar, intoxicating smell of leather as soon as you open one of the Jaguar's doors. Inside, the veneer finish of the centre console, dashboard and door cappings is equally inviting but alas, the XJS instrument cluster housing those four barrel gauges for volts, petrol, oil pressure and water temperature, looks as awful as it did back in 1975.

The Porsche instruments, grouped together in a large oval display, are considerably clearer (even if the temperature gauge is still partially obscured by the rim of the wheel) and thus far easier to assimilate at a glance. The same goes for the 944's minor controls – they're clearly labelled and simple to use. The Jaguar also fares well under this heading but details such as the awkward headlamp switch are the kind of thing you just wouldn't find inside the German car.

If the Jaguar's interior exudes a gentlemen's clubroom type atmosphere, the 944 (or at least that of our test car) could hardly be more different, being furnished in such a brash purple and white colour scheme it almost offended the eye.

However for fit and finish, the 944 shows how it should be done – it looks and feels what it is, a quality made product that will still be in fine shape and rattle free long after the warranty period has run out. The Jaguar is screwed together well too these days and constructed from high quality materials as befitting its station in life. It's certainly a more impressively made product than the early examples of its kind but whether it will stay the course in as good condition as the Porsche over the coming years is something that remains to be seen.

Yet for overall refinement, there's no dispute: it's the Jaguar that wins. Though the 944 qualifies as a suave and civilised machine, it has to give best to the XJS where ride suppleness and noise absorption are concerned. Due to some wind roar and road noise at speed, it's not quite as refined in these areas as the Jaguar.

However it still comes across as a high quality, precision piece of engineering. The engine is a delight and the car drives smoothly, not to say effortlessly over long distances. A shame then, that the Porsche's less than perfect suppression of road rumble is so evident.

Both cars come well stocked with a lengthy standard equipment list. Still though, even at this price level, there's plenty of extras that can be added on, cruise control, sunroof and full leather trim being just tree of the items available on the list.

Specify these three from your neighbourhood Porsche dealer and you won't get much change from £3300. For the Jaguar, the bill is much cheaper – around £2000 – simply because the price of factory leather trim is around

XJS is fast, luxurious and thanks to suspension work, more fun in uprated 3.6-guise

■ TWIN TEST   Jaguar XJS 3.6, Porsche 944 ■

one sixth of the price! In each case, the cost of an electric metal sunroof is about £1000 and cruise control £400. A limited slip differential is standard on the XJS but a £725 extra on the German car.

With the XJS, air conditioning is factory-fitted and once mastered, the complex system works well. Porsche charges £1647 for full climate control. Is this money well spent? We're not so sure, for the standard heating system is pretty sophisticated, splitting hot/cold air as required, providing good through flow ventilation independent of the four-speed fan and having such refinements as a quick defrost facility for cold winter mornings.

COSTS
PORSCHE ●●
JAGUAR ●●

There's no doubt that the cost of having a 944 serviced and repaired by one of the UK's 31 official Porsche Centres won't be cheap by any means. Insurance, at group nine, is also pretty frightening and the initial purchase price of some £25,000 before options, isn't a bargain either.

And yet the 944 isn't perhaps the extravagance it seems. True, it's risen inexorably in price from the moment it first appeared in Britain back in 1982. Indeed, when we first tested it in 1983, it was under £15,000 so in the space of some five years, it's gone by no less than £10,000! Furthermore, during that time, it's remained hardly altered at all. Compare an original car with one of the new ones and aside from the restyled dashboard, you'd be hard pressed to tell the difference.

But now the good news. Expensive though it undoubtedly is, the 944 still represents a sound financial investment. After all, few cars hold their value so convincingly as a Porsche (especially so now that the prices of new ones are so high). And it's this exceptionally strong resale factor that, in the end, should theoretically offset some of those high running costs.

It scores under economy too, offering a 30mpg potential with careful driving. We saw 22mpg overall from our test car however, with over 20mpg on offer even under performance testing. Over the same course, the Jaguar provided figures of 20.8 and 16.5mpg respectively – not as frugal as the Porsche but reasonable nonetheless for a big, high performance car driven enthusiastically.

On its twin tanks, the Jaguar thus has a potential fuel range of well over 400 miles. The more economical Porsche can manage the same feat on its single, smaller tank.

Economy aside, the Jaguar should be roughly on par with the Porsche as far as overall running costs are concerned. It's in the same insurance bracket for instance and parts and option prices, though expensive, are generally comparable.

Perhaps its biggest asset is purchase price. The XJS is listed at over £1200 cheaper than the Porsche and this must obviously have an important bearing on the situation. Discounts aren't likely on either car. But while the Jaguar is clearly the hands-down winner for value for money while new, long term, we wonder how well it will stand up on the secondhand market.

Servicing sees the Jaguar with an important advantage, needing major attention every 15,000 miles to the Porsche's 12,000-mile intervals (both cars also requiring check-ups at each half way point). That the Jaguar dealer network is over three times the size of Porsche's approved operation must also count in the British car's favour.

But for warranty cover, it's the Porsche all the way. Alongside the German car's multi year package covering mechanical components, paint and anti corrosion, the Jaguar's single year guarantee looks pretty feeble by comparison. There's no rust protection policy either.

VERDICT
JAGUAR ●●●●
PORSCHE ●●●●

At the finishing post, it's a win on points for the Jaguar, not a knock out victory as some predicted at the outset.

Was it right, they said, that we were going to compare six cylinder Jaguar against four cylinder Porsche? Surely, the argument continued, they're not the same kind of car, each appealing to its own individual and quite different style of customer. There was even a shout of 'unfair' from one camp when this test was suggested.

We don't agree. The impartial buyer with around £25,000 to spend on something fast and glamorous is surely going to look at both cars: the 944 is perhaps the standard setter in the class, the XJS recently rejuvenated to make it more of a match for the smaller sporting Porsche.

You can make a case for giving either one the vote. If you want a car that won't depreciate, that exudes class and quality and which stands as one of life's acknowledged status symbols – go for the 944.

If, on the other hand, you want a prestige product that gives you even higher performance, usable interior space for four plus greater value for money, then turn towards the XJS.

Choosing between the two isn't easy and frustratingly, neither car is perfect. The Porsche, though beautifully built and refined to drive, is exceptionally expensive for its size and capabilities. The Jaguar offers more for less but we're not convinced about the handling modifications. The car is very fast and improved over earlier XJSs but feels nervous when driven hard.

In the final analysis, the Jaguar's healthy price advantage, superior performance and space are factors that are hard to ignore. Its attractive wood and leather interior treatment appeals too. It's a big car, the XJS, less nimble perhaps than the Porsche and not as sexy looking but still, for all that, a highly desirable proposition. The 944 is good, but for the money, the Jaguar is better.

HOW THE CARS COMPARE

CAR	JAGUAR XJS 3.6	PORSCHE 944
PRICE	£23,500	£24,771
Other models	1 coupe 1 convertible	2 coupes
Price span	£23,500-£31,000	£24,771-£36,874

PERFORMANCE

Max in 5th (mph)	136	130
Max in 4th (mph)	117	112
Max in 3rd (mph)	84	82
Max in 2nd (mph)	59	57
Max in 1st (mph)	33	33
0-30 (sec)	2.4	2.6
0-40 (sec)	3.6	4.1
0-50 (sec)	5.1	5.8
0-60 (sec)	7.1	8.0
0-70 (sec)	9.2	10.6
0-80 (sec)	11.8	13.7
0-90 (sec)	15.3	18.2
0-100 (sec)	20.0	23.6
0-400 metres (sec)	15.6	16.1
Terminal speed (mph)	89	86
30-50 in 3rd/4th/5th(sec)	4.8/6.9/10.3	4.9/7.4/11.3
40-60 in 3rd/4th/5th(sec)	4.5/7.0/10.9	4.8/7.1/10.9
50-70 in 3rd/4th/5th(sec)	4.3/7.0/11.1	5.0/7.2/11.0
60-80 in 3rd/4th/5th(sec)	4.9/7.0/11.9	5.8/7.4/12.4

SPECIFICATIONS

Cylinders/capacity (cc)	6/3590	4/2479
Bore x stroke (mm)	91 x 92	100 x 79
Valve gear	dohc	ohc
Compression ratio	9.6:1	10.2:1
Fuel system	injection	injection
Power/rpm (bhp)	221/5100	160/5900
Torque/rpm (lbs/ft)	249/4000	155/4500
Steering	PA/rack/pin	PA/rack/pin
Turning circle (ft)/(Turns)	42.7 (2.5)	34.0 (3.6)
Brakes	Di(v)/Di	Di(v)/Di(v)
Suspension – front	I/Wi/C/AR	I/McP/AR
rear	I/Wi/C/RA/AR	I/STA/Tor/AR
Tyres	235/60VR15	195/65VR15

COSTS

Test mpg	16.5-24.0	20.7-24.4
Govt mpg City/56/75	18.6/37.1/32.0	22.4/41.5/32.8
Tank galls (grade)	20.0 (4)	17.6 (4)
Major service miles (hrs)	15,000(4.45)	12,000(3.6)
Parts costs (fitting hours)		
Front wing	£198.00 (N/A)	£194.14 (2.4)
Front bumper	£365.90 (N/A)	£69.33 (0.8)
Headlamp unit	£69.50 (0.3)	£35.13 (0.7)
Rear light lens	£39.75 (0.25)	£47.56 (0.6)
Front brake pads	£34.00 (0.55)	£36.83 (0.7)
Shock absorber	£25.00 (0.55)	£99.99 (1.2)
Windscreen	£137.00 (4.3)	£291.81 (3.5)
Exhaust system	£294.90 (2.2)	£385.76 (1.9)
Clutch unit	£137.90 (3.5)	£177.16 (7.8)
Alternator	£180.00 (0.45)	£266.70 (1.4)
Insurance group	9	9
Warranty	12/UL	24/UL
Anti-rust	no	10 yrs
Paint	no	3 yrs

EQUIPMENT

Automatic transmission	£740	£980
Power steering	yes	yes
Alloy wheels	yes	yes
Anti lock brakes	N/A	£2022
Cruise control	£425	£384
Air conditioning	yes	£1647
Electric windows	yes	yes
Sunroof	£1115	£1007

DIMENSIONS

Front headroom (ins)	36	36
Front legroom (ins)	35-42	34.5-42
Rear headroom (ins)	36	32.5
Rear kneeroom (ins)	18-24.5	18-26
Length (ins)	187.6	165.3
Wheelbase (ins)	102.0	94.5
Height (ins)	49.7	50.2
Overall width (ins)	70.6	68.3
Kerb weight (cwt)	33.0	24.8
Boot capacity (cu ft)	13.3	4.8/7.7
Drag factor	0.40	0.35

KEY. Valve gear: ohc, overhead camshaft; dohc, double overhead camshaft. **Steering:** rack/pin, rack and pinion; PA, power assistance. **Brakes:** Di(v), ventilated discs; Di, discs. **Suspension:** I, independent; AR, anti roll bar; C, coil springs; McP, McPherson struts; RA, radius arms; STA, semi-trailing arms; Tor, torsion bar springs; Wi, wishbones

XJ-S V12 Convertible

AUTOCAR TEST EXTRA

The E-type is dead, long live the XJ-S convertible. Fresh-air fiends can once more indulge their passion in the grand style and enjoy Jaguar's V12 performance and handling to match

Price £36,000 **Top Speed** 144mph **0-60** 8.0secs **MPG** 13.8
For Performance, hood mechanism **Against** Economy

Chassis refinement and handling are up to usual high Jaguar standards

MENTION THE E-TYPE ROADSTER TO any motorist with soul and they will smile. One of *the* classic British sports cars: stunning in appearance, muscular in performance and a full convertible. Production ceased in 1974 and we had to wait patiently for 14 years for the next factory-built full convertible Jaguar, the XJ-S V12 convertible.

It replaces the V12 Cabriolet which has been the S-range flagship since its launch in 1985 and in the region of 2000 were sold worldwide last year. Of course there is a big difference between a cabriolet with its strengthening roll over bar and a convertible which has to have a completely re-engineered body to replace the strengthening lost by removing the roof.

Jaguar has spent three years developing the V12 convertible in peace and quiet, as the bulk of the work was overshadowed by the XJ40 project. Not all the work was done in-house, however, as Jaguar — probably quite sensibly — liaised very closely with Karmann of West ▶

TEST EXTRA

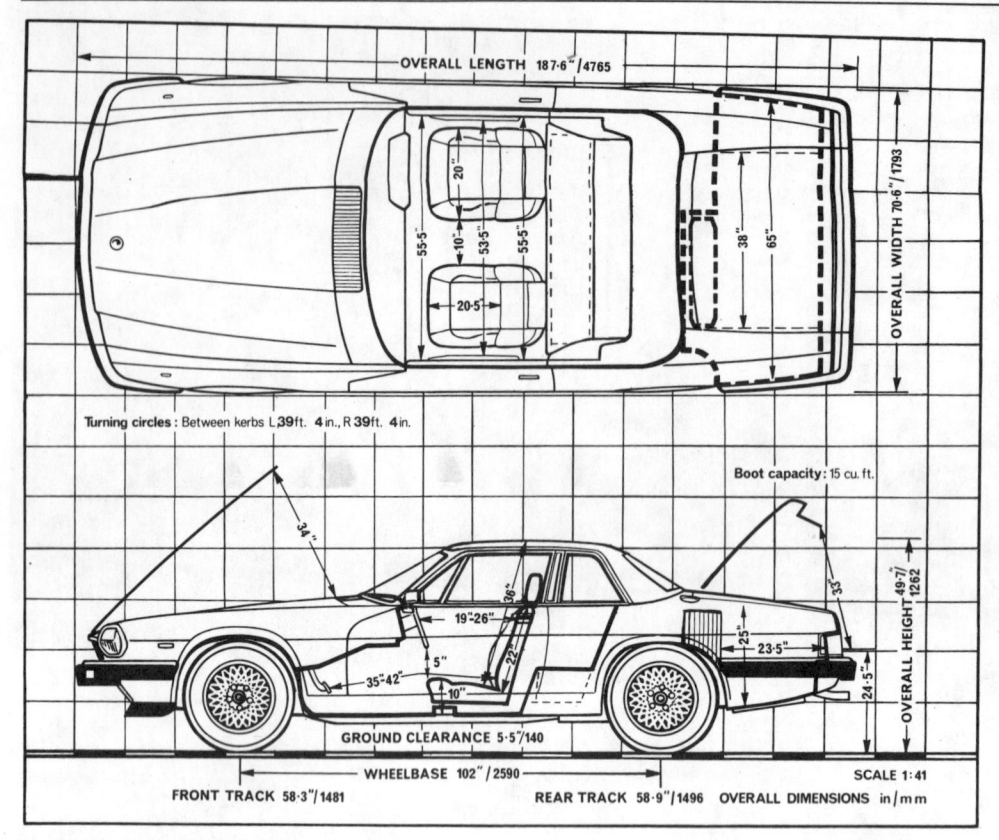

Model: JAGUAR XJ-S V12 CONVERTIBLE

PRODUCED BY:
Jaguar plc
Browns Lane
Allesley
Coventry CV5 9DR

SPECIFICATION

ENGINE
Longways, front, rear-wheel drive. Head/block al. alloy/al. alloy. 12 cylinders in 60deg V.
Bore 90mm, **stroke** 70mm, **capacity** 5345cc.
Valve gear sohc per bank, 2 valves per cylinder.
Compression ratio 12.5 to 1. Electronic ignition. Lucas digital electronic fuel injection.
Max power 291bhp (PS-DIN) (217kW ISO) at 5500rpm. **Max torque** 317lb ft (430 Nm) at 3000rpm.

TRANSMISSION
3-speed automatic.

Gear	Ratio	mph/1000rpm
Top	1.00	26.2
2nd	1.48	17.7
1st	2.48	10.6

Final drive ratio 2.88: limited-slip differential.

SUSPENSION
Front, independent, double wishbones, anti-dive geometry, coil springs, telescopic dampers, anti-roll bar.
Rear, independent, lower transverse wishbones, driveshaft upper links, radius arm, coil springs, telescopic dampers.

STEERING
Rack and pinion, power assistance. Steering wheel diameter 16ins, 2.6 turns lock to lock.

BRAKES
Front 11.2ins (284mm) dia ventilated discs. **Rear** 10.4ins (264mm) dia discs. Antilock standard. Vacuum servo.

WHEELS
Sports alloy 6½ins rims. Pirelli P600 tyres, size 235/60VR15.

PERFORMANCE

MAXIMUM SPEEDS

Gear	mph	km/h	rpm
Top (Mean)	144	232	5500
(Best)	146	235	5570
2nd	115	185	6500
1st	69	111	6500

ACCELERATION FROM REST

True mph	Time (secs)	Speedo mph
30	3.3	32
40	4.8	43
50	6.3	54
60	8.0	64
70	10.4	75
80	12.9	86
90	16.0	97
100	20.4	107
110	26.0	118
120	32.7	128
130	—	138
140	—	149

Standing ¼-mile: 16.3secs, 91mph
Standing km: 29.2secs, 115mph

IN EACH GEAR

mph	Top	2nd	1st
0-20	—	—	1.9
10-30	—	—	2.5
20-40	—	—	3.0
30-50	—	—	3.1
40-60	—	5.0	3.4
50-70	—	5.1	—
60-80	—	5.2	—
70-90	—	5.7	—
80-100	—	6.7	—
90-110	10.9	8.8	—
100-120	12.3	—	—

FUEL CONSUMPTION
Overall mpg: 13.8 (20.5 litres/100km)
Grade of fuel: Premium, 4-star (97 RM)
Fuel tank: 18 Imp galls (82 litres)
Mileage recorder: 1.3 per cent short

BRAKING
Fade (from 91mph in neutral)
Pedal load for 0.5g stops in lb
start/end

	start/end		start/end
1	30-35	6	30-38
2	30-35	7	35-40
3	30-35	8	35-40
4	30-35	9	30-35
5	30-35	10	30-35

Response (from 30mph in neutral)

Load	g	Distance
10lb	0.10	301ft
20lb	0.35	86ft
30lb	0.58	52ft
40lb	0.90	33ft
50lb	1.10	27ft
Parking brake	0.35	86ft

WEIGHT
Kerb 4055lb/1835kg
Distribution % F/R 54/46
Test 4425lb/2002kg
Max payload 442lb/200kg

COSTS

Prices
Total (in GB)	£36,000.00
Road tax, delivery, no plates	£428.21
Total on the Road	£36,428.21
Insurance group	OA
EXTRAS (fitted to test car)	
Total as tested	£36,428.21

SERVICE & PARTS

Change	7500	interval miles 15,000	30,000
Engine oil	Yes	Yes	Yes
Oil filter	Yes	Yes	Yes
Gearbox oil	No	No	Yes
Spark plugs	No	Yes	Yes
Air cleaner	No	No	Yes
Total cost	£161.11	£000.63	£004.47

(Labour cost £36.80 an hour inc VAT)

PARTS COST (inc VAT)
Brake pads (2 wheels) front	£39.10
Brake pads (2 wheels) rear	£28.52
Exhaust complete	£475.20
Tyre — each (typical)	£217.29
Windscreen	£160.65
Headlamp unit	£87.98
Front wing	£250.70
Rear bumper	£381.23

WARRANTY
12 months/unlimited mileage

EQUIPMENT
Automatic	•
Five speed	N/A
Limited slip differential	•
Power steering	•
Steering rake adjustment	N/A
Head restraints front	•
Heated seats	•
Height adjustment	N/A
Seat back recline	•
Cruise control	•
Electric windows F/R	•
Headlamp wash/wipe	•
Central locking	•
Radio/cassette	•
Aerial	•
Speakers	•

• Standard N/A Not available

TEST CONDITIONS
Wind	4-10mph
Temperature	6deg C (43deg F)
Barometer	996mbar (29.4ins Hg)
Humidity	68 per cent
Surface	dry asphalt
Test distance	543 miles

Figures taken at 1485 miles by our own staff at the Lotus Group proving ground at Millbrook and on the Continent.

All *Autocar* test results are subject to world copyright and may not be reproduced without the Editor's written permission.

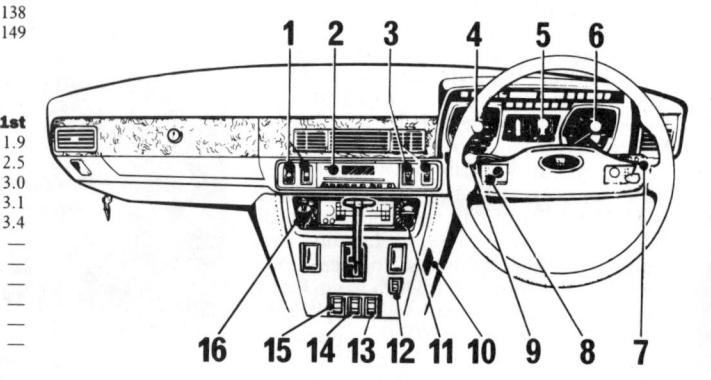

1 Hazard warning lights, rear screen demist; **2** Trip computer; **3** Interior lights; **4** Speedometer; **5** Water temperature, oil pressure, fuel and battery voltage vauges; **6** Revcounter; **7** Windscreen wiper stalk; **8** Headlights, front/rear fog lamps; **9** Indicators, dip/main beam; **10** Lumbar adjustment, heated front seat; **11** Air conditioning; **12** Cruise control; **13** Offside door window; **14** Electric hood switch; **15** Nearside door window; **16** Air conditioning temperature control

JAGUAR XJ-S CONVERTIBLE

Germany, acknowledged expert in the field. Karmann not only helped with the basic design of the hood and its electro-hydraulic mechanism but also with the design of press tooling and jig assembly manufacture.

Jaguar has always paid particular attention to chassis refinement and this remains the case with the V12 convertible. The basic problem with convertibles is how to achieve a high standard of dynamic refinement with minimal body vibration and deflection over varying surfaces without the rigidity normally provided by a roof.

The bodyshell of the V12 convertible is significantly different to that of the XJ-S coupé but they share the same underframe. There are 108 completely new panels and 48 modified panels which, in total, account for one third of the total pressings.

The main differences are found in the rear wings, rear saddle panel behind the hood, windscreen, header and A-posts and the doors which have frameless windows without quarterlights. Rigidity has been added to the shell by strengthening around the central transmission tunnel, front and rear bulkheads and the rear floor area. In addition, steel tubes have been fitted inside the inner sills and A-posts to optimise torsional rigidity.

The hood is a very substantial unit which puts the E-type arrangement to shame. It is filled with sound-deadening material, fully lined, has tinted and heated glass rear window and is electrically operated. Undo the two securing catches attached to the leading edge of the hood frame — and you won't break your nails doing it — press the rocker switch in the centre console and the hood folds smoothly away — all the rear three-quarter windows dissappear into the bodywork. The process is simple, efficient and takes about 12 secs.

Raising the hood is just as easy and perhaps the most impressive aspect of the whole affair is the way the locating lugs on the leading edge of the hood mate exactly with the holes in the top of the A-posts. The fit and finish is very impressive. To ensure longevity, Jaguar has put the hood and mechanism through rigorous testing which included raising and lowering the hood 8000 times — the equivalent of one hood operation a day for 22 years.

Jaguar engineers have also made the hood operation idiot-proof. It cannot be operated unless the handbrake is on and the gear selector in Park. They also seem to have overcome the one major problem when driving convertibles; wind intrusion. Even with the hood and windows down it is possible to drive the V12 convertible at 70mph without excessive buffeting. This has been achieved by altering the rake and shape of the windscreen after exhaustive wind-tunnel tests at MIRA.

Under the bonnet there is no change. The engine is the classic super-smooth normally aspirated two-valves-per-cylinder 5345cc fuel-injected V12 which has appeared in Jaguar in carburettored form since 1972 and in fuel-injected form since 1975. It produces maximum power of 291bhp at 5500rpm and a hefty 317lb ft of peak torque at 3000rpm.

One result of the extensive wind-tunnel work is a Cd figure of 0.39 — compared to 0.38 for the coupé — which is not bad for a full convertible and goes some way towards explaining the more than acceptable mean maximum speed of 144mph. This corresponds to 5500rpm in top and is bang on peak power, indicating that the V12 convertible is ideally geared in top. We recorded a wind-assisted best of 146mph which equates to 5570rpm.

Taking the indicated 6500rpm red-line as maximum engine revs, in-gear maxima are 115 and 69mph in second and first gears respectively. The convertible feels positively sedate off the line from a standing start, but it is deceptive. It is not particularly impressive at 3.3secs to 30 mph, but in the main this is due to the three-speed automatic transmission. The 0-60mph time of 8.0secs is acceptable, but the V12 convertible really starts to show its mettle at the top-end. It reaches 100mph in 20.4secs, covers the standing quarter-mile 16.3secs at a terminal speed of 91mph and the standing kilometre in an impressive 29.2secs at 115mph.

The in-gear incremental times show the Jaguar V12 unit to be extremely flexible with plenty of mid-range torque which endows the convertible with a 50-70mph incremental time of 5.1secs in second. Maximum kick-down speeds — important to know when gauging over-taking — are 35mph for second to first and 87mph for third to second.

The V12 engine is not a particularly economical unit. We recorded an overall figure of 13.8mpg which is comparable to the 14.5mpg returned by the V12 cabriolet. Of limited compensation, however, is an 18 gallon tank which will give the convertible a theoretical cruising range of 270 miles.

As far as refinement is concerned, the XJ-S V12 is very accomplished for a convertible. With the hood in place and the windows up, high speed cruising reveals slight wind intrusion through the seal between the top of the front window and the hood — probably due to suction of the frameless windows — but, unlike many convertibles, there was no hood buffeting. With the hood down there is an obvious increase in wind noise, road and tyre roar and exhaust, but Jaguar has managed to keep wind intrusion to a minimum.

Slip into the contoured, suportive sports seat and the interior will be familiar to any XJ-S driver: Clear and informative analogue instruments, reach-adjustable steering wheel, column stalks for wash/wipe and the indicators and the other ancillary controls operated by facia or centre console switches. The fly-off handbrake is to the side of the driver's seat.

The V12 convertible has an impressive level of standard equipment including electric windows, central locking, electric hood, electric lumbar adjustment heated seats, air conditioning, cruise control, headlamp wash/wipe and a removable four-speaker radio/cassette player. The air conditioning is particularly impressive with the hood down as it can supply plenty of hot air to the footwells.

Rear visibility is good even with the hood up thanks to the heated glass rear window, but care has to be taken when reversing with the hood down. When folded it protrudes above the body line, putting the rear three-quarters of the car out of the driver's sight line.

Three years of intense development and attention to detail have really paid off for Jaguar. The XJ-S V12 convertible is a very complete and accomplished tourer in the true sense of the word. It is capable of taking two people and luggage over long distances in great comfort. At £36,000 it may be the most expensive production Jaguar to date, but we feel it is worth every penny. ■

Operating the V12's hood takes about 12 secs. For safety reasons it cannot be used unless the gear selector is in Park and the handbrake on. Hood has a tinted and heated glass rear window, and it folds into the bodywork. Wind intrusion slight

KOENIG XJ-S

Some people never lose their auto enthusiasm

STORY & PHOTOS BY JOHN LAMM

BRAND COOPER GOT the message when he went sailing off Mulholland Drive in his Ferrari 308. In the hospital with four crushed vertebrae and a smashed right leg, Cooper came to the conclusion it was time to give up on really fast cars and go with something a bit more conservative. In the two years it took to recuperate from the accident, he settled on what new car he wanted, a Jaguar XJ-S. Though still viewing his upcoming driving through the hindsight of his Ferrari incident, Cooper wasn't going to be content with just a regular XJ-S . . .

Mulholland Drive runs along the top of the Santa Monica mountains, which divide Los Angeles from the San Fernando Valley, home of the infamous valley girls. It's no secret this twisting bit of pavement has been famous for its street racers. The usual image is of young kids with lowered, highly polished sedans and sports cars milling around on a section of the road called "the pits," waiting for a challenge. However, the main challenge the last few years has come from the police.

Cooper lives not far from Mulholland and knows the road very well. He's been driving it for fun for 30 years. First on motorcycles, later in a Cadillac-powered Allard and various Ferrari coupes. Cooper was not only a founding member of the Sports Car Club of Southern California, but also the man who designed the organization's badge. And Mulholland, in the days before the populace began building heavily alongside the roadway, was a great place to drive cars.

It still is, but for a different kind of car and driver. Cooper, who admits to being 72 years old, decided to forsake Ferraris in favor of something more civilized. He wanted the retired hot rodder's hot rod.

Although Cooper was fairly satisfied with the stock XJ-S, it wasn't long before he wanted something a bit more from the English car. He came across a photo of one of the Jaguar coupes as modified by Koenig in Germany. *Road & Track* readers will remember the firm from stories about its modified cars, including the second quickest Porsche Turbo in the magazine's most recent World's Fastest Cars competition.

Koenig cars are always head-turners. Doing his work on Porsches, Ferraris and Jaguars, Koenig always creates an interesting variation on the production theme. Sometimes they are beautiful, almost-organic shapes that make the car look faster, while other times they are more than a bit heavy-handed. When Cooper saw the Jag, he liked it and tracked down a distributor of the parts in Los Angeles.

On went a hood scoop. And a front air dam that's meant to increase both downforce and the interest of folks seeing the big cat as it approaches. They might hear it too, because with only about 5 in. of ground clearance under the nose, driveways must be approached carefully.

There's a neat upsweep to the back of the roofline that matches the rear lip spoiler on the back of the car. But what makes the Koenig Jaguar look most outrageous are the rear fender flares. The bulges are good-looking and not overly imposing from a direct side view, complete with the Testarossa-style vents, though in the case of the Jaguar they are nonfunctional. It's the direct front or rear view of the car where the width of the flares is most dramatic. The view is especially effective from the back as you watch the XJ-S drive away. It looks just plain nasty, with about two extra feet added to the car's width at the rear. Cooper admits that because of the extra-wide rear end, he has to be careful when he parks it to avoid nipping one of the Jag's rear fenders.

Though all the added pieces were done in plastic (and shipped to Los Angeles for installation), the quality of the work looks first-rate. On lesser kits, you can see the waviness of the fiberglass as a contrast to the smooth look of the metal. Not so on the Koenig Jag. There is a still more outrageous whale-tail rear deck that originally came with the body kit. Cooper thought it was a bit much. The strong red was enough for him.

With the exception of a Nardi wood-rim steering wheel and a radar detector, the interior of the Jaguar is stock, which is fine with me. I've long admired Jag interiors, particularly the leather seats in the 2+2.

Best reason for buying a Jaguar XJ-S is, of course, the V-12 engine. At 5.3 liters and fitted with Michael May's 11.5:1 compression ratio heads, the fuel injected engine is one of the most satisfying and luxurious in the business. With a single camshaft per cylinder head, this V-12 creates 262 bhp at 5000 rpm and 290 lb-ft of torque at 3000. With that power comes a 0–60-mph time around 8.5 seconds and a top speed that's just a nudge over 140 mph. Just as important, perhaps, you get a chance to sense and hear all 12 cylinders ahead of you.

Koenig offers three levels of modification to increase the engine's power and amplify acceleration. Cooper's choice was fairly conservative in this area, because he wanted no complications getting a smog certificate for the engine. He chose the least complicated of the changes, forgoing the other two, the 390-bhp stage two or the next step up, the 430. Either one of these thumpers raises the cost spectacularly, including shipping the engine to Germany. (They do sound like fun, though, don't they?). Even with this "lesser choice," we guess our subject's car is capable of doing 0–60 mph in the 7.5-sec range or lower, with the top speed close to 150 mph. This is certainly enough for Mulholland Drive.

Although the XJ-SC with the 6-cylinder engine is available with a 5-speed manual gearbox, the V-12 version comes only with the 3-speed automatic. It's not state-of-the-art, perhaps, but works well with ample torque provided by the 12-cylinder engine.

The standard underpinnings of the XJ-S are designed to offer good balance of ride and handling. Both front and back have independent suspension, with a pair of unequal-length A-arms in front, while the rear uses a lower arm, the halfshaft and trailing arms. Both the shocks and coil springs are paired in back. In addition to the Koenig conversion came stiffer springs, different shackles and Koni shocks with tighter valving.

Stock rubber on the production XJ-S is Pirelli P6, size P215/70VR-15. Not big enough for Koenig. On went 235/55VR-15s P6s for the front wheels, with Cooper choosing 285/50VR-16s P5s at the back, all mounted on the rather classic-looking BBS modular wheels with their deep-set, gold-painted lacework centers.

The remainder of the chassis is stock, with a brake combination that includes big vented discs at the front and inboard discs at the back. Steering is Jaguar's rack and pinion, with power assist.

The combination of power, steering and sophistication is what Cooper seems to like best about his Koenig Jaguar XJ-S coupe. With those fat tires there's little chance he's going to have problems keeping the V-12's torque and horsepower under control. As for being a head-turner,

Cooper admits the Jaguar is a bit too much but worth the trouble for a chance to have this Mulholland racer.

And that suits Brand Cooper just fine. Here's a man who had a Bugatti Type 35B as a teenager, had a small incident with it in which his passenger, actress Carole Lombard, cut her face and ended up suing the high schooler for $100,000. His father took the Bugatti back, Cooper explains, and two owners later, the 35B accidentally was driven off a high cliff near Big Sur, its owner inside the car. There were other cars Cooper talks about having—a Brescia Bugatti, Chrysler Imperials, Lincolns, a Porsche Carrera, a Shelby Mustang—all of which are depicted in a collection of 1/43-scale models.

The cost of the Koenig Jaguar? Cooper grins and admits that eventually he stopped counting. But, when pressed, he pegs it around $87,000. He is more than satisfied.

U.S. SPECIFICATIONS

Price	$87,000
Curb weight, lb	4030
Wheelbase, in.	102.0
Track, front/rear	na
Length	191.7
Width	est 94.5
Height	47.8
Fuel capacity, U.S. gal.	24.0

ENGINE

Type	sohc V-12
Bore x stroke, in./mm	3.54 x 2.76/90.0 x 70.0
Displacement, cu in./cc	326/5343
Compression ratio	11.5:1
Bhp @ rpm, SAE net	est 280 @ 5000
Torque @ rpm, lb-ft	est 300 @ 3000
Fuel injection	Bosch-Lucas L-Jetronic

DRIVETRAIN

Transmission	3-sp automatic
Gear ratios: 3rd (1.00)	2.88:1
2nd (1.48)	4.26:1
1st (2.48)	7.14:1
Final-drive ratio	2.88:1

CHASSIS & BODY

Layout	front engine/rear drive
Brake system	11.1-in. vented discs front, 10.3-in. discs rear; vacuum assist
Wheels	cast alloy; 15 x 10 front, 16 x 12 rear
Tires	Pirelli P6, 235/55VR-15 front; Pirelli P5, 285/50VR-16 rear
Steering type	rack & pinion, power assisted
Turns, lock-to-lock	3.3
Suspension, front/rear: unequal-length A-arms, coil springs, tube shocks, anti-roll bar/lower A-arms, fixed-length halfshafts, trailing arms, dual coil springs & tube shocks, anti-roll bar	

PERFORMANCE

0–60 mph, sec	est 7.5
Standing ¼ mile, sec @ mph	est 15.5 @ 95.0
Top speed, mph	est 150

na means information is not available

Tom's cat

XJR-S turns in with agility, allowing slides to be held or corrected with ease

Not even leather seats with red piping can rescue the outdated interior

Blessed with the intuitive insight of one who'd taken 48 hours and 1000 miles to reach a working relationship with the obvious, I suggested 11 months ago that what JaguarSport's XJR-S needed to sustain a credible existence in the modern world was a 6-litre engine. Groggy from the impact of heavy hints launched by Jaguar's PR machine, it wasn't a hard conclusion to reach but, after driving the new £45,500 6-litre XJR-S, I'm left in no doubt that it was the right one.

It's now possible to see the original XJR-S in its true light, shielded from the over-exposure of hype by the sense of perspective the faster and sharper 6-litre car brings. Too standard and too soft; despite the tauter suspension and steering — the 5.3 wasn't the real McCoy. Those who'd expected something tangible and exciting from the Jaguar-Walkinshaw partnership, especially in the light of Mr Walkinshaw's previous and much respected high-performance interpretations of V12 Jaguars, must have wondered what all the fuss was about and why anyone should want to pay £38,500 for an XJ-S wearing the equivalent of water-based transfer tatoos. The be-spoilered Jaguar looked tough all right but, in hot water, the image simply floated away.

Those who weren't expecting too much, however, weren't disappointed. By sending the XJ-S to the health club instead of the hormone bank, JaguarSport avoided the risk of altering the big two-seater's big-hearted and benign character. It was an XJ-S that had worked out with a few light weights and donned a sharper tracksuit, nothing more. Clearly enough people considered this to be a valid exercise. All 500 5.3s were sold and customer feedback suggested general satisfaction with just a few dissenters — and their beef was over the styling. Encouraged by this response, Jaguar has set the target for 600 6-litres next year.

For its second stab at making a more sporting version of a tried and tested but ageing grand tourer, JaguarSport has laid the curling tongs to one side and, in administering the testosterone, secured the services of a large syringe. The new car represents 'Phase 2' and, this time, it means business.

Those who know Jaguar's V12 know that its capacity can be expanded to 7 litres and that 500bhp isn't asking too much. In the light of this, JaguarSport's 5993cc and 318bhp (just 20bhp up on the 5.3) looks lamentably lame but, to be fair, it was never the company's intention to produce a high specific output. Considered to be more important was flexibility and, for this, you need torque. Here, the new 6-litre engine makes a bigger impression, peaking at 3750rpm with a thumping 362lb ft and maintaining more than 300lb ft from 2000rpm. Now the torque curve looks like a single line rendering of Ayer's Rock.

The extra capacity is the result of lengthening the stroke from 70 to 78.5mm. New pistons, liners and crank also contribute to the changes. In addition to this, the standard three-speed GM400 automatic transmission has been recalibrated for quicker and more responsive kickdown. As far as the chassis is concerned, stiffer coil springs and Bilstein gas-filled dampers have been employed to keep things more firmly in check.

Visually, the new XJR-S looks much as before with well-considered spoilers, flared sills and elegant Speedline alloy wheels. These are now of 16 rather than 15ins diameter, however, and wear broader Dunlop rubber: 225/50 at the front and 245/55 at the back. Two sets of square-section twin exhaust pipes are the most obvious aesthetic clue, though a zero brightwork option and a new range of colours (including black and British Racing Green) have been added to the cosmetic catalogue. So much for the beautification. The proof of this XJR-S is in the driving.

The Jaguar delivers its greater urge with characteristic finesse but more feeling. Its mid-range punch is still silk-gloved but now there's a greater sense of the iron fist behind it, the engine hauling strongly, cleanly and evenly from as little as 1000rpm; there's no mistaking the extra gruffness in its voice, though. From 3500rpm, the exhaust note cuts loose as the engine crests the torque curve, then growls discreetly until the GM400 kills the cresendo. A genuine 130mph can be held with an alluring sense of ease, only now there's something worthwhile left when you floor the accelerator.

With a top speed of around 150mph and 0-60mph acceleration in the 7 to 7.5secs bracket, the standard XJ-S is anything but slow, but, subjectively at least, the 6-litre car goes harder. Jaguar reckons to have recorded 162mph round the Nardo circuit in southern Italy and claimed a "conservative" maxi- ▶

> 'There's a greater sense of ease about the Jaguar's work but you'll look in vain for the thrill that sets your neck hairs bristling'

THE UPS AND DOWNS OF JAGUAR'S V12

THE 6-LITRE XJR-S WAS QUICK, but not *that* quick. When we processed the performance figures, we were dismayed to discover that not only did they fail to meet Jaguar's 'conservative' claims, but also fell well short of those returned by the original XJ-S HE in 1982.

That car, the first recipient of the high-compression version of Jaguar's 5.3-litre V12 (299bhp) with its unique May Fireball heads, had a top speed of 153mph and accelerated from rest to 60mph in a rousing 6.5secs. By comparison, the 6-litre car is all through at 150mph, some 8mph short of Jaguar's claim, and gets from zero to 60mph in 7.1secs — 0.6secs adrift of the 6.5secs claim which, intriguingly, is no quicker than the original HE's actual time.

Now it's just conceivable that *Autocar*'s test HE was an especially good one — one of those cars which, for whatever reason, performs beyond even its maker's expectations. This notion is largely borne out by the figures for *Motor*'s 1981 HE which boasted a similar top speed (152.4mph) but a more plausible 7.5secs for the 0-60mph sprint.

Even so, that's about as fast as a standard XJ-S has ever gone. In the ensuring years, the XJ-S became both heavier and slower. By the mid-80s, maximum power was down to 291bhp and although neither magazine tested a regular V12 coupe in the period, Autocar's test results for a 1984 Cabriolet (140mph/7.7secs 0-60) and a 1988 Convertible (144mph/8.0secs 0-60) would seem to confirm the erosion of performance.

Towards the end of last year, the V12's power output took another minor tumble to 286bhp with the introduction of Marelli ignition and a consequent drop in compression ratio from 12.6 to 11.5:1 — measures to enable the V12 to run on unleaded petrol. This engine was fitted to a 5.3-litre XJR-S we tested just before the introduction of the 6-litre and the results are interesting to say the least: 146mph and 0-60mph in a barely believable 9.3secs.

Of course, it's possible that the XJR-S was off-colour; for every 'good' XJ-S, there's probably a 'bad' one. What the 6-litre engine does is restore that original sparkle. As for taking on the Porsche 928 S4, that, we fear, will have to wait for 'Phase 3'.

	1982 HE	XJR-S 6.0	XJR-S 5.3
Top speed, mph	153	150	146
0-60mph, secs	6.5	7.1	9.3
0-100mph, secs	15.7	17.7	23.7
30-70mph through gears	5.5	6.6	9.3
Standing ¼	14.9	15.5	17.1
Standing km	26.9	27.8	30.1

of 158mph. But things aren't quite that simple (see boxed story).

The 6-litre XJR-S is a sharper sprinter, too, though the engine's plentiful torque contrives to disguise the fact if you rely too heavily on your senses. There's a greater sense of ease about the Jaguar's work — it feels like a lighter car — but you'll look in vain for the thrill that sends the small hairs on the back of your neck bristling. That said, you can't argue with the figures. Displaying excellent traction off the line and a new vigour more effectively harnessed by the reworked gearbox, the Jaguar surges to 60mph from rest in 7.1secs and on to 100mph in a little more than 17secs. No more embarrassing moments from fast hatchbacks. That said, the XJR-S doesn't exactly find itself trading tenths with Porsche's 928S4.

In hard driving, the 6-litre is both more responsive and more forgiving than the 5.3-litre car it supersedes. Most of the standard car's handling shortcomings — strong understeer in tight turns, roll oversteer on the limit, sloppy suspension control over crests and dips — have been exorcised from a chassis which will never feel lithe but has always promised greater ability than it's been allowed to display.

Despite the handicap of steering which remains too light and reticent to communicate, the Jaguar turns in with almost startling agility and precision for such a big car and is better balanced than before, staying neutral for longer. The tail can be punched a little way out of line and easily held or corrected. In steady state cornering on perfectly dry tarmac, the bite of the squat Dunlops is very impressive and the Jaguar feels stable and secure. In short, the vices have been all but eradicated.

What's remarkable is that the ride has hardly suffered at all. It feels much firmer but very little harsher. The small loss of absorbency is more than compensated for by the huge gains in control. Just as impressive are the brakes. The Jaguar's massive all-round ventilated discs are beyond criticism, capable of hauling the car to a standstill from three-figure velocities without any fade.

The inescapable conclusion to all this is that JaguarSport has finally come up with a car worthy of the name and struck some very clever compromises in the process. Any fears that to make the XJ-S handle it would be necessary to ruin its ride have proved unfounded. Likewise, the additional potency has brought with it commensurate gains in driveability and only a small sacrifice in refinement.

What Walkinshaw and his team haven't been able to do, of course, is turn back the clock. The days when the XJ-S offered unrivalled value for money have long since gone. The V12 remains a masterpiece to be sure, but its charms have to be seen in the context of a dated design. Leather seats with red piping can't rescue the interior, nor spoilers and skirts the styling.

If comparisons are to be made with Porsche's 928S4 — and JaguarSport isn't keen to discourage them — then they have to be made on a dynamic level. And, good as the new XJR-S is, it's no 928. What you have to decide is whether the £10,000 price difference is ample compensation. ■

'A genuine 130mph can be held with an alluring sense of ease, only now there's something worthwhile left in reserve'

Jaguar XJ-S Convertible

The perfect Sunday afternoon diversion

by Ron Grable
PHOTOGRAPHY BY CHARLIE RATHBUN

Imagine if you will a lazy Sunday afternoon. Clear blue sky, temperatures in the 70s, and a nice, cool breeze. You have absolutely nothing to do. You've finished the Sunday paper and you're sitting with your feet up, when you suddenly realize that the perfect car in which to do nothing is waiting downstairs.

A Jaguar. Moreover, a killer-red Jaguar—convertible. Consider the ramifications. You can get out that seldom-used ascot from the dresser drawer, along with the tweed cap and pipe. You can search for small country lanes and trundle along with the wind in your hair, taking your afternoon tea and cakes in some lovely, old inn. This laid-back attitude represents, in part, the general ambiance of convertiblism. Convertibles offer open-air motoring at the flick of a switch, and when that top is down, the driving experience is completely altered. When it's cold, you're cold, ditto hot. But convertiblism in total is much more than the sum of its parts.

Take any given trip in a conventional car, which, for the purposes of this discussion, is anything with a fixed top. Then redo the same route with a convertible, and it'll seem like a different trip entirely. You'll remember the smells of the countryside, the damp air of freshly irrigated fields, the hot sun, the roar of a swollen river, and the glow of a soft summer sunset. When you stop, people are friendlier, and you find yourself in conversation with locals who wouldn't otherwise offer the time of day. In a convertible, you experience your travels; in a closed car, you only pass through.

The '89 XJ-S Convertible is the first Jaguar convertible offered by the company since 1974. Jaguar has sold a long list of drop-tops in the U.S., starting with the SS-100 of the late '30s. The XK-120, 140, 150, and E-Type all had a significant portion of their sales as convertibles, but the last E-type rolled off the Coventry assembly line in 1974. This convertible has been developed over a three-year period, and Jaguar solicited the help of Karmann, of West Germany, to engineer the folding soft top. The results are excellent, and lowering and raising the top is as simple as with any we've ever tested (and can be done without leaving the driver's seat). Cycle time from full up to full down is 12 sec, and the system is electric/hydraulic-powered with the motors located in the rear stowage area. An interesting feature of the top is the use of glass (tinted and heated) for the rear window. The Karmann engineers found a method to retract the glass among the folds of the top, and it eliminates the melted look of rear windows found in most convertibles. And then there are the obvious security advantages.

The new XJ-S Convertible was developed by Jag engineers specifically as a drop-top. That's important because of the inherent difficulties of trying to adapt a coupe body into a convertible. The underlying problem with convertibles is to achieve adequate torsional rigidity. If a designer simply cuts the top off a coupe body structure, the result is poor torsional rigidity. As an illustration of what happens, take a cardboard roll from the center of a roll of toilet paper, grasp it in both hands, and twist it. If you twist hard enough, of course, it will crumple and fail (fold up). The resistance it offers to this twisting force is its torsional rigidity. Now cut out an elliptically shaped area from the roll (roughly the size of a silver dollar), and again twist the roll. There will be a dramatic decrease in the torsional rigidity, which is exactly the problem faced by engineers who want to build a convertible.

Torsional rigidity is important, because the chassis locates all the suspension and mounting points for the engine, driveline, etc. If the chassis is flexing and twisting in response to road inputs, then so is the suspension, destroying all the accuracy needed to locate the suspension and determine the handling dynamics of the vehicle. In addition, the twisting of the body shell will show up as squeaks, rattles, and other ills.

Jaguar decided to design a whole new body structure for its convert, and its approach was three-phased, led by Jim Randle, director of product engineering. Phase one was dynamic finite element modeling (DFEM), where the engineers wrote computer programs to analyze their designs for shake and vibration behavior. This allowed them to check their designs without actually building a vehicle. Phase two was a prototype vehicle, based on the DFEM, which was mounted on a "paddle rig"

and subjected to vibratory inputs of certain frequency ranges (representing what would be encountered while driving, in the range of 5-30 Hz). The third phase used complete prototype vehicles tested over various road surfaces to fine-tune the engine and driveline mountings, aimed at optimizing noise isolation and reducing body response to road inputs. Interestingly enough, the final phase achieved a 40% reduction in response (at the body resonant frequency) over the undeveloped prototype, which tells us that seat-of-the-pants engineering is still able to do some things a computer can't.

The new body shell was based on the XJ-S Coupe underframe. The engineers strengthened (compared to the coupe) the shell in the transmission tunnel area, both front and rear bulkheads, and rear floor area. In addition, a structure of steel tubing was fitted inside the sill area (the area beneath the doors when closed), and also inside the "A" posts. There are 108 completely different panels, and 48 modified panels in the convertible, compared to the coupe. All the strengthening and the motors to raise and lower the top add approximately 175 kg (nearly 400 lb) to the curb weight of the convertible.

As an example of Jaguar's attention to detail for this convertible, the door glass was redesigned as frameless, and as a check of durability, a machine slammed the door closed, pushing on the top of the glass, 100,000 times. That's the equivalent of eight times per day for 34 *years*.

New this year on the XJ-S Coupe is Teves ABS, which is also fitted to the convertible. The Teves system works with the Girling 4-wheel disc brakes found on all XJ-S models, and it has a special yaw control refinement for better braking control on surfaces with uneven friction. For example, winter roads that have been plowed and then gone through a couple cy-

The XJ-S Convertible was designed by Jaguar engineers specifically as a drop-top

cles of melting and refreezing, ending up with the center of the road dry asphalt and the edge snowy or icy. Normally, ABS systems will sense the traction available to the wheel on the low friction surface and limit the stopping power of the vehicle to that wheel. This prevents yaw disturbances but drastically reduces stopping power. The Teves system, after initially reducing braking to the lowest friction wheel, *slowly* increases braking power up to the high traction wheel, giving the driver enough time to maintain control.

We were able to try this system when we came across some water running down the curb, flooding out into the street so that the right-side tires were in water while the left sides were on nice dry concrete. Stopping from 100 mph in full ABS resulted in some yaw, but it was progressive enough to allow ample time to add the correct amount of opposite steering to counteract the yaw mo-

tion. The Teves system uses hydraulic boost for the power braking assist, which is an advantage over the more traditional vacuum booster systems in case of engine failure. The hydraulic boost pump is electrically driven, so the electric motor provides power to the hydraulic pump for braking boost even with a dead engine.

Also new this year is a power lumbar adjustment on the front seats, and both driver and passenger seats are heated to warm the tush on those cold, foggy London mornings. And just as the seats warm the tush, the silky, smooth V-12 warms the heart. It's basically carryover, with the exception of a change to Marelli ignition (no changes in either power or torque). The original single-cam cylinder head design remains, with the combustion chamber designed by Swiss engineer Michael May. This particular combustion chamber shape allows high compression ratios (11.5:1 this year) and lean mixtures. This engine is now rated at 262 hp at 5000 rpm with a pavement-wrinkling 290 lb-ft of torque at 3000 rpm, and is the same engine used in the successful IMSA Castrol Oil Jaguars and the '87 World Sports Prototype Champion Silk Cut Jaguar team—impeccable credentials, indeed.

The transmission, on the other hand, is from the world of luxury and convenience, and soothes the cat with buttery-smooth shifts. Unfortunately, the XJ-S Convertible is available only with the 3-speed automatic. Although delivering good and positive shifts, the transmission is no doubt responsible for the rather lethargic 10.3-sec 0-60 and the 17.7-sec/83.3-mph quarter mile times.

Suspension is carryover from last year, fully independent front and rear. The wheels were widened to 6.5 in. and carry 235/60VR15 Pirelli P600s. The increased tire size offers slight increases in handling performance, but probably more important, excellent ride quality, which is the primary goal of the Jaguar engineers for this vehicle.

Driving the Jag Convertible is a study in smooth, refined transport. It soaks up larger-amplitude bumps without notice and even manages to damp out the rippled surface irregularities better than most luxury class vehicles. The interior of the Jag speaks of quiet luxury, with real walnut burl, selected so the grain of the wood is coordinated from door panel to dash and into the other door. The wood is polished and finished beautifully and is a real luxury statement by Jaguar. Soft leather covers the doors, seats, and dash and color-matches the walnut perfectly.

Just as the seats warm the tush, the silky, smooth V-12 warms the heart

The steering column telescopes, with fore-aft, rake, and lumbar adjustments provided for the seats. We found the head room restricting for our 6-ft 1-in. height and had to recline the seats more than we wanted to get adequate head room. Thinking back to our last drive in an XJ-S Coupe, we remember slightly more aft adjustment to the seat, which is possibly due to the restriction of the convertible top mechanisms buried in the stowage area.

We like the Jag better with the top down. Head room is no longer a problem, and the combination of windscreen rake angle and door windows does an excellent job of creating a dead air space for passengers, even at speeds near 100 mph. It's possible to listen to the radio and even use the climate control to maintain a semblance of heating/cooling as needed. With the top up, there's some wind noise emanating from the door-window to top seal, but otherwise the top is quite efficient in sealing out the outside world.

Limit handling is not the Jaguar's forte. It's at its best driven at 70-80% of its limit, where it's smooth and

TECH DATA
SPECIFICATIONS & PERFORMANCE

Jaguar XJ-S Convertible

GENERAL
Make and model	Jaguar XJ-S Convertible
Importer	Jaguar Cars Inc., Leonia, N.J.
Body style	2-door, 2-passenger
Drivetrain layout	Front engine, rear drive
Base price	$57,000
Price as tested	$57,000
Options included	None
Typical market competition	Mercedes 300SL, Cadillac Allanté

DIMENSIONS
Wheelbase, in./mm	102.0/2590
Track, f/r, in./mm	58.6/59.2/ 1488/1504
Length, in./mm	191.7/4869
Width, in./mm	70.6/1793
Height, in./mm	49.4/1255
Ground clearance, in./mm	4.5/140
Manufacturer's curb weight, lb	4190
Weight distribution, f/r, %	56/44
Cargo capacity, cu ft	13.6
Fuel capacity, gal.	21.6
Power/weight ratio, lb/hp	16.0

PERFORMANCE AND TEST DATA
Acceleration, sec
0-30 mph	4.1
0-40 mph	6.0
0-50 mph	8.0
0-60 mph	10.3
0-70 mph	13.1
0-80 mph	16.9

Standing quarter mile,
sec @ mph	17.7 @ 83.3

Braking, ft
30-0 mph	33
60-0 mph	129

Handling
Lateral acceleration, g	0.80
Speed through 600-ft slalom, mph	61.1

Speedometer error, mph
Indicated	Actual
30	30
40	40
50	50
60	60

Interior noise, dBA
Steady 60 mph in top gear	70

FUEL ECONOMY
EPA, city/hwy., mpg	12/16
Est. range, city/hwy., miles	259/346

ENGINE
Type	V-12, liquid cooled, alloy block and heads
Bore x stroke, in./mm	3.54 x 2.76/ 90.0 x 70.0
Displacement, ci/cc	326/5344
Compression ratio	11.5:1
Valve gear	SOHC, 2 valves/cylinder
Fuel/induction system	Multi point EFI
Horsepower, hp @ rpm, SAE net	262 @ 5000
Torque, lb/ft @ rpm, SAE net	290 @ 3000
Horsepower/liter	49.0
Redline, rpm	6500
Recommended fuel	Unleaded premium

DRIVELINE
Transmission type	3-speed auto.
Gear ratios (1st)	2.50:1
(2nd)	1.50:1
(3rd)	1.00:1
Axle ratio	2.88:1
Final drive ratio	2.88:1
Engine rpm, 60 mph in top gear	2300

CHASSIS
Suspension
Front	Upper and lower control arms, coil springs, anti-roll bar
Rear	Upper and lower control arms with halfshaft as upper arm, trailing arms, coil springs

Steering
Type	Rack and pinion, power assist
Ratio	16.0:1
Turns, lock to lock	2.7
Turning circle, ft	39.4

Brakes
Front, type/dia., in	Vented discs/11.1
Rear, type/dia., in	Discs/10.3
Anti-lock	Standard

Wheels and tires
Wheel size, in.	15 x 6.5
Wheel type/material	Cast alloy
Tire size	235/60VR15
Tire mfr. and model	Pirelli P600

INSTRUMENTATION
Instruments	0-160-mph speedo; odo; trip odo; 0-7000-rpm tach; coolant temp; oil press; fuel level; volts; digital clock
Warning lamps	Oil press; batt; brakes; coolant temp; fuel level; hazard; headlamps; ABS

---

predictable and can cover miles in a big hurry. The wonderful V-12 will pull the car off the corners quickly, so the technique that works best is to use the brakes to slow the car down from speed approaching the corner, select a line through the corner that gets the Jag positioned for the exit, then sit back and enjoy the powerful engine as it hustles you down to the next corner.

Rushing along on the perfect back road in this manner with the top down on a nice summer evening can be pretty close to convertible nirvana and will, for some, justify the dent in the wallet the XJ-S demands. At a suggested retail of $57,000, it's pretty pricey for a 2-seat convertible, but when you offer the right product, apparently price isn't much of a concern in this market segment. In the first week the XJ-S convertible was available, over 300 were snapped up by eager buyers. The $57,000 buys lots of content, as the salesman will tell you, but there are some things that don't appear on the window sticker. One of them is a service-on-site roadside assistance program that takes care of any failure, including flat tires or an empty gas tank. If it happens more than 50 miles from home, you get motel, food, and even alternate transportation costs. Ya can't beat that.

Some of the other specifics not listed on the window sticker are prestige, quality, luxury, and those warm summer nights with the top down on Pacific Coast Highway—or your personal equivalent. **MT**

Second Opinion

The XJ-S is a little like a tuxedo: Wearing either one, you cut a dashing, jauntily formal profile that makes you feel like a million bucks. There are times when nothing else comes close. For day-to-day use, though, I feel a little out of place in such attire, be it automotive or haberdashery. Sure, you can wear a tux to run over to the hardware store, but people tend to look at you funny, and you have to give the paint-mixing machine a wide berth.

The same goes for the Jag. Subjecting it to the drudgery of daily use seems somehow disrespectful. Letting the unwashed fling their doors into the Jag's tender flanks would be a tragedy I'm not sure I could bear.

Maybe if I traveled in better circles, I wouldn't have this problem. Please send money. —*Jeff Karr*

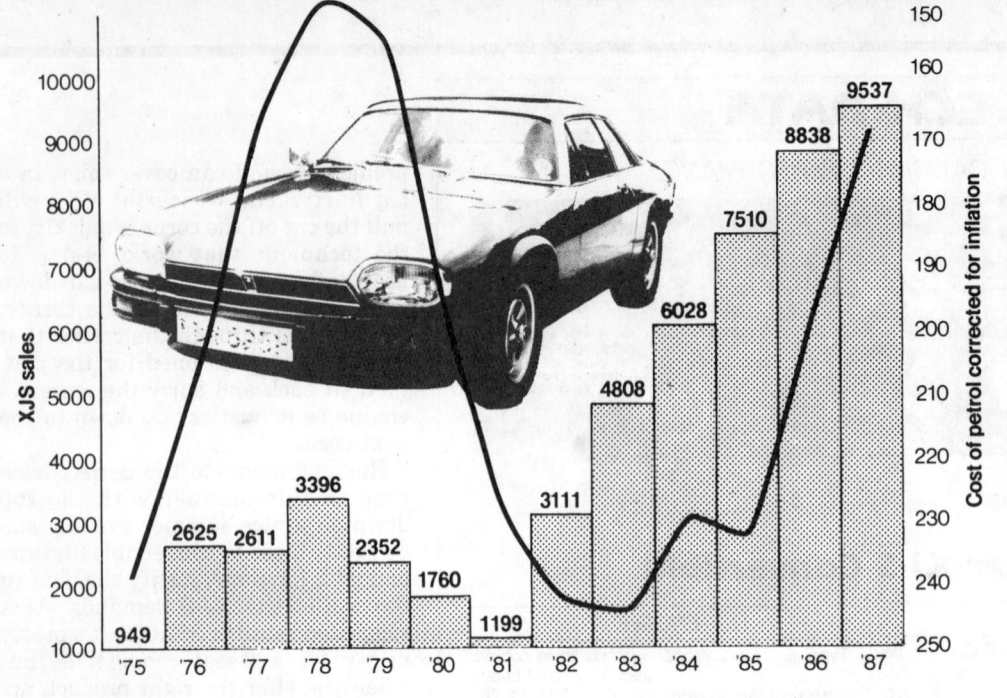

The fall and rise of the XJ-S

Born into a hostile world, the XJ-S started life as a car nobody loved. Now it's selling like hot cakes. What happened? John Simister explains

Were the Jaguar XJ-S a person, its parents would be smiling indulgently and calling it a late developer. The bad times – the times their offspring was a sloven, once even a dropout – are forgotten. It has all come good.

A dropout? Yes, in 1980. That was the year that Jaguar stopped making the XJ-S for a few months. The unsold stocks, achieved despite cutting production to the bone, were embarrassing. Petrol prices had soared and the XJ-S – all 5.3 litres, 34cwt and 14mpg of it – didn't fit into the order of things any more. Jaguar sold 1760 XJ-Ss in 1980, but made only 1057; sales actually bottomed out the following year, with 1199 cars finding new owners from a production of 1292.

But look at it now. The XJ-S, as a breed, sold 9537 units last year. This year it should crack the 10,000 mark, a year in which the design celebrates its 14th birthday. What *has* happened?

The Jaguar got off to an inauspicious start. That controversial flying-buttress styling did not endear the Jaguar to its target customers, even allowing that it was never intended to be an E-Type replacement. Its position in the market was unclear to those used to what Jaguars had been until then, a lack of clarity which reflected exactly the outlook of the British Leyland management then inflicted on Browns Lane.

Billed as a luxurious grand tourer, a sort of cut-price Aston Martin, the XJ-S was paradoxically designed to appeal to what BL perceived to be a new desire for functionalism, à la Porsche. So there was less leather than expected, and instead of wood on the facia there was sombre black vinyl. Was this what the market wanted?

And did the market want, in the wake of the Yom Kippur war and the threat of petrol rationing, a car which would only manage 14mpg? Even if it did, and bought XJ-Ss in affluent hordes, it would soon realise that the car woefully lacked both build quality and reliability. This did the image no good at all, which meant that residual values provided a shock come trade-in time.

Then John (later Sir John) Egan came in, bringing with him his now famous purge on poor quaililty. This, and some rethinking about the XJ-S's role in life, resulted in 1981's XJ-S HE. With its high-compression, squish-inducing cylinder heads designed by Swiss engineer Michael May, the HE proved dramatically less profligate with fuel – over 20mpg became possible with a sympathetic right foot. And, inside, dark tones became light, leather replaced vinyl on the door casings and rear quarter trims, and burr elm veneer adorned the facia and door trim waist rails. At last, it looked like a Jaguar should.

And sales soared, at a much greater rate than they had previously declined. Falling petrol prices helped (the graph shows those for the UK, using as our 'inflation' factor the rise in XJ-S V12 coupé prices over the years) but the worldwide XJ-S sales recovery pre-empted them. Its upward trend began while UK petrol rises were still levelling out.

It was overseas sales that benefited the most. Hitherto the mix had been 50-50 home market and export; by 1987, export sales – that is, sales to markets less likely to be sympathetic to any problems Jaguar might have – were three-quarters of the total. And those 7083 XJ-Ss earned a lot of foreign revenue, especially as 79 per cent were V12s.

The other 21 per cent, of course, were powered by the six-cylinder, 3.6-litre AJ6 engine introduced in 1983. The arrival of this cheaper XJ-S (available with manual transmission, unlike the V12 version) ensured the continued rise of the sales graph without altering its slope significantly. The UK market, in fact, currently accounts for just under half of XJ-S 3.6 output, that world-wide sales figure of 21 per cent 3.6s reflecting the burgeoning love affair that the Americans (the most numerous of XJ-S buyers) have with the V12.

After the 3.6 came Cabriolet versions of both cars, broadening the XJ-S's appeal further and helping the sales graph to climb. The slowing of that graph's rate of ascent last year shows how timely is the arrival of the Convertible – a V12-only car – which replaces both cabriolets. Jaguar reckon there's little market for a 3.6 convertible at present.

Truly, the XJ-S has flowered late. But, with sales still some way from levelling off, there's little chance of the bloom wilting for a few years yet. Drive an original XJ-S from 1975, though, and there's one thing that will strike you hard – it really won't feel much different from the current version.

The concept always was good.